Library of
Davidson College

Progress in Mathematics
Volume 104

Series Editors
J. Oesterlé
A. Weinstein

Jeffrey Adams
Dan Barbasch
David A. Vogan, Jr.

The Langlands Classification and Irreducible Characters for Real Reductive Groups

1992

Birkhäuser
Boston • Basel • Berlin

Jeffrey Adams
Department of Mathematics
University of Maryland
College Park, MD 20742

Dan Barbasch
Department of Mathematics
Cornell University
Ithaca, NY 14853

David A. Vogan, Jr.
Department of Mathematics
Massachusetts Institute of Technology
Cambridge, MA 02139

Library of Congress Cataloging-in-Publication Data

Adams, Jeffrey,
    The Langlands classification and irreducible characters for real
reductive groups / J. Adams, D. Barbasch, D. Vogan, Jr.
      p.   cm. -- (Progress in mathematics : v. 104)
    Includes index.
    ISBN 0-8176-3634-X (alk. paper). -- ISBN 3-7643-3634-X (alk. paper)
    1. Representation of groups. 2. Algebraic geometry. I. Adams,
Jeffrey. II. Barbasch, D. (Dan), 1951- . III. Vogan, David A.,
1954- . IV. Series: Progress in mathematics (Boston, Mass.) :
vol. 104 .
QA176. L36 1992   91-48265
512 '.2--dc20   CIP

Printed on acid-free paper.
© Birkhäuser Boston, 1992.

Copyright is not claimed for works of U.S. Government employees.
All rights reserved. No part of this publication may be reproduced, stored in a retrieval system, or transmitted, in any form or by any means, electronic, mechanical, photo-copying, recording or otherwise, without prior permission of the copyright owner.

Permission to photocopy for internal or personal use, or the internal or personal use of specific clients, is granted by Birkhäuser Boston for libraries and other users registered with the Copyright Clearance Center (CCC), provided that the base fee of $0.00 per copy, plus $0.20 per page is paid directly to CCC, 21 Congress Street, Salem, MA 01970, U.S.A. Special requests should be addressed directly to Birkhäuser Boston, 675 Massachusetts Avenue, Cambridge, MA 02139, U.S.A.

ISBN 0-8176-3634-X
ISBN 3-7643-3634-X

Camera-ready copy prepared by the Authors in TeX.
Printed and bound by Quinn-Woodbine, Woodbine, New Jersey.
Printed in the USA.

9 8 7 6 5 4 3 2 1

Contents

Acknowledgments vii
Index of notation ix
1. Introduction 1
2. Structure theory: real forms 28
3. Structure theory: extended groups and Whittaker models . . . 41
4. Structure theory: L-groups 47
5. Langlands parameters and L-homomorphisms 55
6. Geometric parameters 64
7. Complete geometric parameters and perverse sheaves 82
8. Perverse sheaves on the geometric parameter space 98
9. The Langlands classification for tori 105
10. Covering groups and projective representations 113
11. The Langlands classification without L-groups 120
12. Langlands parameters and Cartan subgroups 139
13. Pairings between Cartan subgroups and the proof
    of Theorem 10.4 147
14. Proof of Propositions 13.6 and 13.8 157
15. Multiplicity formulas for representations 167
16. The translation principle, the Kazhdan-Lusztig algorithm,
    and Theorem 1.24 175
17. Proof of Theorems 16.22 and 16.24 189
18. Strongly stable characters and Theorem 1.29 205
19. Characteristic cycles, micro-packets, and Corollary 1.32 . . 212
20. Characteristic cycles and Harish-Chandra modules 222
21. The classification theorem and Harish-Chandra modules
    for the dual group 234
22. Arthur parameters 239
23. Local geometry of constructible sheaves 248
24. Microlocal geometry of perverse sheaves 252
25. A fixed point formula 266
26. Endoscopic lifting 275
27. Special unipotent representations 295
References 311
Index 315

Acknowledgments

All three authors have been supported in part by grants from the National Science Foundation, most recently DMS-8802586, DMS-9104117, and DMS-9011483. Jeffrey Adams is a Sloan Research Fellow.

It is a great pleasure to thank a few of the many mathematicians who helped us to write a book we barely feel competent to read. Mike Artin and Steve Kleiman led us patiently through the algebraic geometry, answering hundreds of foolish questions as well as several serious ones. Bob MacPherson, Kari Vilonen, Victor Ginsburg, and Jean-Luc Brylinski taught us about perverse sheaves, $\mathcal{D}$-modules, and microlocal geometry. George Lusztig knows everything about reductive groups, and so does Bert Kostant; they have helped us through many treacherous swamps.

The specific form of the results we owe to Martin Andler. He suggested that something like Proposition 1.11 had to be true; this led to the definition of the geometric parameter space, and so made everything else possible. He has also offered many corrections and improvements to the manuscript.

Most of all, we are grateful to Jim Arthur. In papers, in lectures, and in conversation, he has opened up for us some of the mysteries of automorphic representation theory. His gentle and encouraging manner allowed us to persist through long periods when we felt (no doubt correctly) that we were in far over our heads. Working with his ideas has been a privilege mathematically and personally.

Index of notation

We list here some of our principal notation and terminology, with a very brief summary of definitions and references to more complete discussion.

$A_\phi^{loc,alg}$: group of connected components of the centralizer in $^\vee G^{alg}$ of a Langlands parameter ϕ (Definition 5.11). A Langlands parameter ϕ gives a point x of the geometric parameter space $X(^\vee G^\Gamma)$ (Proposition 6.17), which generates an orbit S under $^\vee G$. The $^\vee G^{alg}$-equivariant fundamental group of S based at x (Definition 7.1) is written $A_S^{loc,alg}$ or $A_x^{loc,alg}$; it is naturally isomorphic to $A_\phi^{loc,alg}$ (Lemma 7.5).

$A_\phi^{mic,alg}$: micro-component group for a Langlands parameter ϕ (Definition 23.13). Write S for the orbit on $X(^\vee G^\Gamma)$ corresponding to ϕ, and (x,ν) for a sufficiently generic conormal vector to S at $x \in S$. Then $A_\phi^{mic,alg}$ is the group of connected components of the stablizer of (x,ν) in $^\vee G^{alg}$.

A_ψ^{alg}: group of connected components of the centralizer in $^\vee G^{alg}$ of an Arthur parameter ψ (Definition 21.4). Naturally isomorphic to the micro-component group $A_{\phi_\psi}^{mic,alg}$ for the corresponding Langlands parameter (Proposition 21.9 and Definition 23.7).

Ch($\mathcal{M}$): characteristic cycle for a $\mathcal{D}$-module $\mathcal{M}$ (Definition 19.9).

$d(\xi)$: dimension of the underlying orbit for a geometric parameter ξ ((1.22),(7.10)).

$e(G_\mathbb{R})$, $e(\xi)$: Kottwitz' sign attached to a complete geometric parameter or to the corresponding real form of G (Definition 15.8, Lemma 15.9).

$e(\cdot)$: normalized exponential map $\exp(2\pi i \cdot)$; usually from semisimple elements or canonical flats in $^\vee\mathfrak{g}$ ((6.2), Lemma 6.11).

G: connected complex reductive algebraic group.

G^Γ: weak extended group for G (Definition 2.13).

$(G^\Gamma, \mathcal{W})$: extended group for G (Definition 1.12, Proposition 3.6).

$G(\mathbb{R}, \delta)$: real points of G with real form defined by $\delta \in G^\Gamma - G$ (Definition 2.13).

$^\vee G$: dual group for G (Definition 4.2).

$^\vee G^\Gamma$: weak L-group or E-group for G; an algebraic extension of the Galois group of $\mathbb{C}/\mathbb{R}$ by a dual group (Definition 4.3).

$(^\vee G^\Gamma, \mathcal{D})$: L-group or E-group for G, essentially as defined by Langlands (in Galois form); $\mathcal{D}$ is a class of distinguished splittings.

$(^\vee G^\Gamma, \mathcal{S})$: L-group or E-group for G, as defined in [1] (Definition 4.14). Equivalent to giving an L-group or E-group as in the preceding definition, but the set $\mathcal{S}$ is different from $\mathcal{D}$.

$\mathcal{H}(\mathcal{O})$: Hecke algebra attached to a regular semisimple conjugacy class $\mathcal{O}$ in $^\vee\mathfrak{g}$ ((16.10)).

$KX(\mathcal{O}, ^\vee G^\Gamma)$: Grothendieck group of $^\vee G^{alg}$-equivariant perverse sheaves (or constructible sheaves, or holonomic $\mathcal{D}$-modules) on $X(\mathcal{O}, ^\vee G^\Gamma)$ (Definition 7.13, (7.10)).

$K\Pi^z(\mathcal{O}, G_\mathbb{R})$: Grothendieck group of finite-length canonical projective representations of $G_\mathbb{R}$ of type z and infinitesimal character $\mathcal{O}$ ((15.5)). Without the $\mathcal{O}$, no assumption on infinitesimal character. With $G/\mathbb{R}$, the sum of these over all strong real forms ((15.7)). With $\overline{K}$, infinite sums (still with finite multiplicities) are allowed (Corollary 1.26, (16.12)). With $\overline{K}_f$, these infinite sums are required to include only finitely many representations of each strong real form (Definition 18.5). With $\overline{K}_\mathbb{C}$, complex-linear combinations of irreducible representations are allowed (Proposition 25.1).

$\mathcal{K}\Pi^z(\mathcal{O}, G/\mathbb{R})$: mixed Grothendieck group of finite-length canonical projective representations of strong real forms of G of type z and infinitesimal character $\mathcal{O}$ ((16.12)). (Formally, $\mathcal{K}$ denotes extension of scalars to $\mathbb{Z}[u^{1/2}, u^{-1/2}]$.)

$l^I(\Lambda)$, $l^I(\xi)$: integral length for a standard limit character of regular infinitesimal character, or for the corresponding complete geometric parameter (Definition 16.16).

$L^z(G_\mathbb{R})$: equivalence classes of final standard limit characters of Cartan subgroups of $G_\mathbb{R}$ (Definition 11.13).

$L^z(G/\mathbb{R})$: equivalence classes of final standard limit characters of Cartan subgroups of G^Γ (Definition 12.1).

Lift$_0$: endoscopic lifting from stable virtual (canonical projective representations) representations of a quasisplit endoscopic group to complex virtual (canonical projective) representations of strong real forms of G (Definition 26.18). Without the subscript zero, its (technically more convenient) natural extension to "σ_H-stable" virtual representations of strong real forms of H.

$\mathcal{M}^z(G_\mathbb{R})$, $\mathcal{M}^z(G/\mathbb{R})$: category of (canonical projective of type z) representations of a real reductive group $G_\mathbb{R}$, or of strong real forms of

G^Γ ((15.2) and (15.7)).

$M(\Lambda)$, $M(\xi)$: standard limit representation attached to a limit character Λ, or to the corresponding complete geometric parameter ξ (Definition 11.2, (15.7)).

$\mathcal{O}$: semisimple $^\vee G$-conjugacy class in $^\vee \mathfrak{g}$; parametrizes an infinitesimal character for $\mathfrak{g}$. Inserted in other notation, it restricts attention to objects living on that orbit, or having that infinitesimal character ((6.10), Proposition 6.16, Lemma 15.4).

$SS(\mathcal{M})$: characteristic variety (or singular support) of the $\mathcal{D}$-module $\mathcal{M}$ (Definition 19.9).

T^Γ: Cartan subgroup of G^Γ (Definition 12.1).

$X(^\vee G^\Gamma)$: set of geometric parameters for the E-group $^\vee G^\Gamma$ (Definition 6.9). With $\mathcal{O}$, the subset of infinitesimal character $\mathcal{O}$ (an algebraic variety).

Λ: limit character of a Cartan subgroup of a real reductive group; parameter for a standard representation (Definitions 11.2, 11.6, 11.10, 11.13; Definition 12.1).

$\chi_\phi^{mic}(\pi)(\sigma)$: trace at $\sigma \in A_\phi^{mic,alg}$ of the representation $\tau_\phi^{mic,alg}$ (Definition 23.13). Without the argument σ, the dimension of this representation, which is equal to the multiplicity of the conormal bundle to the orbit S_ϕ in the characteristic cycle of the perverse sheaf attached to π (Definition 19.13, Theorem 23.8).

η_ϕ^{loc}: strongly stable standard formal virtual representation of strong real forms of G attached to a Langlands parameter ϕ (Definition 18.13). We may replace ϕ by the corresponding $^\vee G$-orbit S on the geometric parameter space $X(^\vee G^\Gamma)$.

η_ϕ^{mic}, η_S^{mic}: strongly stable formal virtual representation of strong real forms of G attached to a Langlands parameter ϕ or to the corresponding orbit S on the geometric parameter space $X(^\vee G^\Gamma)$ (Corollary 19.16, Corollary 1.32).

ξ: complete geometric parameter $(S_\xi, \mathcal{V}_\xi)$. Here S_ξ is an orbit of $^\vee G$ on the geometric parameter space $X(^\vee G^\Gamma)$, and $\mathcal{V}_\xi$ is an irreducible equivariant local system on S_ξ (Definition 1.17, Definition 7.6).

$\pi(\xi)$: irreducible representation corresponding to the complete geometric parameter ξ (Theorem 10.4, Definition 10.8, and (1.19)). This is a representation of $G(\mathbb{R}, \delta(\xi))^{can}$.

$\pi(\Lambda)$: Langlands quotient of the standard representation $M(\Lambda)$ (Definition 11.2(e), Theorem 11.14).

$\Pi^z(G/\mathbb{R})$: equivalence classes of pairs (π, δ), with δ a strong real form (for a fixed extended group G^Γ) and π an irreducible representation of $G(\mathbb{R},\delta)^{can}$ of type z (Definition 10.3, Definition 1.14). This set is in one-to-one correspondence with $\Xi^z(G/\mathbb{R})$.

$\Pi^z(G/\mathbb{R})_\phi$: the L-packet attached to the Langlands parameter ϕ; that is, the subset of $\Pi^z(G/\mathbb{R})$ parametrized by complete Langlands parameters of the form (ϕ,τ), or by complete geometric parameters supported on the orbit attached to ϕ (Definition 10.8). Includes at most finitely many representations of each strong real form.

$\Pi^z(G/\mathbb{R})_\phi^{mic}$: the micro-packet attached to the Langlands parameter ϕ (Definition 19.15); the set of irreducible constituents of the stable virtual representation η_ϕ^{mic}. Contains the L-packet.

$\Xi(^\vee G^\Gamma)$: complete geometric parameters; that is, pairs $\xi = (S, \mathcal{V})$ with S an orbit of $^\vee G$ on the geometric parameter space $X(^\vee G^\Gamma)$, and $\mathcal{V}$ an irreducible $^\vee G^{alg}$-equivariant local system on S (Definition 7.6). Equivalently, complete Langlands parameters; that is, equivalence classes of pairs (ϕ, τ), with ϕ a Langlands parameter and τ an irreducible representation of $A_\phi^{loc,alg}$ (Definition 5.11). When $^\vee G^\Gamma$ has the structure of an E-group with second invariant z, we may write instead $\Xi^z(G/\mathbb{R})$.

$\Phi(^\vee G^\Gamma)$: equivalence classes of Langlands parameters; that is, $^\vee G$-conjugacy classes of quasiadmissible homomorphisms ϕ from $W_\mathbb{R}$ to $^\vee G^\Gamma$ (Definition 5.2). Equivalently, equivalence classes of geometric parameters; that is, $^\vee G$-orbits on $X(^\vee G^\Gamma)$ (Definition 7.6). When $^\vee G^\Gamma$ has the structure of an E-group with second invariant z, we write instead $\Phi^z(G/\mathbb{R})$.

$\Psi(^\vee G^\Gamma)$: equivalence classes of Arthur parameters; that is, $^\vee G$-conjugacy classes of homomorphisms from $W_\mathbb{R} \times SL(2,\mathbb{C})$ satisfying some additional conditions (Definition 21.4). We sometimes write $\Psi^z(G/\mathbb{R})$ as for Langlands parameters.

$\tau_\phi^{mic}(\pi)$: representation (possibly reducible) of $A_\phi^{mic,alg}$ attached to an irreducible representation π of a strong real form of G (Definition 23.13). Non-zero if and only if $\pi \in \Pi^z(G/\mathbb{R})_\phi^{mic}$.

$\Theta(\eta)$: distribution character of the virtual representation η ((18.1)). When the argument includes a strong real form δ, η is a (locally finite) virtual representation of all strong real forms, and $\Theta(\eta,\delta)$ is the character of those defined on the group $G(\mathbb{R},\delta)^{can}$.

$\Theta_{SR}(\eta)$: smooth function on the strongly regular elements obtained by restricting the character of η.

1. Introduction

In [2] and [3], Arthur has formulated a number of conjectures about automorphic forms. These conjectures would have profound consequences for the unitary representation theory of the group $G(\mathbb{R})$ of real points of a connected reductive algebraic group G defined over $\mathbb{R}$. Our purpose in this book is to establish a few of these local consequences. In order to do that, we have been led to combine the ideas of Langlands and Shelstad (concerning dual groups and endoscopy) with those of Kazhdan and Lusztig (concerning the fine structure of irreducible representations).

We will recall Arthur's conjectures in detail in Chapters 22 and 26, but for the moment it is enough to understand their general shape. We begin by recalling the form of the Langlands classification. Define

$$\Pi(G(\mathbb{R})) \supset \Pi_{\mathrm{unit}}(G(\mathbb{R})) \supset \Pi_{\mathrm{temp}}(G(\mathbb{R})) \tag{1.1}$$

to be the set of equivalence classes of irreducible admissible (respectively unitary or tempered) representations of $G(\mathbb{R})$. Now define

$$\Phi(G(\mathbb{R})) \supset \Phi_{\mathrm{temp}}(G(\mathbb{R})) \tag{1.2}$$

to be the set of Langlands parameters for irreducible admissible (respectively tempered) representations of $G(\mathbb{R})$ (see [34], [10], [1], Chapter 5, and Definition 22.3). To each $\phi \in \Phi(G(\mathbb{R}))$, Langlands attaches a finite set $\Pi_\phi \subset \Pi(G(\mathbb{R}))$, called an *L-packet of representations*. The L-packets Π_ϕ partition $\Pi(G(\mathbb{R}))$. If $\phi \in \Phi_{\mathrm{temp}}(G(\mathbb{R}))$, then the representations in Π_ϕ are all tempered, and in this way one gets also a partition of $\Pi_{\mathrm{temp}}(G(\mathbb{R}))$.

Now the classification of the unitary representations of $G(\mathbb{R})$ is one of the most interesting unsolved problems in harmonic analysis. Langlands' results immediately suggest that one should look for a set between $\Phi(G(\mathbb{R}))$ and $\Phi_{\mathrm{temp}}(G(\mathbb{R}))$ parametrizing exactly the unitary representations. Unfortunately, nothing quite so complete is possible: Knapp has found examples in which some members of the set Π_ϕ are unitary and some are not.

The next most interesting possibility is to describe a set of parameters giving rise to a large (but incomplete) family of unitary representations. This is the local aim of Arthur's conjectures. A little more precisely, Arthur defines a new set

$$\Psi(G/\mathbb{R}) \qquad (1.3)(a)$$

of parameters (Definition 22.4). (We write $G/\mathbb{R}$ rather than $G(\mathbb{R})$ because Arthur's parameters depend only on an inner class of real forms, and not on one particular real form.) Now assume that $G(\mathbb{R})$ is quasisplit. Then Arthur defines an inclusion

$$\Psi(G/\mathbb{R}) \hookrightarrow \Phi(G(\mathbb{R})), \quad \psi \mapsto \phi_\psi. \qquad (1.3)(b)$$

Write $\Phi_{\text{Arthur}}(G(\mathbb{R}))$ for the image of this inclusion. Then

$$\Phi(G(\mathbb{R})) \supset \Phi_{\text{Arthur}}(G(\mathbb{R})) \supset \Phi_{\text{temp}}(G(\mathbb{R})). \qquad (1.3)(c)$$

Roughly speaking, Arthur proposes that $\Psi(G/\mathbb{R})$ should parametrize all the unitary representations of $G(\mathbb{R})$ that are of interest for global applications. More specifically, he proposed the following problems (still for $G(\mathbb{R})$ quasisplit at first).

Problem A. Associate to each parameter $\psi \in \Psi(G/\mathbb{R})$ a finite set $\Pi_\psi \subset \Pi(G(\mathbb{R}))$. This set (which we might call an *Arthur packet*) should contain the L-packet Π_{ϕ_ψ} (cf. (1.3)(b)) and should have other nice properties, some of which are specified below.

The Arthur packet will not in general turn out to be a union of L-packets; so we cannot hope to define it simply by attaching some additional Langlands parameters to ψ.

Associated to each Arthur parameter is a certain finite group A_ψ (Definition 21.4).

Problem B. Associate to each $\pi \in \Pi_\psi$ a non-zero finite-dimensional representation $\tau_\psi(\pi)$ of A_ψ.

Problem C. Show that the distribution on $G(\mathbb{R})$

$$\sum_{\pi \in \Pi_\psi} (\epsilon_\pi \dim(\tau_\psi(\pi))) \Theta(\pi)$$

is a stable distribution in the sense of Langlands and Shelstad ([35], [48]). Here $\epsilon_\pi = \pm 1$ is also to be defined.

Problem D. Prove analogues of Shelstad's theorems on lifting tempered characters (cf. [48]) in this setting.

Problem E. Extend the definition of Π_ψ to non-quasisplit G, in a manner consistent with appropriate generalizations of Problems B, C, and D.

1. Introduction

Problem F. Show that every representation $\pi \in \Pi_\psi$ is unitary.

We give here complete solutions of problems A, B, C, D, and E. Our methods offer no information about Problem F. (In that direction the best results are those of [5], where Problem F is solved for complex classical groups.)

The central idea of the proofs is by now a familiar one in the representation theory of reductive groups. It is to describe the representations of $G(\mathbb{R})$ in terms of an appropriate geometry on an L-group. So let $^\vee G$ be the (complex reductive) dual group of G, and $^\vee G^\Gamma$ the (Galois form of the) L-group attached to the real form $G(\mathbb{R})$. The L-group is a complex Lie group, and we have a short exact sequence

$$1 \to {}^\vee G \to {}^\vee G^\Gamma \to \operatorname{Gal}(\mathbb{C}/\mathbb{R}) \to 1 \qquad (1.4)(a)$$

(The complete definition of the L-group is recalled in Chapter 4.) We also need the Weil group $W_\mathbb{R}$ of $\mathbb{C}/\mathbb{R}$; this is a real Lie group, and there is a short exact sequence

$$1 \to \mathbb{C}^\times \to W_\mathbb{R} \to \operatorname{Gal}(\mathbb{C}/\mathbb{R}) \to 1. \qquad (1.4)(b)$$

(The Weil group is not a complex Lie group because the action of the Galois group on $\mathbb{C}^\times$ is the non-trivial one, which does not preserve the complex structure.)

Definition 1.5 ([34], [10]). A *quasiadmissible homomorphism* ϕ from $W_\mathbb{R}$ to $^\vee G^\Gamma$ is a continuous group homomorphism satisfying

(a) ϕ respects the homomorphisms to $\operatorname{Gal}(\mathbb{C}/\mathbb{R})$ defined by (1.4); and
(b) $\phi(\mathbb{C}^\times)$ consists of semisimple elements of $^\vee G$.

(Langlands' notion of "admissible homomorphism" includes an additional "relevance" hypothesis on ϕ, which will not concern us. This additional hypothesis is empty if $G(\mathbb{R})$ is quasisplit.) Define

$$P(^\vee G^\Gamma) = \{\, \phi : W_\mathbb{R} \to {}^\vee G^\Gamma \mid \phi \text{ is quasiadmissible}\,\}.$$

Clearly $^\vee G$ acts on $P(^\vee G^\Gamma)$ by conjugation on the range of a homomorphism, and we define

$$\Phi(G/\mathbb{R}) = \{\, {}^\vee G \text{ orbits on } P(^\vee G^\Gamma)\,\}.$$

(If $G(\mathbb{R})$ is quasisplit, this is precisely the set of parameters in (1.2). In general Langlands omits the "irrelevant" orbits.)

Now a homomorphism ϕ is determined by the value of its differential on a basis of the real Lie algebra of $\mathbb{C}^\times$, together with its value at a single specified element of the non-identity component of $W_\mathbb{R}$; that is, by an element of the complex manifold ${}^\vee\mathfrak{g} \times {}^\vee\mathfrak{g} \times {}^\vee G^\Gamma$. The conditions (a) and (b) of Definition 1.5 amount to requiring the first two factors to be semisimple, and the third to lie in the non-identity component. Requiring that these elements define a group homomorphism imposes a finite number of complex-analytic relations, such as commutativity of the first two factors. Pursuing this analysis, we will prove in Chapter 5

Proposition 1.6. *Suppose ${}^\vee G^\Gamma$ is an L-group. The set $P({}^\vee G^\Gamma)$ of quasiadmissible homomorphisms from $W_\mathbb{R}$ into ${}^\vee G^\Gamma$ may be identified with the set of pairs (y, λ) satisfying the following conditions:*
a) $y \in {}^\vee G^\Gamma - {}^\vee G$, and $\lambda \in {}^\vee\mathfrak{g}$ is a semisimple element;
b) $y^2 = \exp(2\pi i \lambda)$; and
c) $[\lambda, \mathrm{Ad}(y)\lambda] = 0$.

The Langlands classification described after (1.2) is thus already geometric: L-packets are parametrized by the orbits of a reductive group acting on a topological space. Subsequent work of Langlands and Shelstad supports the importance of this geometry. For example, one can interpret some of the results of [48] as saying that the L-packet Π_ϕ may be parametrized using ${}^\vee G$-equivariant local systems on the ${}^\vee G$ orbit of ϕ.

By analogy with the theory created by Kazhdan-Lusztig and Beilinson-Bernstein in [28] and [8], one might hope that information about irreducible characters is encoded by perverse sheaves on the closures of ${}^\vee G$-orbits on $P({}^\vee G^\Gamma)$. Unfortunately, it turns out that the orbits are already closed, so these perverse sheaves are nothing but the local systems mentioned above. On the other hand, one can often parametrize the orbits of several rather different group actions using the same parameters; so we sought a different space with a ${}^\vee G$ action, having the same set of orbits as $P({}^\vee G^\Gamma)$, but with a more interesting geometry.

In order to define our new space, we need some simple structure theory for reductive groups. (This will be applied in a moment to ${}^\vee G$.)

Definition 1.7. Suppose H is a complex reductive group, with Lie algebra $\mathfrak{h}$, and $\lambda \in \mathfrak{h}$ is a semisimple element. Set

$$\mathfrak{h}(\lambda)_n = \{ \mu \in \mathfrak{h} \mid [\lambda, \mu] = n\mu \} \quad (n \in \mathbb{Z}) \tag{1.7)(a}$$

$$\mathfrak{n}(\lambda) = \sum_{n=1,2,\ldots} \mathfrak{h}(\lambda)_n \tag{1.7)(b}$$

$$e(\lambda) = \exp(2\pi i \lambda) \in H. \tag{1.7)(c}$$

1. Introduction

The *canonical flat through* λ is the affine subspace

$$\mathcal{F}(\lambda) = \lambda + \mathfrak{n}(\lambda) \subset \mathfrak{h}. \qquad (1.7)(d)$$

We will see in Chapter 6 that the canonical flats partition the semisimple elements of $\mathfrak{h}$ — in fact they partition each conjugacy class — and that the map e is constant on each canonical flat. If Λ is a canonical flat, we may therefore write

$$e(\Lambda) = \exp(2\pi i \lambda) \qquad \text{(any } \lambda \in \Lambda\text{)}. \qquad (1.7)(e)$$

Finally, write $\mathcal{F}(\mathfrak{h})$ for the set of all canonical flats in $\mathfrak{h}$.

Definition 1.8 Suppose $^\vee G^\Gamma$ is the L-group of a real reductive group (cf. (1.4)). The *geometric parameter space for* $^\vee G^\Gamma$ is the set

$$X = X(^\vee G^\Gamma) = \{\, (y,\Lambda) \mid y \in {^\vee G^\Gamma} - {^\vee G},\ \Lambda \in \mathcal{F}(^\vee\mathfrak{g}),\ y^2 = e(\Lambda) \,\}.$$

This is our proposed substitute for Langlands' space $P(^\vee G^\Gamma)$. The set $\mathcal{F}(^\vee\mathfrak{g})$ is difficult to topologize nicely, as one can see already for $SL(2)$; this difficulty is inherited by X. To make use of geometric methods we will always restrict to the subspaces appearing in the following lemma.

Lemma 1.9 (cf. Proposition 6.16 below). *In the setting of Definition 1.8, fix a single orbit $\mathcal{O}$ of $^\vee G$ on the semisimple elements of $^\vee\mathfrak{g}$, and set*

$$X(\mathcal{O}, {^\vee G^\Gamma}) = \{\, (y,\Lambda) \in X \mid \Lambda \subset \mathcal{O} \,\}.$$

Then $X(\mathcal{O}, {^\vee G^\Gamma})$ has in a natural way the structure of a smooth complex algebraic variety, on which $^\vee G$ acts with a finite number of orbits.

The variety $X(\mathcal{O}, {^\vee G^\Gamma})$ need not be connected or equidimensional, but this will cause no difficulties. We topologize X by making the subsets $X(\mathcal{O}, {^\vee G^\Gamma})$ open and closed. (It seems likely that a more subtle topology will be important for harmonic analysis, as soon as continuous families of representations are involved.)

Of course the first problem is to check that the original Langlands classification still holds.

Proposition 1.10. (cf. Proposition 6.17 below). *Suppose $^\vee G^\Gamma$ is an L-group. Then there is a natural $^\vee G$-equivariant map*

$$p : P(^\vee G^\Gamma) \to X(^\vee G^\Gamma), \quad p(y,\lambda) = (y, \mathcal{F}(\lambda))$$

inducing a bijection on the level of $^\vee G$-orbits. The fibers of p are principal homogeneous spaces for unipotent algebraic groups. More precisely, suppose $x = p(\phi)$. Then the isotropy group $^\vee G_\phi$ is a Levi subgroup of $^\vee G_x$.

This proposition shows that (always locally over $\mathbb{R}$!) the geometric parameter space X shares all the formal properties of $P(^\vee G^\Gamma)$ needed for the Langlands classification. In particular, if $G(\mathbb{R})$ is quasisplit, L-packets in $\Pi(G(\mathbb{R}))$ are parametrized precisely by $^\vee G$-orbits on X. What has changed is that the orbits on the new space X are not closed; so the first new question to consider is the meaning of the closure relation.

Proposition 1.11. *Suppose $G(\mathbb{R})$ is quasisplit. Let ϕ, $\phi' \in \Phi(G/\mathbb{R})$ be two Langlands parameters, and S, $S' \subset X$ the corresponding $^\vee G$-orbits. Then the following conditions are equivalent:*
(i) S is contained in the closure of S'.
(ii) there are irreducible representations $\pi \in \Pi_\phi$ and $\pi' \in \Pi_{\phi'}$ with the property that π' is a composition factor of the standard representation of which π is the unique quotient.
If ϕ is a tempered parameter, then the orbit S is open in the variety $X(\mathcal{O}, ^\vee G^\Gamma)$ containing it (cf. Lemma 1.9).

(In the interest of mathematical honesty, we should admit that this result is included only for expository purposes; we will not give a complete proof. That (ii) implies (i) (even for $G(\mathbb{R})$ not quasisplit) follows from Corollary 1.25(b) and (7.11). The other implication in the quasisplit case can be established by a subtle and not very interesting trick. The last assertion follows from Proposition 22.9(b) (applied to an Arthur parameter with trivial $SL(2)$ part).)

Proposition 1.11 suggests the possibility of a deeper relationship between irreducible representations and the geometry of orbit closures on X. To make the cleanest statements, we need to introduce some auxiliary ideas. (These have not been emphasized in the existing literature on the Langlands classification, because they reflect phenomena over $\mathbb{R}$ that are non-existent or uninteresting globally.) The reader should assume at first that G is adjoint. In that case the notion of "strong real form" introduced below amounts to the usual notion of real form, and the "algebraic universal covering" of $^\vee G$ is trivial.

Definition 1.12. Suppose G is a complex connected reductive algebraic group. An *extended group* for $G/\mathbb{R}$ is a pair $(G^\Gamma, \mathcal{W})$, subject to the following conditions.
(a) G^Γ is a real Lie group containing G as a subgroup of index two, and every element of $G^\Gamma - G$ acts on G (by conjugation) by antiholomorphic automorphisms.

(b) $\mathcal{W}$ is a G-conjugacy class of triples (δ, N, χ), with
  (1) The element δ belongs to $G^\Gamma - G$, and $\delta^2 \in Z(G)$ has finite order. (Write $\sigma = \sigma(\delta)$ for the conjugation action of δ on G, and $G(\mathbb{R})$ or $G(\mathbb{R}, \delta)$ for the fixed points of σ; this is a real form of G.)
  (2) $N \subset G$ is a maximal unipotent subgroup, and δ normalizes N. (Then N is defined over $\mathbb{R}$; write $N(\mathbb{R}) = N(\mathbb{R}, \delta)$ for the subgroup of real points.)
  (3) The element χ is a one-dimensional non-degenerate unitary character of $N(\mathbb{R})$. (Here "non-degenerate" means non-trivial on each simple restricted root subgroup of N.)

We will discuss this definition in more detail in Chapters 2 and 3. For now it suffices to know that each inner class of real forms of G gives rise to an extended group. The groups $G(\mathbb{R})$ appearing in the definition are quasisplit (because of (b)(2)) and the pair $(N(\mathbb{R}), \chi)$ is the set of data needed to define a Whittaker model for $G(\mathbb{R})$.

Definition 1.13. Suppose $(G^\Gamma, \mathcal{W})$ is an extended group. A *strong real form* of $(G^\Gamma, \mathcal{W})$ (briefly, of G) is an element $\delta \in G^\Gamma - G$ such that $\delta^2 \in Z(G)$ has finite order. Given such a δ, we write $\sigma = \sigma(\delta)$ for its conjugation action on G, and

$$G(\mathbb{R}) = G(\mathbb{R}, \delta)$$

for the fixed points of σ. Two strong real forms δ and δ' are called *equivalent* if they are conjugate by G; we write $\delta \sim \delta'$. (The elements δ of Definition 1.12 constitute a single equivalence class of strong real forms, but in general there will be many others.)

The usual notion of a real form can be described as an antiholomorphic involution σ of G. Two such are equivalent if they differ by the conjugation action of G. This is exactly the same as our definition if G is adjoint. The various groups $G(\mathbb{R}, \delta)$ (for δ a strong real form of $(G^\Gamma, \mathcal{W})$) constitute exactly one inner class of real forms of G.

Definition 1.14. Suppose $(G^\Gamma, \mathcal{W})$ is an extended group. A *representation of a strong real form* of $(G^\Gamma, \mathcal{W})$ (briefly, of G) is a pair (π, δ), subject to
(a) δ is a strong real form of $(G^\Gamma, \mathcal{W})$ (Definition 1.13); and
(b) π is an admissible representation of $G(\mathbb{R}, \delta)$.

Two such representations (π, δ) and (π', δ') are said to be *(infinitesimally) equivalent* if there is an element $g \in G$ such that $g\delta g^{-1} = \delta'$, and $\pi \circ \mathrm{Ad}(g^{-1})$ is (infinitesimally) equivalent to π'. (In particular, this is possible only if the strong real forms are equivalent.) Finally, define

$$\Pi(G^\Gamma, \mathcal{W}) = \Pi(G/\mathbb{R})$$

to be the set of (infinitesimal) equivalence classes of irreducible representations of strong real forms of G.

Lemma 1.15. *Suppose $(G^\Gamma, \mathcal{W})$ is an extended group for G (Definition 1.12). Choose representatives $\{\delta_s \mid s \in \Sigma\}$ for the equivalence classes of strong real forms of G (Definition 1.13). Then the natural map from left to right induces a bijection*

$$\coprod_{s \in \Sigma} \Pi(G(\mathbb{R}, \delta_s)) \simeq \Pi(G/\mathbb{R})$$

(Definition 1.14; the set on the left is a disjoint union).

This lemma is an immediate consequence of the definitions; we will give the argument in Chapter 2.

The set $\Pi(G/\mathbb{R})$ is the set of representations we wish to parametrize. To do so requires one more definition on the geometric side.

Definition 1.16. Suppose ${}^\vee G^\Gamma$ is the L-group of the inner class of real forms represented by the extended group G^Γ (cf. (1.4) and Definition 1.13). The *algebraic universal covering* ${}^\vee G^{alg}$ is the projective limit of all the finite coverings of ${}^\vee G$. This is a pro-algebraic group, of which each finite-dimensional representation factors to some finite cover of ${}^\vee G$.

With the algebraic universal covering in hand, we can define a complete set of geometric parameters for representations.

Definition 1.17. Suppose G is a connected reductive algebraic group endowed with an inner class of real forms, and ${}^\vee G^\Gamma$ is a corresponding L-group for G. A *complete geometric parameter* for G is a pair

$$\xi = (S, \mathcal{V}),$$

where

(a) S is an orbit of ${}^\vee G$ on $X({}^\vee G^\Gamma)$ (Definition 1.8); and
(b) $\mathcal{V}$ is an irreducible ${}^\vee \widetilde{G}$-equivariant local system on S, for some finite covering ${}^\vee \widetilde{G}$ of ${}^\vee G$.

We may write $(S_\xi, \mathcal{V}_\xi)$ to emphasize the dependence on ξ. In (b), it is equivalent to require $\mathcal{V}$ to be ${}^\vee G^{alg}$-equivariant. Write $\Xi(G/\mathbb{R})$ for the set of all complete geometric parameters.

1. Introduction

A slightly different formulation of this definition is sometimes helpful. Fix a $^\vee G$-orbit S on X, and a point $x \in S$. Write $^\vee G_x^{alg}$ for the stabilizer of x in $^\vee G^{alg}$, and define

$$A_S^{loc,alg} = {^\vee G_x^{alg}} / \left({^\vee G_x^{alg}}\right)_0$$

for its (pro-finite) component group. We call $A_S^{loc,alg}$ the *equivariant fundamental group of S*; like a fundamental group, it is defined only up to inner automorphism (because of its dependence on x). Representations of $A_S^{loc,alg}$ classify equivariant local systems on S, so we may also define a complete geometric parameter for G as a pair

$$\xi = (S, \tau),$$

where

(a) S is an orbit of $^\vee G$ on $X(^\vee G^\Gamma)$; and
(b) τ is an irreducible representation of $A_S^{loc,alg}$.

Again we may write (S_ξ, τ_ξ).

Theorem 1.18. *Suppose $(G^\Gamma, \mathcal{W})$ is an extended group for G (Definition 1.12), and $^\vee G^\Gamma$ is an L-group for the corresponding inner class of real forms. Then there is a natural bijection between the set $\Pi(G/\mathbb{R})$ of equivalence classes of irreducible representations of strong real forms of G (Definition 1.14), and the set $\Xi(G/\mathbb{R})$ of complete geometric parameters for G (Definition 1.17). In this parametrization, the set of representations of a fixed real form $G(\mathbb{R})$ corresponding to complete geometric parameters supported on a single orbit is precisely the L-packet for $G(\mathbb{R})$ attached to that orbit (Proposition 1.10).*

As we remarked after Proposition 1.6, one can find results of this nature in [48].

For each complete geometric parameter ξ, we define (using Theorem 1.18 and Definition 1.14)

$$(\pi(\xi), \delta(\xi)) = \text{ some irreducible representation parametrized by } \xi$$
$$(1.19)(a)$$

$$M(\xi) = \text{ standard representation with Langlands quotient } \pi(\xi).$$
$$(1.19)(b)$$

As a natural setting in which to study character theory, we will also use

$$K\Pi(G/\mathbb{R}) = \text{ free } \mathbb{Z}\text{-module with basis } \Pi(G/\mathbb{R}). \quad (1.19)(c)$$

We will sometimes call this the *lattice of virtual characters*. One can think of it as a Grothendieck group of an appropriate category of representations of strong real forms. In particular, any such representation ρ has a well-defined image

$$[\rho] \in K\Pi(G/\mathbb{R}).$$

By abuse of notation, we will usually drop the brackets, writing for example $M(\xi) \in K\Pi(G/\mathbb{R})$. (All of these definitions are discussed in somewhat more depth in Chapters 11 and 15.)

In order to write character formulas, we will also need a slight variant on the notation of (1.19)(a). Fix a strong real form δ of G, and a complete geometric parameter ξ. By the proof of Lemma 1.15, there is at most one irreducible representation π of $G(\mathbb{R}, \delta)$ so that (π, δ) is equivalent to $(\pi(\xi), \delta(\xi))$. We define

$$\pi(\xi, \delta) = \pi. \qquad (1.19(d)$$

If no such representation π exists, then we define

$$\pi(\xi, \delta) = 0. \qquad (1.19)(e)$$

Similarly we can define $M(\xi, \delta)$.

Lemma 1.20 (Langlands — see [54], [56]). *The (image in $K\Pi(G/\mathbb{R})$ of the) set*

$$\{ M(\xi) \mid \xi \in \Xi(G/\mathbb{R}) \}$$

is a basis for $K\Pi(G/\mathbb{R})$.

Because the standard representations of real groups are fairly well understood, it is natural to try to describe the irreducible representations in terms of the standard ones. On the level of character theory, this means relating the two bases $\{\pi(\xi)\}$ and $\{M(\xi)\}$ of $K\Pi(G/\mathbb{R})$:

$$M(\xi) = \sum_{\gamma \in \Xi} m_r(\gamma, \xi) \pi(\gamma). \qquad (1.21)$$

(The subscript r stands for "representation-theoretic," and is included to distinguish this matrix from an analogous one to be introduced in Definition 1.22.) Here the *multiplicity matrix* $m_r(\gamma, \xi)$ is what we want. The Kazhdan-Lusztig conjectures (now proved) provide a way to compute the multiplicity matrix, and a geometric interpretation of it —

the "Beilinson-Bernstein picture" of [8]. Unfortunately, this geometric interpretation is more complicated than one would like in the case of non-integral infinitesimal character, and it has some fairly serious technical shortcomings in the case of singular infinitesimal character. (What one has to do is compute first at nonsingular infinitesimal character, then apply the "translation principle." The translation principle can introduce substantial cancellations, which are not easy to understand in the Beilinson-Bernstein picture.) We have therefore sought a somewhat different geometric interpretation of the multiplicity matrix. Here are the ingredients. (A more detailed discussion appears in Chapter 7.)

Definition 1.22. Suppose Y is a complex algebraic variety on which the pro-algebraic group H acts with finitely many orbits. Define

$$\mathcal{C}(Y, H) = \text{category of } H\text{-equivariant constructible sheaves on } Y.$$
$$(1.22)(a)$$

$$\mathcal{P}(Y, H) = \text{category of } H\text{-equivariant perverse sheaves on } Y.$$
$$(1.22)(b)$$

(For the definition of perverse sheaves we refer to [9]. The definition of H-equivariant requires some care in the perverse case; see [38], section 0, or [39], (1.9.1) for the case of connected H.) Each of these categories is abelian, and every object has finite length. (One does not ordinarily expect the latter property in a category of constructible sheaves; it is a consequence of the strong assumption about the group action.) The simple objects in the two categories may be parametrized in exactly the same way: by the set of pairs

$$\xi = (S_\xi, \mathcal{V}_\xi) = (S, \mathcal{V}) \qquad (1.22)(c)$$

with S an orbit of H on Y, and $\mathcal{V}$ an irreducible H-equivariant local system on S. The set of all such pairs will be written $\Xi(Y, H)$, the set of *complete geometric parameters for H acting on Y*. Just as in Definition 1.17, we may formulate this definition in terms of the equivariant fundamental group

$$A_S^{loc} = H_y/(H_y)_0 \qquad (y \in S)$$

and its representations. We write $\mu(\xi)$ for the irreducible constructible sheaf corresponding to ξ (the extension of ξ by zero), and $P(\xi)$ for the irreducible perverse sheaf (the "intermediate extension" of ξ — cf. [9], Definition 1.4.22).

The Grothendieck groups of the two categories $\mathcal{P}(Y,H)$ and $\mathcal{C}(Y,H)$ are naturally isomorphic (by the map sending a perverse sheaf to the alternating sum of its cohomology sheaves, which are constructible). Write $K(Y,H)$ for this free abelian group. The two sets $\{P(\xi) \mid \xi \in \Xi\}$ and $\{\mu(\xi) \mid \xi \in \Xi\}$ are obviously bases of their respective Grothendieck groups, but they are *not* identified by the isomorphism. Write $d(\xi)$ for the dimension of the underlying orbit S_ξ. We can write in $K(Y,H)$

$$\mu(\xi) = (-1)^{d(\xi)} \sum_{\gamma \in \Xi(Y,H)} m_g(\gamma,\xi) P(\gamma) \qquad (1.22)(d)$$

with $m_g(\gamma,\xi)$ an integer. (The subscript g stands for "geometric.") In this formula, it follows easily from the definitions that

$$m_g(\xi,\xi) = 1, \qquad m_g(\gamma,\xi) \neq 0 \text{ only if } S_\gamma \subset (\overline{S_\xi} - S_\xi) \quad (\gamma \neq \xi).$$
$$(1.22)(e)$$

The matrix $m_g(\gamma,\xi)$ is essentially the matrix relating our two bases of $K(Y,H)$. It is clearly analogous to (1.21). In each case, we have a relationship between something uncomplicated (the standard representations, or the extensions by zero) and something interesting (the irreducible representations, or the simple perverse sheaves). One can expect the matrix m to contain interesting information, and to be difficult to compute explicitly.

Definition 1.23. In the setting of Definition 1.8, define

$$\mathcal{C}(X(^\vee G^\Gamma), {}^\vee G^{alg})$$

to be the direct sum over semisimple orbits $\mathcal{O} \subset {}^\vee\mathfrak{g}$ of the categories $\mathcal{C}(X(\mathcal{O}, {}^\vee G^\Gamma), {}^\vee G^{alg})$ of Definition 1.22. The objects of this category are called (by a slight abuse of terminology) ${}^\vee G^{alg}$-*equivariant constructible sheaves on* X. Similarly we define

$$\mathcal{P}(X(^\vee G^\Gamma), {}^\vee G^{alg}),$$

the ${}^\vee G^{alg}$-*equivariant perverse sheaves on* X. The irreducible objects in either category are parametrized by $\Xi(G/\mathbb{R})$ (cf. Definition 1.22), and we write

$$KX(^\vee G^\Gamma)$$

for their common Grothendieck group. We write $\mu(\xi)$ and $P(\xi)$ for the irreducible objects constructed in Definition 1.22, or their images in $KX(^{\vee}G^{\Gamma})$. These satisfy (1.22)(d) and (e).

Since Theorem 1.18 tells us that the two Grothendieck groups $KX(^{\vee}G^{\Gamma})$ and $K\Pi(G/\mathbb{R})$ have bases in natural one-to-one correspondence, it is natural to look for a functorial relationship between a category of representations of strong real forms of G, and one of the geometric categories of Definition 1.23. We do not know what form such a relationship should take, or how one might hope to establish it directly. What we are able to establish is a formal relationship on the level of Grothendieck groups. This will be sufficient for studying character theory.

Theorem 1.24 *Suppose $(G^{\Gamma}, \mathcal{W})$ is an extended group for G (Definition 1.12), and $^{\vee}G^{\Gamma}$ is an L-group for the corresponding inner class of real forms. Then there is a natural perfect pairing*

$$<,>: K\Pi(G/\mathbb{R}) \times KX(^{\vee}G^{\Gamma}) \to \mathbb{Z}$$

between the Grothendieck group of the category of finite length representations of strong real forms of G, and that of $^{\vee}G^{alg}$-equivariant (constructible or perverse) sheaves on X (cf. (1.19) and Definition 1.23). This pairing is defined on the level of basis vectors by

$$<M(\xi), \mu(\xi')> = e(G(\mathbb{R}, \delta(\xi)))\delta_{\xi,\xi'}.$$

Here we use the notation of (1.19) and Definition 1.22. The group $G(\mathbb{R}, \delta(\xi))$ is the real form represented by $M(\xi)$; the constant $e(G(\mathbb{R})) = \pm 1$ is the one defined in [32] (see also Definition 15.8), and the last δ is a Kronecker delta. In terms of the other bases of (1.19) and Definition 1.23, we have

$$<\pi(\xi), P(\xi')> = e(G(\mathbb{R}, \delta(\xi)))(-1)^{d(\xi)}\delta_{\xi,\xi'}.$$

The content of this theorem is in the equivalence of the two possible definitions of the pairing. We will deduce it from the main result of [56]. As an indication of what the theorem says, here are three simple reformulations.

Corollary 1.25.
a) The matrices m_r and m_g of (1.21) and Definition 1.22(d) are essentially inverse transposes of each other:

$$\sum_\gamma (-1)^{d(\gamma)} m_r(\gamma, \xi) m_g(\gamma, \xi') = (-1)^{d(\xi)} \delta_{\xi,\xi'}.$$

b) The multiplicity of the irreducible representation $\pi(\gamma)$ in the standard representation $M(\xi)$ is up to a sign the multiplicity of the local system $\mathcal{V}_\xi$ in the restriction to S_ξ of the Euler characteristic of the perverse sheaf $P(\gamma)$:

$$m_r(\gamma, \xi) = (-1)^{d(\gamma)-d(\xi)} \sum_i (-1)^i (\text{multiplicity of } \mathcal{V}_\xi \text{ in } H^i P(\gamma)\mid_{S_\xi}).$$

c) The coefficient of the standard representation $M(\gamma)$ in the expression of the irreducible representation $\pi(\xi)$ is equal to $(-1)^{d(\gamma)-d(\xi)}$ times the multiplicity of the perverse sheaf $P(\xi)$ in the expression of $\mu(\gamma)[-d(\gamma)]$.

Here part (c) refers to the expansion of $\pi(\xi)$ in the Grothendieck group as a linear combination of standard representations (cf. Lemma 1.20); and similarly for $\mu(\gamma)$.

Another way to think of Theorem 1.24 is this.

Corollary 1.26. *In the setting of Theorem 1.24, write*

$$\overline{K} = \overline{K}\Pi(G/\mathbb{R})$$

for the set of (possibly infinite) integer combinations of irreducible representations of strong real forms of G. Then $\overline{K}$ may be identified with the space of $\mathbb{Z}$-linear functionals on the Grothendieck group $KX(^\vee G^\Gamma)$:

$$\overline{K}\Pi(G/\mathbb{R}) \simeq \text{Hom}_{\mathbb{Z}}(KX(^\vee G^\Gamma), \mathbb{Z}).$$

In this identification,
a) the standard representation $M(\xi)$ of $G(\mathbb{R}, \delta(\xi))$ corresponds to $e(G(\mathbb{R}, \delta(\xi)))$ times the linear functional "multiplicity of $\mathcal{V}_\xi$ in the restriction to S_ξ of the constructible sheaf C;" and
b) the irreducible representation $\pi(\xi)$ of $G(\mathbb{R}, \delta(\xi))$ corresponds to $e(G(\mathbb{R}, \delta(\xi)))(-1)^{d(\xi)}$ times the linear functional "multiplicity of $P(\xi)$ as a composition factor of the perverse sheaf Q."

Here in (a) we are interpreting $KX({}^\vee G^\Gamma)$ as the Grothendieck group of constructible sheaves, and in (b) as the Grothendieck group of perverse sheaves.

We call elements of $\overline{K}\Pi(G/\mathbb{R})$ *formal virtual characters of strong real forms of G*.

In order to bring Langlands' notion of stability into this picture, we must first reformulate it slightly.

Definition 1.27. In the setting of Definition 1.14 and (1.19), suppose

$$\eta = \sum_{\xi \in \Xi} n(\xi)(\pi(\xi), \delta(\xi))$$

is a formal virtual character. We say that η is *locally finite* if for each strong real form δ there are only finitely many ξ with $n(\xi) \neq 0$ and $\delta(\xi)$ equivalent to δ. Suppose that η is locally finite, and that δ is a strong real form of G. There is a finite set $\pi_1, \ldots, \pi_r$ of inequivalent irreducible representations of $G(\mathbb{R}, \delta)$ so that each (π_j, δ) is equivalent to some $(\pi(\xi_j), \delta(\xi_j))$ with $n(\xi_j) \neq 0$. Each of these representations has a character $\Theta(\pi_j)$, a generalized function on $G(\mathbb{R}, \delta)$; and we define

$$\Theta(\eta, \delta) = \sum_j n(\xi_j) \Theta(\pi_j),$$

a generalized function on $G(\mathbb{R}, \delta)$. This generalized function has well-defined values at the regular semisimple elements of $G(\mathbb{R}, \delta)$, and these values determine $\Theta(\eta, \delta)$. In the notation of (1.19)(d,e), we can write

$$\Theta(\eta, \delta) = \sum_\xi n(\xi) \Theta(\pi(\xi, \delta)).$$

We say that η is *strongly stable* if it is locally finite, and the following condition is satisfied. Suppose δ and δ' are strong real forms of G, and $g \in G(\mathbb{R}, \delta) \cap G(\mathbb{R}, \delta')$ is a strongly regular semsimple element. Then

$$\Theta(\eta, \delta)(g) = \Theta(\eta, \delta')(g).$$

A necessary condition for η to be strongly stable is that each $\Theta(\eta, \delta)$ should be stable in Langlands' sense. Conversely, Shelstad's results in [48] imply that if Θ is a stable finite integer combination of characters on a real form $G(\mathbb{R}, \delta)$, then there is a strongly stable η with $\Theta = \Theta(\eta, \delta)$.

Corollary 1.26 gives a geometric interpretation of formal virtual characters. We can now give a geometric interpretation of the notion of stability.

Definition 1.28. In the setting of Definition 1.22, fix an H-orbit $S \subset Y$, and a point $y \in S$. For a constructible sheaf C on Y, write C_y for the stalk of C at y, a finite-dimensional vector space. The map

$$\chi_S^{loc} : \mathrm{Ob}\,\mathcal{C}(Y, H) \to \mathbb{N}, \quad \chi_S^{loc}(C) = \dim(C_y)$$

is independent of the choice of y in S. It is additive for short exact sequences, and so defines a $\mathbb{Z}$-linear map

$$\chi_S^{loc} : K(Y, H) \to \mathbb{Z},$$

the *local multiplicity along S*. If we regard $K(Y, H)$ as a Grothendieck group of perverse sheaves, then the formula for χ_S^{loc} on a perverse sheaf P is

$$\chi_S^{loc}(P) = \sum (-1)^i \dim(H^i P)_y.$$

Any $\mathbb{Z}$-linear functional η on $K(Y, H)$ is called *geometrically stable* if it is in the $\mathbb{Z}$-span of the various χ_S^{loc}.

In the setting of Definition 1.23, a $\mathbb{Z}$-linear functional η on $K(Y, H)$ is called *geometrically stable* if its restriction to each summand $K(X(\mathcal{O}), {}^\vee G^\Gamma)$ is geometrically stable, and vanishes for all but finitely many $\mathcal{O}$.

Theorem 1.29. *In the identification of Corollary 1.26, the strongly stable formal virtual characters correspond precisely to the geometrically stable linear functionals.*

This is an immediate consequence of Corollary 1.26 and Shelstad's description of stable characters in [47]. (It is less easy to give a geometric description of the stable characters on a single real form of G, even a quasisplit one.)

In a sense Arthur's conjectures concern the search for interesting new stable characters. We have now formulated that problem geometrically, but the formulation alone offers little help. The only obvious geometrically stable linear functionals are the χ_S^{loc}. For S corresponding to an L-packet by Proposition 1.10, the corresponding strongly stable formal virtual character is essentially the sum of all the standard representations attached to the L-packet. This sum is stable and interesting, but not new, and not what is needed for Arthur's conjectures. To continue, we need a different construction of geometrically stable linear functionals on $K(Y,H)$.

Definition 1.30. Suppose Y is a smooth complex algebraic variety on which the pro-algebraic group H acts with finitely many orbits. To each orbit S we associate its conormal bundle

$$T_S^*(Y) \subset T^*(Y);$$

this is an H-invariant smooth Lagrangian subvariety of the cotangent bundle. Attached to every H-equivariant perverse sheaf P on Y is a *characteristic cycle*

$$\mathrm{Ch}(P) = \sum_S \chi_S^{mic}(P) \overline{T_S^*(Y)}.$$

Here the coefficients $\chi_S^{mic}(P)$ are non-negative integers, equal to zero unless S is contained in the support of P. One way to construct $\mathrm{Ch}(P)$ is through the Riemann-Hilbert correspondence ([12]): the category of H-equivariant perverse sheaves on Y is equivalent to the category of H-equivariant regular holonomic D-modules on Y, and the characteristic cycle of a D-module is fairly easy to define (see for example [16] or [24]). The functions χ_S^{mic} are additive for short exact sequences, and so define $\mathbb{Z}$-linear functionals

$$\chi_S^{mic} : K(Y,H) \to \mathbb{Z},$$

the *microlocal multiplicity along S*.

Theorem 1.31 (Kashiwara — see [23], [24], Theorem 6.3.1, or [16], Theorem 8.2.) *The linear functionals χ_S^{mic} of Definition 1.30 are geometrically stable. More precisely, for every H-orbit S' such that $\overline{S'} \supset S$ there is an integer $c(S,S')$ so that for every H-equivariant perverse sheaf on Y,*

$$\chi_S^{mic}(P) = \sum_{S'} c(S,S') \chi_{S'}^{loc}(P).$$

Here $\chi_{S'}^{loc}$ is defined in Definition 1.28.

In fact Kashiwara's interest was in an inverted form of this relationship, expressing $\chi_{S'}^{loc}$ in terms of the various $\chi_{S'}^{mic}$. (The invertibility of the matrix $c(S,S')$ is an immediate consequence of the facts that $c(S,S) = (-1)^{\dim S}$, and that $c(S,S') \neq 0$ only if $\overline{S'} \supset S$.) We could therefore have defined geometrically stable in terms of the linear functionals χ_S^{mic}.

Perhaps the main difficulty in Theorem 1.31 is the definition of the matrix $c(S, S')$. That definition is due independently to Macpherson in [41]. Although the D-module approach to characteristic cycles is intuitively very simple, it entails some great technical problems (notably that of lifting [44]). We will therefore find it convenient to use a geometric definition of χ_S^{mic} due to MacPherson (see (24.10) and Definition 24.11 below). With this definition, Theorem 1.31 has a very simple proof due to MacPherson; we reproduce it at the end of Chapter 24.

The matrix $c(S, S')$ and its inverse have been extensively studied from several points of view (see for example the references in [16]). If $S \neq S'$ is contained in the smooth part of $\overline{S'}$, then $c(S, S') = 0$. Nevertheless (and in contrast with the multiplicity matrices of (1.22)(d)) there is no algorithm known for computing it in all the cases of interest to us.

Corollary 1.32. *Suppose $(G^\Gamma, \mathcal{W})$ is an extended group for G (Definition 1.12), and $^\vee G^\Gamma$ is an L-group for the corresponding inner class of real forms. Fix an orbit S of $^\vee G$ on $X(^\vee G^\Gamma)$ (Definition 1.8) (or, equivalently, an L-packet for the quasisplit form of G (Proposition 1.10)). Then the linear functional χ_S^{mic} on $KX(^\vee G^\Gamma)$ (Definition 1.30) corresponds via Corollary 1.26 to a strongly stable formal virtual representation η_S^{mic}. The irreducible representations of strong real forms occurring in η_S^{mic} are those for which the corresponding perverse sheaf P has $\chi_S^{mic}(P) \neq 0$. This includes all perverse sheaves attached to the orbit S itself, and certain sheaves attached to orbits S' containing S in their closures. With notation as in (1.19) and Definition 1.27, the corresponding stable distribution on $G(\mathbb{R}, \delta)$ is*

$$\Theta(\eta_S^{mic}, \delta) = e(G(\mathbb{R}, \delta)) \sum_{\xi' \in \Xi} (-1)^{d(\xi') - \dim S} \chi_S^{mic}(P(\xi')) \Theta(\pi(\xi', \delta)).$$

In terms of standard representations, this distribution may be expressed as

$$\Theta(\eta_S^{mic}, \delta) = e(G(\mathbb{R}, \delta))(-1)^{\dim S} \sum_{\xi' \in \Xi} c(S, S_{\xi'}) \Theta(M(\xi', \delta)).$$

The set $\{\eta_S^{mic}\}$ (as S varies) is a basis of the lattice of strongly stable formal virtual representations.

(Recall that the tempered representations correspond to open orbits; in that case χ_S^{mic} is equal to $(-1)^{\dim S} \chi_S^{loc}$, and we get nothing new.)

As the second formula of Corollary 1.32 shows, obtaining explicit character formulas for η_S^{mic} amounts to computing the matrix $c(S, S')$ in Theorem 1.31.

1. Introduction

To approach Arthur's conjectures, we need an extension of some of the notation in Definitions 1.22 and 1.30.

Definition 1.33. Suppose Y is a smooth complex algebraic variety on which the pro-algebraic group H acts with finitely many orbits. Fix a point y belonging to an H-orbit $S \subset Y$, and write $T^*_{S,y}(Y)$ for the conormal bundle at y to the orbit S. (This is a subspace of the cotangent space at y, having dimension equal to the codimension of S in Y.) The isotropy group H_y acts linearly on $T^*_{S,y}(Y)$; so for any $\nu \in T^*_{S,y}(Y)$, the isotropy group $H_{y,\nu}$ is a pro-algebraic subgroup of H_y. We can therefore form the pro-finite component group $A_{y,\nu} = H_{y,\nu}/(H_{y,\nu})_0$. This family of groups will be locally constant in the variable ν over most of $T^*_{S,y}(Y)$ (Lemma 24.3 below), so we can define the *equivariant micro-fundamental group* A^{mic}_S to be $A_{y,\nu}$ for generic ν.

Attached to every H-equivariant perverse sheaf P on Y is a representation $\tau^{mic}_S(P)$ of A^{mic}_S (Theorem 24.8 and Corollary 24.9 below), of dimension equal to $\chi^{mic}_S(P)$ (Definition 1.30).

The differences between Langlands' original conjectures and those of Arthur amount geometrically to the difference between local geometry (on the orbits S) and microlocal geometry (on the union of the conormal bundles $T^*_S(Y)$); formally, to the difference between the equivariant fundamental group A^{loc}_S and the equivariant micro-fundamental group A^{mic}_S. (Here we are writing Y for the geometric parameter space $X(\mathcal{O}, {}^\vee G^\Gamma)$ containing S.) Notice that if S is open, then $T^*_{S,y}(Y)$ is zero, and A^{mic}_S coincides with the equivariant fundamental group A^{loc}_S. Because tempered representations correspond to open orbits in Theorem 1.18 (Proposition 1.11), we see why the Langlands theory is so effective for tempered representations: the local and the microlocal geometry coincide.

More generally, an L-packet is called *generic* if some irreducible representation in it admits a Whittaker model (see (3.11) and (14.11)–(14.14) below. Here we must understand L-packets as extending over all strong real forms of G.) The generic L-packets (in fact the individual generic representations) were explicitly determined in [30] and [52]. Using those results and Proposition 1.11, it is not difficult to show that an L-packet is generic if and only if the corresponding orbit S is open; that is, if and only if S coincides with the conormal bundle $T^*_S(Y)$. We believe that this collapsing of microlocal to local geometry explains why representation theory should be simpler for generic L-packets.

The existence of the representation $\tau^{mic}_S(P)$ in Definition 1.33 is well-known but quite subtle. We will construct it, following Goresky and MacPherson in [17], using a pair of small spaces $J \supset K$ that reflect the local nature of the singularity of the orbit stratification of Y along S. The representation $\tau_S^{mic}(P)$ will be the hypercohomology of the pair (J, K) with coefficients in P. By a "purity" theorem of Goresky-MacPherson and Kashiwara-Schapira ([17], [26]; see also Chapter 24 below) this hypercohomology is non-zero in only one degree.

Our approach to Arthur's conjectures is now fairly straightforward. Arthur attaches to his parameter ψ a Langlands parameter ϕ_ψ, and thus (by Proposition 1.10) an orbit S_ψ. We define Π_ψ to consist of all those representations appearing in $\eta_{S_\psi}^{mic}$; that is, representations for which the corresponding perverse sheaf has the conormal bundle of S_ψ in its characteristic cycle. We will show (Corollary 27.13) that this agrees with the previous definition of Barbasch and Vogan (in the case of "unipotent" parameters) in terms of primitive ideals. It follows from Proposition 22.9 that Arthur's group A_ψ is isomorphic to a quotient of the equivariant micro-fundamental group $A_{S_\psi}^{mic,alg}$. The difference arises only from our use of the algebraic universal covering of $^\vee G$, and for these local purposes our choice seems preferable. We therefore define

$$A_\psi^{alg} = A_{S_\psi}^{mic,alg} \qquad (1.34)(a)$$

Definition 1.33 provides a representation $\tau_{S_\psi}^{mic}(P)$ of $A_{S_\psi}^{mic,alg}$, of dimension equal to $\chi_{S_\psi}^{mic}(P)$. Now Problems A, B, C, and E are resolved as special cases of Corollary 1.32 and the preceding definitions. In particular, the representation $\tau_\psi(\pi)$ of Problem B is defined to be

$$\tau_\psi = \tau_{S_\psi}^{mic}(P(\pi)), \qquad (1.34)(b)$$

with $P(\pi)$ the irreducible perverse sheaf corresponding to π (Theorem 1.24).

Arthur's Problem C identifies one interesting linear combination of the representations in Π_ψ, using the dimensions of the representations $\tau_\psi(\pi)$. If we use instead other character values, we can immediately define several more. Fix an element $\sigma \in A_\psi^{alg}$, and consider the complex-valued linear functional on equivariant perverse sheaves given by

$$P \mapsto \operatorname{tr}[\tau_{S_\psi}^{mic}(P)](\sigma) \qquad (1.34)(c)$$

By Corollary 1.26, this linear functional corresponds to a complex formal virtual representation $\eta_\psi(\sigma)$; that is, to a formal sum with complex coefficients of representations of various strong real forms of G. Just as in Corollary 1.32, we can write this virtual representation on a single strong real form δ as

$$\Theta(\eta_\psi(\sigma),\delta) = e(G(\mathbb{R},\delta)) \sum_{\pi\in\Pi(G(\mathbb{E},\delta))_\psi} (-1)^{d(\pi)-\dim S_\psi} \operatorname{tr} \tau_\psi(\pi)(\sigma)\Theta(\pi).$$

(1.34)(d)

(More details may be found in Definition 26.8.) Arthur's Problem D asks for a description of the (complex formal virtual) character $\eta_\psi(\sigma)$ in terms of stable characters like $\eta_\psi(1)$ on smaller groups.

To see how this might be possible, we need a long digression about Langlands' functoriality principle. This principle concerns relationships between the representations of real forms of G and those of a smaller reductive group H. The simplest way that such relationships arise is when G and H are equipped with fixed real forms, and $H(\mathbb{R}) \subset G(\mathbb{R})$. In that case we have functors of induction (carrying representations of $H(\mathbb{R})$ to representations of $G(\mathbb{R})$) and restriction (carrying representations of $G(\mathbb{R})$ to representations of $H(\mathbb{R})$). Except in a few very special cases (for example, when G/H is symmetric and we restrict attention to the trivial representation of $H(\mathbb{R})$) these functors are poorly behaved on irreducible representations, and offer little insight into their structure.

A more interesting situation arises when H is a Levi subgroup of a real parabolic subgroup $P = HN$ of G. Then we have the functor of parabolic induction, which carries irreducible representations of $H(\mathbb{R})$ to finite-length representations of $G(\mathbb{R})$. (There are also various "Jacquet functors," analogous to restriction, carrying irreducible representations of $G(\mathbb{R})$ to finite-length (sometimes only virtual) representations of $H(\mathbb{R})$.)

The parabolic induction functors provide the basic model for Langlands functoriality. They exhibit a number of important features of functoriality in general, of which we will mention two. First, they are most simply defined on the level of virtual representations. The reason is that we want to go directly from representations of $H(\mathbb{R})$ to representations of $G(\mathbb{R})$. The definition of parabolic induction requires the choice of a parabolic subgroup P with Levi subgroup H. Different choices of P lead to inequivalent representations, but to the same virtual representations. If we work with virtual representations, we may therefore suppress the dependence on P.

The second feature is actually hidden within the first. To get independence of P even on the level of virtual representations, we must normalize parabolic induction using certain "ρ-shifts." The definition of these shifts requires the extraction of a square root of a character of H (on the top exterior power of the Lie algebra of N). This character happens to be real-valued on $H(\mathbb{R})$, so the square root more or less exists on $H(\mathbb{R})$. (The problem of square roots of -1 can be swept under the rug.) Nevertheless it is clear that (linear) coverings of H are waiting in the wings.

To get correspondences from representations of a small group H to those of a larger group G that behave like parabolic induction, we will use Theorem 1.24. It is therefore natural to begin with extended groups $(G^\Gamma, \mathcal{W})$ and $(H^\Gamma, \mathcal{W}_H)$, and corresponding L-groups ${}^\vee G^\Gamma$ and ${}^\vee H^\Gamma$. Adopting the suggestion from the preceding paragraph that we should seek only a correspondence of virtual representations, we find that we want something like a $\mathbb{Z}$-linear map

$$\epsilon_* : K\Pi(H/\mathbb{R}) \to K\Pi(G/\mathbb{R}). \tag{1.35}(a)$$

According to Theorem 1.24 (compare Corollary 1.26), such a map is more or less the same as the transpose of a $\mathbb{Z}$-linear map

$$\epsilon^* : KX({}^\vee G^\Gamma) \to KX({}^\vee H^\Gamma). \tag{1.35}(b)$$

(The "more or less" refers only to issues of finiteness — the difference between K and $\overline{K}$ in Corollary 1.26.) Recall now that the Grothendieck groups in (1.35)(b) are built from equivariant constructible sheaves on geometric parameter spaces. Bearing in mind that H is supposed to be smaller than G, we find that a natural source for a map like (1.35)(b) is the pullback of constructible sheaves by an equivariant morphism of varieties. It is therefore natural to seek a morphism of pro-algebraic groups

$$\epsilon_\bullet : {}^\vee H^{alg} \to {}^\vee G^{alg}, \tag{1.35}(c)$$

and a compatible morphism of geometric parameter spaces

$$X(\epsilon) : X({}^\vee H^\Gamma) \to X({}^\vee G^\Gamma). \tag{1.35}(d)$$

Because the geometric parameter spaces are constructed from the L-groups, this suggests finally the definition at the heart of the functoriality principle.

Definition 1.36 (cf. Definitions 5.1 and 26.3 below). Suppose $(G^\Gamma, \mathcal{W})$ and $(H^\Gamma, \mathcal{W}_H)$ are extended groups (Definition 1.12), with corresponding L-groups ${}^\vee G^\Gamma$ and ${}^\vee H^\Gamma$. An *L-homomorphism* is a morphism

$$\epsilon : {}^\vee H^\Gamma \to {}^\vee G^\Gamma$$

respecting the homomorphisms to Gal($\mathbb{C}/\mathbb{R}$) (cf. (1.4)(a)).

Proposition 1.37 (Corollary 6.21 and Proposition 7.18 below). *Suppose $(G^\Gamma, \mathcal{W})$ and $(H^\Gamma, \mathcal{W}_H)$ are extended groups with corresponding L-groups $^\vee G^\Gamma$ and $^\vee H^\Gamma$, and $\epsilon: {}^\vee H^\Gamma \to {}^\vee G^\Gamma$ is an L-homomorphism.*

(a) *The restriction of ϵ to the identity component $^\vee H$ induces a morphism of pro-algebraic groups*

$$\epsilon_\bullet : {}^\vee H^{alg} \to {}^\vee G^{alg}$$

as in (1.35)(c).

(b) *The map ϵ induces a morphism of geometric parameter spaces*

$$X(\epsilon): X({}^\vee H^\Gamma) \to X({}^\vee G^\Gamma)$$

as in (1.35)(d), compatible with the actions of $^\vee H^{alg}$ and $^\vee G^{alg}$ and the morphism $\epsilon_\bullet$ of (a).

(c) *Pullback of constructible sheaves via the morphism $X(\epsilon)$ of (b) defines a $\mathbb{Z}$-linear map*

$$\epsilon^* : KX({}^\vee G^\Gamma) \to KX({}^\vee H^\Gamma)$$

as in (1.35)(b).

(d) *The transpose of the map in (c) is a $\mathbb{Z}$-linear map*

$$\epsilon_* : K\Pi(H/\mathbb{R}) \to \overline{K\Pi}(G/\mathbb{R})$$

(cf. Corollary 1.26), which we call Langlands functoriality. It carries representations of strong real forms of H to formal virtual characters of strong real forms of G.

The most important point about this proposition is that the relationship between H and G is entirely on the dual group side; there may be no homomorphism from H to G dual to ϵ in any sense. (A typical example is provided by the split symplectic group $G = Sp(2n)$, and the split orthogonal group $H = SO(2n)$. The corresponding L-groups are $^\vee G^\Gamma = SO(2n+1) \times \Gamma$ and $^\vee H^\Gamma = SO(2n) \times \Gamma$, so there is an obvious L-homomorphism as in Definition 1.36. For n at least 3, however, any homomorphism from H to G must be trivial.) Similarly, a group homomorphism from H to G need not give rise to an L-homomorphism in general. On the other hand, a real parabolic subgroup $P = HN$ of G *does* provide an L-homomorphism. (Here "real" should be interpreted with respect to one of the special real forms defining the extended group structure on G.) It is not very difficult to show that in this case the Langlands functoriality map of Proposition 1.37 implements the parabolic induction functor discussed earlier; this is implicit in Proposition 26.4.

The primary motivations for studying Langlands functoriality are connected with automorphic representation theory and the trace formula (see for example [2]); we will not discuss them further. However, there are also purely local motivations. One is that the distribution character of the (virtual) representation $\epsilon_*(\pi_H)$ can be expressed in terms of the distribution character of π_H. We will not explain in detail how to do this; but in Proposition 26.4(b) we will solve the closely related problem of calculating ϵ_* in the basis of standard representations.

In the setting of (1.34), we can now refine slightly our formulation of Arthur's Problem D: the goal is to find $\eta_\psi(\sigma)$ in the image of a Langlands functoriality map ϵ_*. Choose an elliptic element

$$\tilde{s} \in {}^\vee G^{alg} \qquad (1.38)(b)$$

representing the class $\sigma \in A_\psi^{alg}$; this is possible by Lemma 26.20. (We may actually choose $\tilde{s}$ to have finite order in every algebraic quotient of ${}^\vee G^{alg}$.) The idea is to use the element $\tilde{s}$ to construct the data (ϵ, H) required for Proposition 1.37. To do this, we begin by defining

$$ {}^\vee H = \text{identity component of centralizer in } {}^\vee G \text{ of } \tilde{s}. \qquad (1.38)(c)$$

We can then define H to be a complex reductive group with dual group ${}^\vee H$. The Arthur parameter ψ defines a Langlands parameter ϕ_ψ, and therefore a geometric parameter x_ψ (Proposition 1.10). As a point of a geometric parameter space, x_ψ is a pair (y_ψ, Λ_ψ) (Definition 1.8), with y_ψ an element of ${}^\vee G^\Gamma - {}^\vee G$ commuting with $\tilde{s}$. Set

$$ {}^\vee H^\Gamma = \text{group generated by } y_\psi \text{ and } {}^\vee H, \qquad (1.38)(d)$$

and let $\epsilon : {}^\vee H^\Gamma \to {}^\vee G^\Gamma$ be the identity map. Because of our choice of $\tilde{s}$, the Arthur parameter ψ takes values in the group ${}^\vee H^\Gamma$; we may write it as ψ_H when we wish to emphasize this.

We are now nearly in the setting of Proposition 1.37. The group ${}^\vee H^\Gamma$ inherits from ${}^\vee G^\Gamma$ a short exact sequence

$$1 \to {}^\vee H \to {}^\vee H^\Gamma \to \text{Gal}(\mathbb{C}/\mathbb{R}) \to 1 \qquad (1.38)(e)$$

as in (1.4)(a). In particular, there is an action of $\text{Gal}(\mathbb{C}/\mathbb{R})$ on the based root datum for ${}^\vee H$ and for H (see Definition 2.10). This gives rise to an extended group structure $(H^\Gamma, \mathcal{W}_H)$ (see Chapter 3). The reductive group H, together with the inner class of real forms defined by the extended group structure, is an *endoscopic group* for G (see Definition 26.15 and (26.17) below). Unfortunately, $^\vee H^\Gamma$ fails to be an L-group for H^Γ (Definition 4.6 below) for two reasons. First, the sequence (1.38)(e) may admit no distinguished splittings. Second, even if such splittings exist, there is no natural way to fix a $^\vee H$-conjugacy class of them. The second of these problems is only a minor nuisance, but the first is somewhat more serious. We will postpone discussing it for a moment in order to formulate a solution to Arthur's Problem D.

Theorem 1.39 (cf. Theorem 26.25 below). *Suppose $(G^\Gamma, \mathcal{W})$ is an extended group with L-group $^\vee G^\Gamma$, ψ is an Arthur parameter for G, and $\sigma \in A_\psi^{alg}$. Following (1.34), define a complex formal virtual character $\eta_\psi(\sigma)$ for strong rational forms of G. Choose an elliptic representative $\tilde{s} \in {}^\vee G^{alg}$ for σ, and define $^\vee H^\Gamma$, ϵ, and $(H, \mathcal{W}_H)$ as in (1.38). Finally, choose any preimage $\tilde{s}_H \in {}^\vee H^{alg}$ for $\tilde{s}$ under the map ϵ_* of (1.35)(c).*

Assume that $^\vee H^\Gamma$ is endowed with the structure of an L-group for $(H, \mathcal{W}_H)$. Then ψ_H may be regarded as an Arthur parameter for H, and $\tilde{s}_H$ represents a class $\sigma_H \in A_{\psi_H}^{alg}$. The complex formal virtual characters of strong rational forms of H and G defined in (1.34) are related by Langlands functoriality (Proposition 1.37) as follows:

$$\eta_\psi(\sigma) = \epsilon_*(\eta_{\psi_H}(\sigma_H)).$$

We will discuss the proof of this result in a moment; first there are some formal issues to address. At (1.34), we asked for a description of $\eta_\psi(\sigma)$ in terms of virtual representations like $\eta_\psi(1)$ on smaller groups. The right side of the formula in Theorem 1.39 involves not $\eta_{\psi_H}(1)$, but rather $\eta_{\psi_H}(\sigma_H)$. The difference is harmless, for the following reason. The element $\tilde{s}_H$ representing σ_H is central in $^\vee H^{alg}$ (by (1.38)(c)). Its image in $^\vee H$ is fixed by the action of Γ on $Z(^\vee H)$ (by (1.38)(d)). Now Lemma 26.12 below shows that for any strong real form δ_H of H there is a non-zero complex number $c = \tau_{univ}(\delta_H)(\tilde{s}_H)$ so that

$$\Theta(\eta_{\psi_H}(\sigma_H), \delta_H) = c\Theta(\eta_{\psi_H}(1), \delta_H). \qquad (1.40)$$

(Here we use the notation of Definition 1.27.) If δ_H is one of the distinguished (quasisplit) strong real forms of H defining the extended group structure, then $c = 1$.

The second formal issue is the one we postponed a moment ago: what happens when $^\vee H^\Gamma$ is not the L-group for H? The answer is implicit in [35]. All of the geometry we have discussed for L-groups can still be carried out on $^\vee H^\Gamma$. The resulting geometric parameter space is slightly different from the one constructed using the L-group of H, and so it should correspond to something slightly different from representations of real forms of H. The right objects turn out to be projective representations. The failure of $^\vee H^\Gamma$ to be an L-group is measured by a certain cocycle (the "second invariant of an E-group" introduced in Definition 4.6 below). This same cocycle defines a class of projective representations of real forms of H (Definition 10.3). There is a version of Theorem 1.18 (Theorem 10.4 below) relating these representations to geometry on $^\vee H^\Gamma$. Once all this extra formalism is assembled, Theorem 1.39 makes sense (and is true) without the hypothesis that $^\vee H^\Gamma$ is an L-group. It is this version that is proved in Chapter 26. (Recall also that the use of projective representations in Langlands functoriality is one of the possibilities suggested by the example of parabolic induction.)

A third formal issue is exactly how much Theorem 1.39 is telling us about distribution characters. In light of the remarks after Proposition 1.37 on the computability of Langlands functoriality (cf. Proposition 26.4(a) and (b)), Theorem 1.39 reduces the calculation of the complex formal virtual characters $\eta_\psi(\sigma)$ to the case $\sigma = 1$. This is the case considered in Corollary 1.32, where we found that it was equivalent to an interesting but (in general) unsolved geometric problem. (Another approach that is sometimes effective is described in the next paragraph.)

A fourth issue in the formulation of Theorem 1.39 is the dependence of the endoscopic group H on the choice $\tilde{s}$ of a representative of σ. Simple examples show that this dependence is very strong: different choices lead to very different endoscopic groups. For example, if the identity component of the group $^\vee G_\psi$ (the centralizer of the Arthur parameter ψ) is not central in $^\vee G$, one can choose a non-central element $\tilde{s}$ to represent $1 \in A_\psi^{alg}$. Theorem 1.39 then computes the stable character $\eta_\psi(1)$ in terms of the same kind of character on a strictly smaller group, bypassing the problem of computing the matrix $c(S, S')$. (More precisely, we are showing how to compute the matrix $c(S, S')$ in the presence of a non-trivial torus action.) The variation of H with the choices should therefore be regarded as a helpful tool, rather than as a weakness of the result.

The last formal issue is that of computing irreducible characters. It is natural to consider the identities (1.34)(d) for fixed ψ and varying σ, and to try to invert them to get formulas for the individual irreducible characters $\Theta(\pi)$ (for $\pi \in \Pi_\psi$) as linear combinations of the formal virtual representations $\eta_\psi(\sigma)$. This was done by Shelstad for tempered representations in [48], and by Barbasch-Vogan for special unipotent representations of complex groups in [7]. In both cases the result is an elementary consequence of two facts peculiar to these cases: the representation $\tau_\psi(\pi)$ of A_ψ^{alg} is irreducible, and the map $\pi \mapsto \tau_\pi$ (from Π_ψ to $\widehat{A}_\psi^{alg}$) is injective. The second fact is certainly not true for Arthur packets in general (see Theorem 27.18 and the remarks after it). It seems likely that the first fails as well; but a counterexample would have to be geometrically rather complicated, and we have not found one. In any case, the identities (1.34)(d) cannot be inverted in general. We do not know whether to expect the existence of a larger set of natural identities that could be inverted.

Here is a sketch of the proof of Theorem 1.39. After unwinding the definition in (1.34) of the virtual representations $\eta_\psi(\sigma)$, and the definition in Proposition 1.37 of the Langlands functoriality map ϵ_*, what must be proved is the following formula. Suppose C is any ${}^\vee G^{alg}$-equivariant constructible complex on the geometric parameter space $X({}^\vee G^\Gamma)$. Then

$$\sum_p (-1)^p \mathrm{tr}\,(\sigma, H^p(J, K; C)) = \sum_q (-1)^q \mathrm{tr}\,(\sigma_H, H^q(J_H, K_H; \epsilon^*(C))). \tag{1.41}$$

Here $J \supset K$ is the pair of spaces arising in the definition of τ_ψ at (1.34); the cohomology is the hypercohomology of the pair with coefficients in the complex C. (By taking C perverse, we could arrange for this cohomology to be zero except in one degree. But even in that case $\epsilon^*(C)$ would not necessarily be perverse, so we would still need the alternating sum on the right.) The objects on the right are defined similarly for H and the Arthur parameter ψ_H. The spaces J and K may be chosen invariant under the action of $\tilde{s}$; then the left side of (1.41) is a Lefschetz number for $\tilde{s}$ acting on $J \supset K$ (an automorphism of finite order). If we recall that ${}^\vee H$ was defined as the centralizer of $\tilde{s}$, it is perhaps not surprising that $J_H \supset K_H$ turns out to be the fixed point set of $\tilde{s}$. We should be able to compute a Lefschetz number in terms of local contributions along this fixed point set; a theorem of Goresky and MacPherson allows us to do this explicitly, leading to (1.41). The answer is so simple because the automorphism $\tilde{s}$ is of finite order. The geometric details are in Chapter 25 (Theorem 25.8).

2. Structure theory: real forms

In this chapter we review the basic facts about real forms of reductive groups. Since our concern is entirely with local problems over $\mathbb{R}$, we have included proofs of several well-known results that are perhaps less familiar to experts on real groups. (The definitions, statements and some of the proofs have nevertheless been constructed with more general fields in mind.)

Suppose G is a connected reductive complex algebraic group. A *real form* of G is an antiholomorphic involutive automorphism

$$\sigma : G \to G. \qquad (2.1)(a)$$

Here "antiholomorphic" means that if f is any algebraic function on G, then the function

$$g \mapsto \overline{f(\sigma g)} \qquad (2.1)(b)$$

is also algebraic. Equivalently, the differential of σ (an automorphism of $\mathfrak{g} = \text{Lie}(G)$) satisfies

$$d\sigma(ix) = -id\sigma(x). \qquad (2.1)(b')$$

"Involutive" means that σ^2 is the identity. The *group of real points of* σ is

$$G(\mathbb{R}, \sigma) = G(\mathbb{R}) = \{\, g \in G \mid \sigma g = g \,\}. \qquad (2.1)(c)$$

Condition $(2.1)(b')$ shows that the Lie algebra $\mathfrak{g}(\mathbb{R})$ determines $d\sigma$, and hence determines σ. We may therefore speak of $G(\mathbb{R})$ as the real form of G without danger of confusion. However, not every real form of the Lie algebra $\mathfrak{g}$ exponentiates to a real form of G. The multiplicative group $G = \mathbb{C}^\times$ has exactly two real forms, given by the automorphisms

$$\sigma_s(z) = \overline{z}, \qquad \sigma_c(z) = \overline{z}^{-1}. \qquad (2.2)(a)$$

The corresponding groups are

$$G(\mathbb{R}, \sigma_s) = \mathbb{R}^\times, \qquad G(\mathbb{R}, \sigma_c) = S^1. \qquad (2.2)(b)$$

2. Structure theory: real forms

The real forms of the Lie algebra $\mathbb{C}$, on the other hand, are parametrized by real lines in $\mathbb{C}$. The corresponding subgroups $\{e^{tz} \mid t \in \mathbb{R}\}$ are not (identity components of) real forms unless z is real or imaginary.

It is worth remarking that this definition is quite restrictive in some slightly surprising ways. For example, the identity component of a disconnected group of real points is *not* usually a group of real points. Thus familiar linear groups like $SO(p,q)_0$ are excluded. This is necessary to get the very clean character formulas discussed in the introduction. (Of course it is a routine matter — though not a trivial one — to relate the representations of $SO(p,q)_0$ to those of $SO(p,q)$, which *is* a group of real points.)

Two real forms σ and σ' are called *equivalent* if there is an element $g \in G$ such that

$$\sigma' = \operatorname{Ad}(g) \circ \sigma \circ \operatorname{Ad}(g^{-1}). \qquad (2.3)(a)$$

We can write this as

$$\sigma' = \operatorname{Ad}(g\sigma(g^{-1})) \circ \sigma. \qquad (2.3)(a')$$

In terms of the groups of real points, it is equivalent to

$$G(\mathbb{R}, \sigma') = gG(\mathbb{R}, \sigma)g^{-1}. \qquad (2.3)(a'')$$

The set of real forms equivalent to a fixed real form σ is therefore a homogeneous space G/H; the isotropy group is

$$H = \{\, g \in G \mid g\sigma(g^{-1}) \in Z(G) \,\}. \qquad (2.3)(b)$$

This isotropy group contains $G(\mathbb{R}, \sigma)Z(G)$, but may be larger. (For example, if $G(\mathbb{R}) = SL(2, \mathbb{R})$, then the group H contains the diagonal matrix with entries $(i, -i)$.) The point of the theory of strong real forms is to find a notion of equivalence for which the corresponding isotropy group is just $G(\mathbb{R})$. This in turn will allow us to formulate results like Lemma 1.15.

This notion of equivalence is already a little more subtle than the one sometimes encountered in the classification of real forms of simple Lie algebras. There one is interested in the question of when the groups of real points of two real forms are isomorphic. The group $SO(4n, \mathbb{C})$ has two isomorphic but inequivalent real forms (the isomorphism class is represented by $SO^*(4n)$) if $n \geq 1$. The involutions are conjugate by the non-identity component of $O(4n, \mathbb{C})$. (There is a unique equivalence class of real forms of $SO(4n+2, \mathbb{C})$ with $G(\mathbb{R})$ isomorphic to $SO^*(4n+2)$.) A less subtle example is provided by the inequivalent real forms $\mathbb{R}^\times \times S^1$ and $S^1 \times \mathbb{R}^\times$ of $\mathbb{C}^\times \times \mathbb{C}^\times$.

Two real forms σ and σ' are said to be *inner* to each other if there is an element $g \in G$ such that

$$\sigma' = Ad(g) \circ \sigma. \tag{2.4}$$

This is an equivalence relation. (If σ is a real form, the automorphism σ' defined by (2.4) will certainly *not* be a real form for arbitrary g.) Because of (2.3)(a'), equivalent real forms are inner to each other. We therefore get an equivalence relation on equivalence classes of real forms. Obviously the relation is trivial if G is abelian. The two isomorphic inequivalent real forms of $SO(4n, \mathbb{C})$ mentioned earlier are inner to each other, and to the real forms $SO(2p, 2q)$. The real forms $SO(2p+1, 2q-1)$ constitute a separate inner class.

Example 2.5. Suppose $G = GL(n, \mathbb{C})$, and σ_c is the compact real form

$$\sigma_c(g) = {}^t \overline{g}^{-1}, \qquad G(\mathbb{R}, \sigma_c) = U(n).$$

For $0 \leq p \leq n$, let g_p be the diagonal matrix with p entries equal to 1 and $n-p$ equal to -1. Set

$$\sigma_p = Ad(g_p) \circ \sigma_c.$$

Then

$$G(\mathbb{R}, \sigma_p) = U(p, q),$$

so all these forms are inner to each other. On the other hand, the split real form

$$\sigma_s(g) = \overline{g}, \qquad G(\mathbb{R}, \sigma_s) = GL(n, \mathbb{R})$$

is not inner to σ_c (as one can see — rather unfairly — by examining their restrictions to the center of G).

We recall that a real form σ is called *quasisplit* if there is a Borel subgroup $B \subset G$ such that

$$\sigma(B) = B. \tag{2.6}$$

Proposition 2.7. *Suppose σ' is a real form of the connected reductive complex algebraic group G. Then there is a unique equivalence* *class of quasisplit real forms of G inner to σ'. Specifically, there is a quasisplit real form σ of G and an element g of G so that*

$$\sigma' = \mathrm{Ad}(g) \circ \sigma.$$

Before giving the proof, we recall the basic structural fact on which it is based.

Proposition 2.8. *Suppose G is a connected reductive complex algebraic group. Fix a Borel subgroup B of G, a maximal torus $T \subset B$, and a set of basis vectors $\{X_\alpha\}$ for the simple root spaces of T in the Lie algebra $\mathfrak{b}$. Let $B', T', \{X_\alpha'\}$ be another set of choices of these objects. Then there is an element $g \in G$ such that*

$$gBg^{-1} = B', \qquad gTg^{-1} = T', \qquad \mathrm{Ad}(g)(\{X_\alpha\}) = \{X_\alpha'\}.$$

The inner automorphism $\mathrm{Ad}(g)$ is uniquely determined by these requirements; that is, any two choices of g differ by $Z(G)$. If we require only the first two conditions, then the coset gT is uniquely determined; and if we require only the first condition, then the coset gB is uniquely determined.

To get a formulation valid over any algebraically closed field k, one need only replace the Lie algebra elements X_α by one-parameter subgroups $x_\alpha : k \to B$.

Proof of Proposition 2.7. Fix $B, T \subset B$, and $\{X_\alpha\}$ as in Proposition 2.8. Then the automorphism σ' carries these objects to others of the same kind:

$$\sigma'(B) = B', \qquad \sigma'(T) = T' \subset B', \qquad d\sigma(X_\alpha) = X_\alpha'. \qquad (2.9)(a)$$

By Proposition 2.8, we can therefore find an element $g \in G$ such that

$$\mathrm{Ad}(g^{-1})(B') = B, \qquad \mathrm{Ad}(g^{-1})(T') = T, \qquad \mathrm{Ad}(g^{-1})(\{X_\alpha'\}) = \{X_\alpha\}. \qquad (2.9)(b)$$

Define $\sigma = \mathrm{Ad}(g^{-1}) \circ \sigma'$; then σ is an antiholomorphic automorphism of G, and

$$\sigma(B) = B, \qquad \sigma(T) = T, \qquad \sigma(\{X_\alpha\}) = \{X_\alpha\}. \qquad (2.9)(c)$$

Furthermore σ^2 is an inner automorphism of G (namely $\mathrm{Ad}(g^{-1}) \circ \mathrm{Ad}(\sigma'(g^{-1}))$) with the properties in (2.9)(c). By Proposition 2.8, such an inner automorphism is necessarily trivial, so σ is an involution. It is therefore a real form. Since σ preserves B, it is quasisplit. By construction it is inner to σ'.

It remains to prove the uniqueness of the equivalence class of σ. So suppose σ'' is another quasisplit real form in the inner class of σ. By definition of quasisplit, this means that there is a Borel subgroup B'' fixed by σ''. The first problem is to find a σ''-stable maximal torus in B''. Write N'' for the unipotent radical of B'', and $\overline{T''} = B''/N''$; this torus inherits a quotient real form $\overline{\sigma''}$, and is isomorphic by the quotient map to any maximal torus of B''. Using any of these isomorphisms, we get a well-defined set of positive roots (characters of $\overline{T''}$). These roots are permuted by the action of $\overline{\sigma''}$ on the character lattice $X^*(\overline{T''})$. Let $\overline{Z}$ be any element of the Lie algebra $\overline{\mathfrak{t}''}$ on which all the positive roots take positive values; then $\overline{Z} + \overline{\sigma''}(\overline{Z})$ has the same property. It follows that any preimage of $\overline{Z} + \overline{\sigma''}(\overline{Z})$ in $\mathfrak{b}''$ is a regular semisimple element. Let $Z \in \mathfrak{b}$ be any preimage of $\overline{Z}$; then $Z + \sigma''(Z)$ is a preimage of $\overline{Z} + \overline{\sigma}(\overline{Z})$, so it is a σ''-fixed regular semisimple element of $\mathfrak{b}''$. Its centralizer in B'' is therefore a σ''-stable maximal torus.

Finally, we want to find a σ''-stable set of basis vectors for the simple root spaces of T'' in B''. We know that σ'' permutes these root spaces by a permutation of order 2. We consider first a pair $\{\alpha, \beta\}$ of distinct simple roots interchanged by σ''. Choose any basis vector X_α for the α root space, and define $X_\beta = \sigma''(X_\alpha)$. Then the relation $(\sigma'')^2 = 1$ forces $X_\alpha = \sigma''(X_\beta)$. Next, suppose the root α is fixed by σ''. Let Y_α be any basis vector for the root space. Then there is a complex number c_α with $\sigma''(Y_\alpha) = c_\alpha Y_\alpha$. Applying σ'' to this relation, and using the fact that σ'' is an antiholomorphic involution, we get $c_\alpha \overline{c_\alpha} = 1$. Let z_α be a square root of c_α; then it follows easily that $X_\alpha = z_\alpha Y_\alpha$ is fixed by σ''.

We have shown how to construct B'', $T'' \subset B''$, and $\{X_\alpha''\}$ as in Proposition 2.8 preserved by σ''. By Proposition 2.8, we can find a $g \in G$ carrying B to B'', etc. It follows that the antiholomorphic involution

$$\widetilde{\sigma''} = \mathrm{Ad}(g) \circ \sigma \circ \mathrm{Ad}(g^{-1})$$

preserves B'', etc. Therefore $(\sigma'')^{-1}\widetilde{\sigma''}$ preserves them as well. But the assumption that σ and σ'' are in the same inner class means that this last automorphism is inner. By Proposition 2.8, it is trivial. So $\sigma'' = \widetilde{\sigma''}$. The right side here is equivalent to σ, as we wished to show. Q.E.D.

Definition 2.10 (see [50]). Suppose G is a complex connected reductive algebraic group, $B \subset G$ is a Borel subgroup, and $T \subset G$ is a maximal torus. The *based root datum for G defined by B and T* is the quadruple

$$\Psi_0(G, B, T) = (X^*(T), \Delta(B,T), X_*(T), \Delta^\vee(B,T)).$$

Here $X^*(T)$ is the lattice of rational characters of T, $\Delta(B,T) \subset X^*(T)$ is the set of simple roots of T in $\mathfrak{b}$, $X_*(T)$ is the lattice of rational one-parameter subgroups of T, and $\Delta^\vee(B,T) \subset X_*(T)$ is the set of simple coroots.

It is a consequence of Proposition 2.8 that any two based root data are canonically isomorphic. Following [33], we use these canonical isomorphisms to define the *based root datum for* G, $\Psi_0(G)$, as the projective limit over all such $T \subset B$:

$$\Psi_0(G) = (X^*, \Delta, X_*, \Delta^\vee) = \varprojlim_{B,T} \Psi_0(G, B, T).$$

The structure of the based root datum consists of the lattice structures on X^* and X_*, the containments $\Delta \subset X^*$ and $\Delta^\vee \subset X_*$, and the perfect pairing

$$\langle,\rangle : X^* \times X_* \to \mathbb{Z}.$$

By an *isomorphism* or *automorphism* of based root data we will understand a map preserving these structures.

Proposition 2.11 (see [50], Corollary 2.14.) *Suppose G is a complex connected reductive algebraic group. Write $\mathrm{Aut}(G)$ for the (complex) group of rational (equivalently, holomorphic) automorphisms of G, and $\mathrm{Aut}(\Psi_0(G))$ for the (discrete) group of automorphisms of the based root datum of G. Then there is a natural short exact sequence*

$$1 \to \mathrm{Int}(G) \to \mathrm{Aut}(G) \xrightarrow{\Psi_0} \mathrm{Aut}(\Psi_0(G)) \to 1.$$

This sequence splits (but not canonically), as follows. Choose a Borel subgroup B of G, a maximal torus $T \subset B$, and a set of basis vectors $\{X_\alpha\}$ for the simple root spaces of T in the Lie algebra $\mathfrak{b}$; and define $\mathrm{Aut}(G, B, T, \{X_\alpha\})$ to be the set of holomorphic automorphisms of G preserving B, T, and $\{X_\alpha\}$ as sets. Then the restriction of Ψ_0 to $\mathrm{Aut}(G, B, T, \{X_\alpha\})$ is an isomorphism.

For the reader's convenience we recall how the map Ψ_0 is defined. Fix $\tau \in \mathrm{Aut}(G)$, and choose any pair $T \subset B$ as in Definition 2.10. (We assume no relationship between τ and B or T.) If λ is any rational character of T, then $\lambda \circ \tau^{-1}$ is a rational character of $\tau(T)$. This defines an isomorphism

$$X^*(T) \to X^*(\tau(T))$$

carrying $\Delta(B,T)$ to $\Delta(\tau(B), \tau(T))$. Similarly, composition with τ carries one-parameter subgroups of T to one-parameter subgroups of $\tau(T)$. Assembling these maps, we get an isomorphism

$$\Psi_0(\tau, B, T) : \Psi_0(G, B, T) \to \Psi_0(G, \tau(B), \tau(T)).$$

Since both range and domain are *canonically* isomorphic to $\Psi_0(G)$, this map provides the automorphism $\Psi_0(\tau)$ that we want.

An automorphism belonging to one of the sets $\mathrm{Aut}(G, B, T, \{X_\alpha\})$ is called *distinguished*.

We have seen that the inner classes of real forms of G are in one-to-one correspondence with the equivalence classes of quasisplit real forms. The quasisplit real forms are easy to classify.

Proposition 2.12. *Suppose G is a complex connected reductive algebraic group. Then the equivalence classes of quasisplit real forms of G (and therefore also the inner classes of all real forms) are in one-to-one correspondence with the involutive automorphisms of $\Psi_0(G)$.*

Proof. Suppose σ is any antiholomorphic automorphism of G; we will show how to define an automorphism $\Psi_0(\sigma)$ of $\Psi_0(G)$. (This is not immediately handled by Proposition 2.11, since σ is not a holomorphic automorphism of G.) Fix a Borel subgroup B of G, and $T \subset B$ a maximal torus. If λ is a holomorphic character of T, then $\lambda \circ \sigma^{-1}$ is an antiholomorphic character of $\sigma(T)$; so $\overline{\lambda \circ \sigma^{-1}}$ is a holomorphic character of $\sigma(T)$. The map sending λ to $\overline{\lambda \circ \sigma^{-1}}$ is an isomorphism

$$X^*(T) \to X^*(\sigma(T)),$$

and it carries $\Delta(B,T)$ to $\Delta(\sigma(B), \sigma(T))$. Continuing in this way, we find that σ induces an isomorphism

$$\Psi_0(\sigma, B, T) : \Psi_0(G, B, T) \to \Psi_0(G, \sigma(B), \sigma(T)).$$

Since both of these objects are canonically isomorphic to $\Psi_0(G)$, we have the automorphism we want. If σ' is another antiholomorphic automorphism of G, then it is clear from the definitions that

$$\Psi_0(\sigma \circ (\sigma')^{-1}) = \Psi_0(\sigma) \circ \Psi_0(\sigma')^{-1}.$$

Here the map on the left is given by Proposition 2.11, and those on the right by the preceding construction. It follows from Proposition 2.11 that $\Psi_0(\sigma) = \Psi_0(\sigma')$ if and only if σ and σ' differ by an inner automorphism. This proves the injectivity of our map from inner classes of real forms to involutive automorphisms of the based root datum.

For the surjectivity, fix an automorphism a of $\Psi_0(G)$ of order 2, and data B, T, and $\{X_\alpha\}$ as in Proposition 2.8. We must find a real form σ of G with $\Psi_0(\sigma) = a$. The theory of Chevalley forms (forms of a reductive group defined and split over any field) shows that there is a real form σ_s of G such that

$$\sigma_s(B) = B, \qquad \sigma_s(T) = T, \qquad \sigma_s(X_\alpha) = X_\alpha,$$

and σ_s induces the identity automorphism of $X^*(T)$ and $X_*(T)$. On the other hand, the splitting of the exact sequence of Proposition 2.11 produces a holomorphic automorphism t_a of G such that

$$t_a(B) = B, \qquad t_a(T) = T, \qquad t_a(X_\alpha) = X_{a(\alpha)},$$

and t_a induces the automorphism a on $X^*(T)$ and $X_*(T)$. The composition $\sigma = t_a \circ \sigma_s$ is an antiholomorphic automorphism of G satisfying

$$\sigma(B) = B, \qquad \sigma(T) = T, \qquad \sigma(X_\alpha) = X_{t(\alpha)},$$

and σ induces the automorphism a on $X^*(T)$ and $X_*(T)$. This implies first of all that $\Psi_0(\sigma) = a$. Since a has order 2, it follows that σ^2 is a holomorphic automorphism of G acting trivially on T and fixing the various X_α. Such an automorphism is trivial by Proposition 2.8; so $\sigma^2 = 1$, and σ is the real form we want. Q.E.D.

We can now describe approximately the context in which we will do representation theory. We fix the connected reductive complex algebraic group G, and an inner class of real forms of G. This inner class is specified by an automorphism of order 2 of the based root datum $\Psi_0(G)$. We need to consider at the same time representations of various real forms of G. It is natural therefore to consider pairs (π, σ), with σ a real form of G (say in the specified inner class) and π a representation of $G(\mathbb{R}, \sigma)$. It is natural to define two such pairs (π, σ) and (π', σ') to be equivalent if there is an element $g \in G$ such that $\sigma' = \text{Ad}(g) \circ \sigma \circ \text{Ad}(g^{-1})$, and $\pi \circ \text{Ad}(g^{-1})$ (which is a representation of $G(\mathbb{R}, \sigma')$) on the space of π) is equivalent to π'. The difficulty with this definition first appears in the example of $SL(2, \mathbb{R})$, discussed after (2.3). If we let π be a holomorphic discrete series representation of $G(\mathbb{R}, \sigma) = SL(2, \mathbb{R})$, and π' the corresponding antiholomorphic discrete series representation, then this definition makes (π, σ) equivalent to (π', σ). Clearly this will lead to inconvenience at least when we try to use the theory to write precise character formulas. The next definition, taken from Definitions 1.12, 1.13, and 1.14 of Chapter 1, provides a way around the problem.

Definition 2.13. Suppose G is a connected reductive complex algebraic group. A *(weak) extended group containing G* is a real Lie group G^Γ subject to the following conditions.

(a1) G^Γ contains G as a subgroup of index two.

(a2) Every element of $G^\Gamma - G$ acts on G (by conjugation) as an antiholomorphic automorphism.

Condition (a1) may be rephrased as follows. Write $\Gamma = \mathrm{Gal}(\mathbb{C}/\mathbb{R})$ for the Galois group.

(a1′) There is a short exact sequence

$$1 \to G \to G^\Gamma \to \Gamma \to 1.$$

A *strong real form* of G^Γ is an element $\delta \in G^\Gamma - G$ such that $\delta^2 \in Z(G)$ has finite order. The *associated real form* for δ is the (antiholomorphic involutive) automorphism $\sigma(\delta)$ of G defined by conjugation by δ:

$$\sigma(\delta)(g) = \delta g \delta^{-1}.$$

The *group of real points of δ* is defined to be the group of real points of $\sigma(\delta)$:

$$G(\mathbb{R}, \delta) = \{\, g \in G \mid \delta g \delta^{-1} = g \,\}.$$

Two strong real forms δ and δ' of G^Γ are called *equivalent* if they are conjugate by G:

$$\delta \sim \delta' \text{ if and only if } \delta' = g \delta g^{-1} \text{ for some } g \in G.$$

Thus the set of strong real forms equivalent to a fixed strong real form δ is a homogeneous space G/H; the isotropy group is

$$H = \{\, g \in G \mid g \delta g^{-1} = \delta \,\} = G(\mathbb{R}, \delta)$$

(cf. (2.3)).

A *representation of a strong real form* of G^Γ is a pair (π, δ), subject to

(a) δ is a strong real form of G^Γ; and
(b) π is an admissible representation of $G(\mathbb{R}, \delta)$.

Two such representations (π, δ) and (π', δ') are said to be *(infinitesimally) equivalent* if there is an element $g \in G$ such that $g\delta g^{-1} = \delta'$, and $\pi \circ \mathrm{Ad}(g^{-1})$ is (infinitesimally) equivalent to π'. (In particular, this is possible only if the strong real forms are equivalent.) Finally, define

$$\Pi(G^\Gamma) = \Pi(G/\mathbb{R})$$

to be the set of (infinitesimal) equivalence classes of irreducible representations of strong real forms of G^Γ. (Here when we use the notation on the right, we must have in mind a particular weak extended group G^Γ.)

Clearly the equivalence of strong real forms implies the equivalence of the associated real forms. We have not formulated the obvious definition of "inner" for strong real forms, since any two strong real forms for the same G^Γ are automatically inner to each other. In order for this to be a reasonable definition, we need to know that every real form is represented by a strong real form of some extended group. This is a consequence of Proposition 2.14 and Corollary 2.16 below.

We pause now to give the proof of Lemma 1.15. (The extra datum $\mathcal{W}$ needed to complete the definition of an extended group plays no rôle in this lemma.) Surjectivity of the map is clear; what must be established is injectivity. So suppose that π and π' are irreducible representations of $G(\mathbb{R}, \delta_s)$ and $G(\mathbb{R}, \delta_{s'})$, respectively, and that the pairs (π, δ_s) and $(\pi', \delta_{s'})$ are equivalent. We must show that $\delta_s = \delta_{s'}$, and that π is equivalent to π'. The hypothesis means that there is an element $g \in G$ such that

$$g\delta_s g^{-1} = \delta_{s'}, \qquad \pi \circ \mathrm{Ad}(g^{-1}) \sim \pi'$$

The first condition says that the strong real forms δ_s and $\delta_{s'}$ are equivalent. Since they belong to a set of representatives for the equivalence classes, they must in fact coincide. Consequently g commutes with δ_s, and so belongs to $G(\mathbb{R}, \delta_s)$. But this means that $\pi \circ \mathrm{Ad}(g^{-1})$ is equivalent to π, the equivalence being implemented by the operator $\pi(g)$ on the space of π. The second condition therefore implies that π' is equivalent to π as representations of $G(\mathbb{R}, \delta_s)$, as we wished to show. Q.E.D.

We conclude this chapter by investigating the possible structures of weak extended groups.

Proposition 2.14. *Suppose G^Γ is a weak extended group. Then the set of real forms of G associated to strong real forms of G^Γ constitutes exactly one inner class of real forms.*

Proof. The conjugation action of any element δ of $G^\Gamma - G$ defines an antiholomorphic automorphism $\sigma(\delta)$ of G. This automorphism preserves $Z(G)$, and in fact its restriction σ_Z to $Z(G)$ is independent of the choice of δ.

The proof of Proposition 2.12 attaches to each element δ of $G^\Gamma - G$ an automorphism $a = \Psi_0(\delta)$ of the based root datum of G. The proof also shows that a is independent of the choice of δ, and that the various conjugation actions $\sigma(\delta)$ give all antiholomorphic automorphisms σ of G such that $\Psi_0(\sigma) = a$. In particular, Proposition 2.12 implies that for any real form σ in the inner class defined by a, there is a $\delta_1 \in G^\Gamma - G$ with $\sigma(\delta_1) = \sigma$. Since σ is an involution, this implies that $\delta_1^2 = z_1 \in Z(G)$. To complete the proof, we must show that our choice of δ_1 can be modified to make z_1 have finite order. We first compute

$$\sigma_Z(z_1) = \delta_1 z \delta_1^{-1} = \delta_1 \delta_1^2 \delta_1^{-1} = \delta_1^2 = z_1.$$

That is, $z_1 \in Z^{\sigma_Z}$. We can now apply the following elementary lemma.

Lemma 2.15. *Suppose Z is a (possibly disconnected) complex reductive abelian algebraic group, and σ is an antiholomorphic involutive automorphism of Z. Put*

$$(1 + \sigma)Z = \{\, z\sigma(z) \mid z \in Z \,\}.$$

Then the quotient group $Z^\sigma/(1 + \sigma)Z$ is finite, and each coset has a representative of finite order in Z.

Of course the quotient in the lemma is a Galois cohomology group.

We can now complete the proof of Proposition 2.14. By Lemma 2.15, there is an element $z_2 \in Z(G)$ such that $z = z_1 z_2 \sigma_Z(z_2)$ has finite order. Set

$$\delta = z_2 \delta_1.$$

Then $\sigma(\delta) = \sigma$, and $\delta^2 = z$ has finite order; so δ is the required strong real form. Q.E.D.

Here is the classification of weak extended groups.

Corollary 2.16. *Suppose G is a connected reductive complex algebraic group.*
 a) *Fix a weak extended group G^Γ for G. Let σ_Z be the antiholomorphic involution of $Z(G)$ defined by the conjugation action of any element*

of $G^\Gamma - G$. We can attach to G^Γ two invariants. The first of these is an involutive automorphism

$$a \in \text{Aut}(\Psi_0(G))$$

of the based root datum of G. The second is a class

$$\bar{z} \in Z(G)^{\sigma_Z}/(1+\sigma_Z)Z(G).$$

b) Suppose G^Γ and $(G^\Gamma)'$ are weak extended groups for G with the same invariants $(a, \bar{z})$. Then the identity map on G extends to an isomorphism from G^Γ to $(G^\Gamma)'$.

c) Suppose $a \in \text{Aut}(\Psi_0(G))$ is an involutive automorphism. Write σ_Z for the antiholomorphic involution of $Z(G)$ defined by the action of any real form σ in the inner class corresponding to a (Proposition 2.12); and suppose

$$\bar{z} \in Z(G)^{\sigma_Z}/(1+\sigma_Z)Z(G).$$

Then there is a weak extended group G^Γ with invariants $(a, \bar{z})$.

Proof. For (a), any element δ of $G^\Gamma - G$ defines by conjugation an antiholomorphic automorphism $\sigma(\delta)$ of G. By Proposition 2.12 and its proof, the corresponding automorphism $a = \Psi_0(\sigma(\delta))$ is independent of the choice of δ. To define $\bar{z}$, fix a quasisplit real form σ_q in the inner class defined by a, and choose $\delta_q \in G^\Gamma - G$ so that

$$\sigma(\delta_q) = \sigma_q$$

(as is possible by Proposition 2.14.) Since $\sigma_q^2 = 1$,

$$\delta_q^2 = z \in Z(G);$$

and using δ_q to compute σ_Z, we see that

$$\sigma_Z(z) = z.$$

By Proposition 2.7, any other choice of σ_q' differs from σ_q by conjugating with an element $\text{Ad}(g)$; so any other δ_q' is of the form

$$\delta_q' = z_1 g \delta_q g^{-1}.$$

We compute immediately that

$$(\delta_q')^2 = z_1 \sigma_Z(z_1) \delta_q^2,$$

so the class $\bar{z}$ is independent of all choices.

For (b), fix a quasisplit real form σ_q of G as in (a), and choose elements $\delta_q \in G^\Gamma - G$, $\delta_q'' \in (G^\Gamma)'$ so that

$$\sigma(\delta_q) = \sigma_q, \qquad \sigma(\delta_q'') = \sigma_q.$$

Write $z = \delta_q^2$, $z' = (\delta_q'')^2$. By the hypothesis on the invariants of the two weak extended groups, there is an element $z_1 \in Z(G)$ so that $z = z' (z_1 \sigma_Z(z_1))$. Set $\delta_q' = z_1 \delta_q''$; then

$$\sigma(\delta_q') = \sigma(\delta_q) = \sigma_q, \qquad \delta_q^2 = (\delta_q')^2 = z. \tag{2.17}(a)$$

Now the group G^Γ is the disjoint union of G and the coset $G\delta_q$; these are multiplied according to the rules

$$(g_1 \delta_q)(g_2 \delta_q) = g_1 \sigma_q(g_2) z, \qquad (g_1 \delta_q)(g_2) = g_1 \sigma_q(g_2) \delta_q \tag{2.17}(b)$$

and the obvious rules for the other two kinds of product. We can define a bijection from G^Γ to $(G^\Gamma)'$ by using the identity on G, and sending $g\delta_q$ to $g\delta_q'$. By (2.17), this bijection is a group homomorphism, proving (b).

For (c), choose a quasisplit real form σ_q of G in the inner class corresponding to a (Proposition 2.12), and a representative $z \in Z(G)$ of the class $\bar{z}$. Define G^Γ to consist of the union of G and the set of formal symbols $g\delta_q$ (topologized as the union of two copies of G). Introduce a multiplication on G^Γ by the rules in (2.17)(b); then it is a simple matter to check that G^Γ is a weak extended group with the desired invariants.

Q.E.D.

3. Structure theory: extended groups and Whittaker models

Part of the goal of the Langlands classification is a parametrization of the representations of real forms of G in terms of L-groups. A difficulty with this goal is that several different pairs (real form, representation) may be isomorphic. The basic example is $(SL(2, \mathbb{R}),$ discrete series), where we may have a holomorphic or an antiholomorphic discrete series representation with the same infinitesimal character. Even though the notion of strong real form allows us to separate these pairs (Lemma 1.15), it still gives no reason to prefer one over another. The L-group parameters we find for these representations (typically local systems of some kind) *do* include a distinguished parameter (a trivial local system). In order to establish a parametrization like Theorem 1.18, we therefore need (roughly speaking) a way to specify a preferred representation in each L-packet.

Langlands' program suggests a way to approach this problem. One can specify a "Whittaker model" for a quasisplit real form $G(\mathbb{R})$ of G (or rather an equivalence class of such models for an equivalence class of quasisplit strong real forms). Two such models must differ by an automorphism of $G(\mathbb{R})$, but not necessarily by an inner automorphism. It is essentially known from [30] that each tempered L-packet for $G(\mathbb{R})$ contains exactly one representation admitting a Whittaker model. (Such a result is expected over any local field, but of course it is unlikely to be established in the absence of a definition of L-packets.) In the case of non-tempered L-packets, often no representation admits a Whittaker model. Nevertheless, there will be exactly one irreducible representation in the L-packet for which the corresponding standard representation admits a Whittaker model. This representation will be taken as the "base point" corresponding to the trivial local system in Theorem 1.18.

Definition 3.1. Suppose $G(\mathbb{R})$ is a quasisplit real form of a complex connected reductive algebraic group. Fix $T \subset B$ Cartan and Borel subgroups of G defined over $\mathbb{R}$, and write $A \subset T$ for the maximal split torus in T. (Thus the identity component of $A(\mathbb{R})$ is an "Iwasawa A" for $G(\mathbb{R})$.) Write N for the unipotent radical of B. List the simple (restricted) roots of A in N as $\{\alpha_1, \ldots, \alpha_l\}$, and write $\mathfrak{g}_{\alpha_j} \subset \mathfrak{n}$ for the corresponding restricted root spaces. Then

$$\mathfrak{n}/[\mathfrak{n},\mathfrak{n}] \simeq \sum_{j=1}^{l} \mathfrak{g}_{\alpha_j}. \qquad (3.1)(a)$$

All the spaces here are defined over $\mathbb{R}$, so we get a corresponding decomposition of $\mathfrak{n}(\mathbb{R})/[\mathfrak{n}(\mathbb{R}),\mathfrak{n}(\mathbb{R})]$. In particular, the space of (one-dimensional) unitary characters of $N(\mathbb{R})$ is isomorphic (by taking differentials) to

$$\sum_{j=1}^{l} i\mathfrak{g}_{\alpha_j}(\mathbb{R})^*, \qquad (3.1)(b)$$

the space of imaginary-valued linear functionals on the simple restricted real root spaces. A unitary character χ of $N(\mathbb{R})$ is called *non-degenerate* if its restriction to each simple restricted root subgroup is non-trivial.

Suppose χ is a non-degenerate unitary character of $N(\mathbb{R})$. Write $\mathbb{C}_\chi$ for the one-dimensional space on which χ acts, and

$$\mathcal{L}_\chi = G(\mathbb{R}) \times_{N(\mathbb{R})} \mathbb{C}_\chi \qquad (3.1)(c)$$

for the corresponding line bundle on $G(\mathbb{R})/N(\mathbb{R})$. The *Whittaker model* $\mathrm{Wh}(\chi)$ *for* $G(\mathbb{R})$ *defined by* χ is the space of smooth sections of $\mathcal{L}_\chi$:

$$\mathrm{Wh}(\chi) \simeq \{\, f \in C^\infty(G(\mathbb{R})) \mid f(gn) = \chi(n)^{-1} f(g) \,\} \qquad (3.1)(d).$$

We make $\mathrm{Wh}(\chi)$ into a smooth representation of $G(\mathbb{R})$ by left translation.

A Hilbert space representation $(\pi, \mathcal{H}_\pi)$ of $G(\mathbb{R})$ is said to *admit a Whittaker model of type* χ if there is a non-zero continuous map

$$\mathcal{H}_\pi^\infty \to \mathrm{Wh}(\chi) \qquad (3.1)(e)$$

respecting the action of $G(\mathbb{R})$. (Equivalently, π should admit a non-zero distribution vector transforming according to the character χ under $N(\mathbb{R})$.)

We consider first the uniqueness of non-degenerate characters. The following lemma is well illustrated by the examples of $GL(2,\mathbb{R})$ (which has up to conjugacy only one kind of Whittaker model) and $SL(2,\mathbb{R})$ (which has two). The element t of (c) below can always be chosen in $T(\mathbb{R})$ in the first case, but not in the second.

3. Structure theory: extended groups and Whittaker models

Lemma 3.2([30], Lemma 6.2.1). *Suppose $G(\mathbb{R})$ is a quasisplit real form of a complex connected reductive algebraic group, and $T \subset B$ are Cartan and Borel subgroups defined over $\mathbb{R}$.*

a) *The decomposition (3.1)(a) of $\mathfrak{n}/[\mathfrak{n},\mathfrak{n}]$ is invariant under $\mathrm{Ad}(B)$. Consequently the adjoint action of $B(\mathbb{R})$ on characters of $N(\mathbb{R})$ preserves the set of non-degenerate unitary characters. In particular, the notion of non-degenerate unitary character of $N(\mathbb{R})$ is independent of the choice of T.*
b) *Suppose $t \in T$. Then the automorphism $\mathrm{Ad}(t)$ of G is defined over $\mathbb{R}$ if and only if it preserves each of the simple real restricted root spaces $\mathfrak{g}_{\alpha_j}(\mathbb{R})$.*
c) *Suppose χ and χ' are non-degenerate unitary characters of $N(\mathbb{R})$. Then there is an element $t \in T$ such that $\mathrm{Ad}(t)$ is defined over $\mathbb{R}$ (so that $\mathrm{Ad}(t)$ defines an automorphism of $N(\mathbb{R})$), and $t \cdot \chi$ is equal to χ' (the action on characters being defined by composition of the character with the inverse of the automorphism of $N(\mathbb{R})$).*
d) *In (c), the coset $tZ(G)$ is uniquely determined by χ and χ'. In particular, χ is conjugate to χ' by $G(\mathbb{R})$ if and only if $t \in T(\mathbb{R})Z(G)$.*

Proof. For (a), $\mathrm{Ad}(N)$ acts trivially on $\mathfrak{n}/[\mathfrak{n},\mathfrak{n}]$; so we need only consider $\mathrm{Ad}(T)$. Because the decomposition arises from the weights of the action of a subtorus of T, the invariance is clear. The second claim follows. For the last, recall that any two choices of $T(\mathbb{R})$ are conjugate by $N(\mathbb{R})$.

For (b), the condition is obviously necessary for $\mathrm{Ad}(t)$ to be defined over $\mathbb{R}$; so suppose that it holds. Write σ for the complex conjugation on G. Because σ preserves T, it must permute the root spaces in $\mathfrak{g}$; we write σ for the corresponding permutation of the roots. Then $\mathrm{Ad}(t)$ is defined over $\mathbb{R}$ if and only if for every root β of T in G, we have

$$\beta(\sigma t) = \overline{(\sigma\beta)(t)}. \qquad (3.3)(a)$$

It suffices to verify this condition for simple roots β. Write α for the restriction of β to A; then the real restricted root space $\mathfrak{g}_\alpha$ consists of elements of the form

$$X + \sigma X \qquad (X \in \mathfrak{g}_\beta).$$

Now

$$\mathrm{Ad}(t)(X + \sigma X) = \beta(t)X + (\sigma\beta(t))\sigma X. \qquad (3.3)(b)$$

It follows that $\operatorname{Ad}(t)$ preserves the real restricted root space if and only if

$$(\sigma\beta(t))\sigma X = \sigma(\beta(t)X).$$

Since σ is conjugate-linear, this is equivalent to (3.3)(a).

For (c), we regard the complexified differential of χ as a character of the Lie algebra n. As such, it has a complex value $d\chi(X) \in \mathbb{C}$ on any $X \in \mathfrak{n}$. The condition that χ be unitary is equivalent to $d\chi$ taking purely imaginary values on each simple restricted real root space. By (3.3)(a), this is equivalent to

$$d\chi(X) = -\overline{d\chi(\sigma X)} \qquad (X \in \mathfrak{g}_\beta) \qquad (3.4)(a)$$

for every simple root β. The non-degeneracy condition is

$$d\chi(X) \neq 0 \qquad (X \in \mathfrak{g}_\beta - 0). \qquad (3.4)(b)$$

Suppose now that χ and χ' are as in (c) of the lemma. Since the simple roots are linearly independent, we can find $t \in T$ with

$$\beta(t) = d\chi(X)/d\chi'(X) \qquad (X \in \mathfrak{g}_\beta - 0). \qquad (3.4)(c)$$

Combining (3.4)(c) with (3.4)(a) gives exactly the condition in (3.3)(a) for $\operatorname{Ad}(t)$ to be defined over $\mathbb{R}$. To compute the action of t on χ, it is enough to evaluate the differential at an element $X \in \mathfrak{g}_\beta$. This is

$$d(t \cdot \chi)(X) = d\chi(\operatorname{Ad}(t^{-1})X) = d\chi(\beta(t^{-1})X) = \beta(t)^{-1}d\chi(X) = d\chi'(X).$$

It follows that $t \cdot \chi = \chi'$, as we wished to show.

For (d), the proof of (c) shows that the choice (3.4)(c) of $\beta(t)$ (for every simple β) is forced by the requirement that $t \cdot \chi = \chi'$. This gives the first assertion. For the second, χ is conjugate to χ' by $G(\mathbb{R})$ if and only if it is conjugate by the normalizer $B(\mathbb{R}) = T(\mathbb{R})N(\mathbb{R})$ of $N(\mathbb{R})$. Since $N(\mathbb{R})$ fixes χ, (d) follows. Q.E.D.

Recall now from Definition 1.12 the notion of an *extended group* $(G^\Gamma, \mathcal{W})$ for G. The *invariants* of the extended group are the automorphism a of the based root datum $\Psi_0(G)$ attached to the underlying weak extended group G^Γ (Corollary 2.16), and the element

$$z = \delta^2 \in Z(G)^{\sigma_Z}. \qquad (3.5)$$

3. Structure theory: extended groups and Whittaker models   45

Here δ is any element of a triple $(\delta, N, \chi) \in \mathcal{W}$ (Definition 1.12). The element z is independent of the choice of δ, since $\mathcal{W}$ is a single conjugacy class under G.

Here is the classsification of extended groups.

Proposition 3.6. *Suppose G is a connected reductive complex algebraic group.*
a) Suppose $(G^\Gamma, \mathcal{W})$ and $\left((G^\Gamma)', \mathcal{W}'\right)$ are extended groups for G with the same invariants (a, z) (cf. (3.5)). Then the identity map on G extends to an isomorphism from G^Γ to $(G^\Gamma)'$ carrying $\mathcal{W}$ to $\mathcal{W}'$. Any two such extensions differ by an inner automorphism of G^Γ from $Z(G)$.
b) Fix a weak extended group G^Γ for G with invariants $(t, \overline{z})$. If $(G^\Gamma, \mathcal{W})$ is an extended group, then its second invariant is a representative for the class of $\overline{z}$. Conversely, if $z \in Z(G)^{\sigma_z}$ is an element of finite order representing the class of $\overline{z}$, then there is an extended group structure on G^Γ with second invariant z.

Proof. For (a), fix $(\delta, N, \chi) \in \mathcal{W}$. Since $\delta^2 = z \in Z(G)$, conjugation by δ defines an antiholomorphic involutive automorphism $\sigma = \sigma(\delta)$ of G, and thus a real form $G(\mathbb{R})$. By condition (b)(2) in Definition 1.12, $G(\mathbb{R})$ is quasisplit. The real forms defined analogously using $\mathcal{W}'$ are also quasisplit, and inner to $G(\mathbb{R})$. By Proposition 2.12, at least one of them must coincide with $G(\mathbb{R})$. That is, we can find $(\delta', N', \chi') \in \mathcal{W}'$ so that

$$\sigma(\delta') = \sigma. \qquad (3.7)(a)$$

Now N and N' are maximal unipotent subgroups of G defined over $\mathbb{R}$. They are therefore conjugate by an element $g \in G(\mathbb{R})$. After replacing (δ', N', χ') by their conjugates by g (which does not change δ', and therefore preserves (3.7)(a)) we may assume that

$$N' = N. \qquad (3.7)(b)$$

Now χ and χ' are non-degenerate unitary characters of $N(\mathbb{R})$. By Lemma 3.2, there is an element $t \in G$ normalizing $N(\mathbb{R})$, carrying χ' to χ, and with $\mathrm{Ad}(t)$ defined over $\mathbb{R}$. This last condition means that

$$\sigma(t) = tw$$

for some $w \in Z(G)$. We finally replace (δ', N', χ') by their conjugates by t. This replaces δ' by $w^{-1}\delta'$, and therefore (since w is central) does not affect (3.7)(a). Condition (3.7)(b) is preserved since t normalizes N. By the choice of t,

$$\chi' = \chi. \qquad (3.7)(c)$$

We can now construct the desired isomorphism between the extended groups as in the proof of Corollary 2.16(b), by sending δ to δ'.

For the uniqueness, any other such isomorphism must send (δ, N, χ) to a G-conjugate $g \cdot (\delta', N, \chi)$ of (δ', N, χ) (since it carries $\mathcal{W}$ to $\mathcal{W}'$). We wish to show that $g \in G(\mathbb{R})Z(G)$. Since the isomorphism is the identity on G, we deduce three facts: the conjugation actions of δ and $g(\delta')g^{-1}$ must agree on G; $gNg^{-1} = N$; and $g \cdot \chi = \chi$. The first of these facts means that $\text{Ad}(g)$ is defined over $\mathbb{R}$. The second fact implies that $g = tn \in B$. Since $\text{Ad}(g)$ is defined over $\mathbb{R}$, it follows that $n \in N(\mathbb{R})$ and that $\text{Ad}(t)$ is defined over $\mathbb{R}$. Because of Lemma 3.2(d), the third fact guarantees that $t \in Z(G)$, as we wished to show.

The first part of (b) is clear from the definitions. For the second part, the argument before (2.17)(a) shows how to find an element δ of $G^\Gamma - G$ with $\sigma(\delta)$ a quasisplit real form of G, and $\delta^2 = z$. Let N be any maximal unipotent subgroup of G defined over $\mathbb{R}$, and let χ be any non-degenerate unitary character of $N(\mathbb{R})$. Then the G-conjugacy class $\mathcal{W}$ of the triple (δ, N, χ) is evidently an extended group structure with second invariant z. Q.E.D.

Corollary 3.8. *Suppose G is a connected complex reductive algebraic group endowed with an inner class of real forms. Then the equivalence classes of extended groups for G are parametrized by elements of finite order in $Z(G)^{\sigma z}$ (cf. Corollary 2.16).*

We have found no use for the wide selection of extended groups provided by Corollary 3.8, and no reason to prefer one choice over another. This should probably be taken as evidence that our definitions are imperfect. The analogous phenomenon for dual groups (Proposition 4.7) has a much clearer rôle, as we will see.

4. Structure theory: L-groups

The based root datum for a complex connected reductive algebraic group (Definition 2.10) characterizes that group up to isomorphism. On the other hand, based root data of reductive groups may be characterized by some simple axioms (see [50]). These axioms are symmetric in X^* and X_*; that is, the quadruple $\Psi_0 = (X^*, \Delta, X_*, \Delta^\vee)$ is the based root datum of a reductive group if and only if $^\vee\Psi_0 = (X_*, \Delta^\vee, X^*, \Delta)$ is as well. We call $^\vee\Psi_0$ the *dual based root datum to* Ψ_0. Notice that the automorphism groups of dual based root data are canonically isomorphic:

$$\operatorname{Aut}(\Psi_0) \simeq \operatorname{Aut}(^\vee\Psi_0). \tag{4.1}$$

Definition 4.2 (see [34] or [10]). Suppose G is a complex connected reductive algebraic group. A *dual group for* G is a complex connected reductive algebraic group $^\vee G$, together with an isomorphism from the dual of the based root datum for G to the based root datum for $^\vee G$:

$$^\vee\Psi_0(G) \simeq \Psi_0(^\vee G).$$

By Proposition 2.8, any two dual groups for G are isomorphic, and the isomorphism is canonical up to inner automorphism of either group.

Definition 4.3 (cf. [1], Definition 10.5). Suppose G is a complex connected reductive algebraic group. A *weak E-group for* G is an algebraic group $^\vee G^\Gamma$ containing a dual group $^\vee G$ for G as a subgroup of index 2. That is, there is a short exact sequence

$$1 \to {}^\vee G \to {}^\vee G^\Gamma \to \Gamma \to 1.$$

Here $\Gamma = \operatorname{Gal}(\mathbb{C}/\mathbb{R})$ as in Definition 2.13.

It will often be convenient for us to use this terminology in a various more specific ways, which we record here though they require Proposition 4.4 below for their formulation. Suppose G^Γ is a weak extended group for G (Definition 2.13). Recall that there is attached to G^Γ an involutive automorphism a of the based root datum of G (Corollary 2.16); of course a really depends only on the inner class of real forms of G defined by G^Γ. A *weak E-group for* G^Γ is a weak E-group for G with second invariant

a (Proposition 4.4(a)). We may also call this a *weak E-group for G and the specified inner class of real forms*, or even a *weak E-group for $G(\mathbb{R})$* (with $G(\mathbb{R})$ a real form in the specified inner class).

Finally, we will occasionally need to use this terminology in a *less specific* way. We will say that a *weak E-group* is an algebraic group H^Γ containing a complex connected reductive algebraic subgroup H of index two. That is, there is a short exact sequence

$$1 \to H \to H^\Gamma \to \Gamma \to 1.$$

Of course every complex connected reductive algebraic group may be regarded as a dual group, so this definition does not enlarge the class of weak E-groups.

The notion of dual group is symmetric, in the sense that if $^\vee G$ is a dual group for G then G is a dual group for $^\vee G$. We emphasize that there is no such symmetry between the notions of weak extended groups and weak E-groups. An E-group is always a complex algebraic group, but an extended group never is (because of condition (a2) in Definition 2.13).

Here is the classification of weak E-groups.

Proposition 4.4. *Suppose G is a connected reductive complex algebraic group.*

a) *Fix a weak E-group $^\vee G^\Gamma$ for G. Let θ_Z be the holomorphic involution of $Z(^\vee G)$ defined by the conjugation action of any element of $^\vee G^\Gamma - {^\vee G}$. We can attach to $^\vee G^\Gamma$ two invariants. The first of these is an involutive automorphism*

$$a \in \mathrm{Aut}(\Psi_0(G))$$

of the based root datum of G. The second is a class

$$\bar{z} \in Z(^\vee G)^{\theta_Z}/(1+\theta_Z)Z(^\vee G).$$

b) *Suppose $^\vee G^\Gamma$ and $(^\vee G^\Gamma)'$ are weak E-groups for G with the same invariants $(a, \bar{z})$. Then any one of the canonical isomorphisms from $^\vee G$ to $^\vee G'$ described after Definition 4.2 extends to an isomorphism from $^\vee G^\Gamma$ to $(^\vee G^\Gamma)'$.*

c) *Suppose $a \in \mathrm{Aut}(\Psi_0(G))$ is an involutive automorphism, and $^\vee G$ is a dual group of G. Write θ_Z for the involution of $Z(^\vee G)$ defined by the action of any automorphism σ corresponding to a (Proposition*

2.11 and (4.1)); and suppose

$$\bar{z} \in Z({}^{\vee}G)^{\theta_Z}/(1+\theta_Z)Z({}^{\vee}G).$$

Then there is a weak E-group ${}^{\vee}G^{\Gamma}$ with invariants $(a, \bar{z})$.

(In order to make sense of the statement in (b), note that by the remarks after Definition 4.2, the centers of any two dual groups of G are canonically isomorphic.)

The proof is a slightly simpler version of the proof of Corollary 2.16, so we omit it. We should record the construction of $\bar{z}$, however. Let ${}^{\vee}\delta$ be any element of ${}^{\vee}G^{\Gamma} - {}^{\vee}G$ such that the conjugation action of ${}^{\vee}\delta$ on ${}^{\vee}G$ is a distinguished automorphism (as defined after Proposition 2.11). (Such elements necessarily exist.) Then

$${}^{\vee}\delta^2 = z \in Z({}^{\vee}G) \tag{4.5}$$

is a representative of $\bar{z}$.

Definition 4.6. Suppose G is a complex connected reductive algebraic group. An *E-group for* G is a pair $({}^{\vee}G^{\Gamma}, \mathcal{D})$, subject to the following conditions.

(a) ${}^{\vee}G^{\Gamma}$ is a weak E-group for G (Definition 4.3).
(b) $\mathcal{D}$ is a ${}^{\vee}G$-conjugacy class of elements of finite order in ${}^{\vee}G^{\Gamma} - {}^{\vee}G$.
(c) Suppose ${}^{\vee}\delta \in \mathcal{D}$. Then conjugation by ${}^{\vee}\delta$ is a distinguished involutive automorphism of ${}^{\vee}G$ (see the definition after Proposition 2.11).

The *invariants* of the E-group are the automorphism a attached to ${}^{\vee}G^{\Gamma}$ by Proposition 4.4, and the element

$$z = {}^{\vee}\delta^2 \in Z({}^{\vee}G)^{\theta_Z};$$

here ${}^{\vee}\delta$ is any element of $\mathcal{D}$.

An *L-group for* G is an E-group whose second invariant is equal to 1. That is, we replace condition (b) above by

(b)' $\mathcal{D}$ is a ${}^{\vee}G$-conjugacy class of elements of order two in ${}^{\vee}G^{\Gamma} - {}^{\vee}G$.

Of course the notion of L-group is due to Langlands (cf. [34], [10]). Just as in Definition 4.3, we can speak of E-groups or L-groups attached to an extended group, to an inner class of real forms, to a single real form, or to nothing at all.

Here is the classification of E-groups.

Proposition 4.7. *Suppose G is a connected reductive complex algebraic group.*

a) Suppose $({}^\vee G^\Gamma, \mathcal{D})$ and $\left(({}^\vee G^\Gamma)', \mathcal{D}'\right)$ are E-groups for G with the same invariants (a, z). Then any one of the canonical isomorphisms from ${}^\vee G$ to ${}^\vee G'$ described after Definition 4.2 extends to an isomorphism from ${}^\vee G^\Gamma$ to $({}^\vee G^\Gamma)'$ carrying $\mathcal{D}$ to $\mathcal{D}'$.

b) In the setting of (a), suppose τ_1 and τ_2 are isomorphisms from ${}^\vee G^\Gamma$ to $({}^\vee G^\Gamma)'$ with the property that $\tau_i |_{{}^\vee G}$ is one of the canonical isomorphisms described after Definition 4.2, and that $\tau_i(\mathcal{D}) = \mathcal{D}'$. Then there is an element $g \in {}^\vee G$ such that $\tau_1 = \tau_2 \circ \mathrm{Ad}(g)$.

c) Fix a weak E-group ${}^\vee G^\Gamma$ for G with invariants $(a, \bar{z})$. If $({}^\vee G^\Gamma, \mathcal{D})$ is an E-group, then its second invariant (Definition 4.6) is a representative for the class of $\bar{z}$. Conversely, if $z \in Z({}^\vee G)^{\theta_z}$ is an element of finite order representing the class of $\bar{z}$, then there is an E-group structure on ${}^\vee G^\Gamma$ with second invariant z.

The argument follows that given for Proposition 3.6, and we leave it to the reader.

Corollary 4.8. *Suppose G is a connected complex reductive algebraic group endowed with an inner class of real forms. Then there is an L-group for G and this class of real forms (Definition 4.6). Any two such L-groups $({}^\vee G^\Gamma, \mathcal{D})$ and $(({}^\vee G^\Gamma)', \mathcal{D}')$ are isomorphic. We may choose this isomorphism τ so that $\tau(\mathcal{D}) = \mathcal{D}'$, and the restriction of τ to ${}^\vee G$ is one of the canonical isomorphisms described after Definition 4.2. This choice of τ is unique up to composition with an inner automorphism from ${}^\vee G$.*

In the course of the proof of Theorem 1.18 (see Definition 13.7 below) it will be convenient to reformulate the definition of E-group. We need a preliminary definition, which will be applied in a moment to a dual group.

Definition 4.9. Suppose G is a complex connected reductive algebraic group, B is a Borel subgroup, and $T \subset B$ is a maximal torus. Write

$$\chi(2^\vee \rho) : \mathbb{C}^\times \to T \qquad (4.9)(a)$$

for the sum of the positive coroots, an element of $X_*(T)$. Set

$$m({}^\vee \rho) = \chi(2^\vee \rho)(i) \in T, \quad z({}^\vee \rho) = m({}^\vee \rho)^2 = \chi(2^\vee \rho)(-1). \qquad (4.9)(b)$$

The familiar fact that the sum of the positive coroots takes the value 2 on each simple root α means that

$$\mathrm{Ad}(m(^\vee\!\rho))(X_\alpha) = i^2 X_\alpha = -X_\alpha, \quad \mathrm{Ad}(z(^\vee\!\rho))(X_\alpha) = X_\alpha \quad (4.9)(c)$$

for any element X_α of the α root space. In particular, $z(^\vee\!\rho) \in Z(G)$; this element is independent of the choice of B and T, so it is preserved by all automorphisms of G.

Suppose θ is an involutive (holomorphic) automorphism of G preserving a Borel subgroup B. Fix a θ-stable maximal torus $T \subset B$ (as is possible). We say that the pair (θ, B) is *large* if θ acts by -1 on each θ-stable simple root space. We say that (θ, B) is *distinguished* if θ acts by $+1$ on each θ-stable simple root space. (Since T is unique up to conjugation by the θ-invariants in B, these definitions do not depend on the choice of T.)

Lemma 4.10. *Suppose G is a complex connected reductive algebraic group, $B \subset G$ is a Borel subgroup, and θ is an involutive holomorphic automorphism of G preserving B. Fix a θ-stable maximal torus $T \subset B$.*
a) The automorphism θ fixes the map $\chi(2^\vee\!\rho)$ of Definition 4.9. In particular, it fixes the element $m(^\vee\!\rho)$; so

$$\theta' = \mathrm{Ad}(m(^\vee\!\rho)) \circ \theta$$

is another involutive automorphism of G preserving B.
b) The pair (θ, B) is large (respectively distinguished) if and only if (θ', B) is distinguished (respectively large).
c) The pair (θ, B) is distinguished if and only if θ is a distinguished automorphism (as defined after Proposition 2.11).

Proof. Parts (a) and (b) are immediate from Definition 4.9 (particularly (4.9)(c)). The "if" in part (c) is also immediate from Definition 4.9. For the "only if," suppose (θ, B) is distinguished in the sense of Definition 4.9, and T is a θ-stable maximal torus in B. To see that θ is distinguished in the sense of Chapter 2, we must find a set of simple root vectors $\{X_\alpha\}$ permuted by θ. Begin with any set $\{X'_\alpha\}$ of simple root vectors. If α is fixed by θ, then the assumption that (θ, B) is distinguished means that $\theta X'_\alpha = X'_\alpha$. For such roots α we define $X_\alpha = X'_\alpha$. The remaining simple roots occur in pairs interchanged by θ; by choosing one representative of each pair, we can list them as $\{\beta_1, \ldots, \beta_r, \theta\beta_1, \ldots, \theta\beta_r\}$. We now define

$$X_{\beta_i} = X'_{\beta_i}, \qquad X_{\theta\beta_i} = \theta X'_{\beta_i}.$$

Then $\theta \in \mathrm{Aut}(G, B, T, \{X_\alpha\})$, as we wished to show. Q.E.D.

Lemma 4.11. *Suppose G is a complex connected reductive algebraic group, and θ is a distinguished involutive holomorphic automorphism of G. Then any two θ-stable Borel subgroups of G are conjugate by the invariants of θ in G.*

Proof. The proof requires a few ideas from the theory of Cartan involutions, and we only sketch it. Write K for the group of fixed points of θ in G, and K_0 for its identity component. Suppose B and B' are θ-stable. Choose θ-stable tori T and T' in B and B'; then $T \cap K_0$ and $T' \cap K_0$ are maximal tori in K_0, contained in Borel subgroups $B \cap K_0$ and $B' \cap K_0$. By conjugacy of maximal tori and Borel subgroups in the reductive group K_0, we may assume (after replacing B' by a conjugate under K_0) that they coincide:

$$T \cap K_0 = T' \cap K_0, \qquad B \cap K_0 = B' \cap K_0.$$

Now $T \cap K_0$ contains regular elements of G (for example in the image of the map $\chi(2^\vee \rho)$), so the identity component of its centralizer is exactly T. It follows that $T = T'$.

Up until now we have used only the assumption that θ is an involutive automorphism. A root of T in G is B-simple if and only if its restriction to $T \cap K_0$ is simple for $B \cap K_0$; this is a consequence of the fact that (θ, B) is distinguished (Definition 4.9 and Lemma 4.10(c)); and similarly for B'. Using this fact, we see that the simple roots of T in B and in B' coincide, so $B = B'$. Q.E.D.

Using these two lemmas, we can formulate a first approximation to our new definition of E-groups.

Definition 4.12. Suppose G is a complex connected reductive algebraic group. An *E-group for G* is a pair $(^\vee G^\Gamma, \mathcal{E})$, subject to the following conditions.

(a) $^\vee G^\Gamma$ is a weak extended group for G (Definition 4.3).
(b) $\mathcal{E}$ is a $^\vee G$-conjugacy class of pairs $(^\vee \delta, {}^d B)$, with $^\vee \delta$ an element of finite order in $^\vee G^\Gamma - {}^\vee G$, and $^d B$ a Borel subgroup of $^\vee G$.
(c) Suppose $(^\vee \delta, {}^d B) \in \mathcal{E}$. Then conjugation by $^\vee \delta$ is a distinguished involutive automorphism of $^\vee G$ (see the definition after Proposition 2.11) preserving $^d B$.

To see that Definition 4.12 is equivalent to Definition 4.6, suppose first that $(^\vee G^\Gamma, \mathcal{E})$ is an E-group in the sense of Definition 4.12. Set

$$\mathcal{D} = \{ {}^\vee \delta \mid (^\vee \delta, {}^d B) \in \mathcal{E} \}. \qquad (4.13)(a)$$

Then $({}^\vee G^\Gamma, \mathcal{D})$ is an E-group in the sense of Definition 4.6. Conversely, suppose $({}^\vee G^\Gamma, \mathcal{D})$ is an E-group in the sense of Definition 4.6. Set

$$\mathcal{E} = \{ ({}^\vee\delta, {}^dB) \mid {}^\vee\delta \in \mathcal{D}, {}^dB \text{ a } {}^\vee\delta\text{-invariant Borel subgroup of }{}^\vee G \}. \tag{4.13}(b)$$

Lemma 4.11 implies that $\mathcal{E}$ is a single ${}^\vee G$-conjugacy class, so $({}^\vee G^\Gamma, \mathcal{E})$ is an E-group in the sense of Definition 4.12.

It is now a simple matter to pass to our final definition of E-groups.

Definition 4.14 ([1], Definition 10.5). Suppose G is a complex connected reductive algebraic group. An *E-group for* G is a pair $({}^\vee G^\Gamma, \mathcal{S})$, subject to the following conditions.

(a) ${}^\vee G^\Gamma$ is a weak E-group for G (Definition 4.3).
(b) $\mathcal{S}$ is a ${}^\vee G$-conjugacy class of pairs $({}^\vee\delta, {}^dB)$, with ${}^\vee\delta$ an element of finite order in ${}^\vee G^\Gamma - {}^\vee G$, and ${}^dB$ a Borel subgroup of ${}^\vee G$.
(c) Suppose $({}^\vee\delta, {}^dB) \in \mathcal{S}$. Then conjugation by ${}^\vee\delta$ defines an involutive automorphism θ of ${}^\vee G$ preserving ${}^dB$. The pair $(\theta, {}^dB)$ is large (Definition 4.9).

The *invariants* of the E-group are the automorphism a attached to ${}^\vee G^\Gamma$ by Proposition 4.5, and the element

$$z = z(\rho){}^\vee\delta^2 \in Z({}^\vee G)^{\theta z};$$

here $({}^\vee\delta, {}^dB)$ is any element of $\mathcal{S}$, and $z(\rho) \in Z({}^\vee G)$ is the distinguished element described in Definition 4.9.

An *L-group for* G is an E-group whose second invariant is equal to 1. That is, the elements ${}^\vee\delta$ appearing in (b) are required to satisfy ${}^\vee\delta^2 = z(\rho)$.

We recall from section 9 of [1] the proof that Definitions 4.12 and 4.14 are equivalent. Suppose that $({}^\vee G^\Gamma, \mathcal{E})$ is an E-group in the sense of Definition 4.12. Fix $({}^\vee\delta, {}^dB) \in \mathcal{E}$, and write θ for the automorphism of ${}^\vee G$ defined by conjugation by ${}^\vee\delta$. Choose a θ-stable maximal torus ${}^dT \subset {}^dB$, and define $m(\rho) \in {}^dT$ as in Definition 4.9. Set

$${}^\vee\delta' = m(\rho)^{-1}{}^\vee\delta; \tag{4.15}(a)$$

the corresponding automorphism of ${}^\vee G$ is

$$\theta' = \text{Ad}(m(\rho))\theta. \tag{4.15}(b)$$

Changing the choice of ${}^dT$ changes ${}^\vee\delta'$ only by conjugation by ${}^dB$; so the conjugacy class

$$\mathcal{S} = \mathrm{Ad}(^\vee G)(^\vee\delta', {}^d B) \qquad (4.15)(c)$$

is well-defined. By hypothesis, the pair $(\theta, {}^d B)$ is distinguished, so by (4.15)(b) and Lemma 4.10 the pair $(\theta', {}^d B)$ is large. Consequently $(^\vee G^\Gamma, \mathcal{S})$ is an E-group in the sense of Definition 4.14. This argument may be reversed without difficulty, proving the equivalence of the two definitions.

5. Langlands parameters and L-homomorphisms

The (local) goal of Langlands' theory of L-groups is the description of representations of real forms of G in terms of an L-group of G. In this chapter we begin our detailed analysis of the parameters that will appear in this description, without as yet explaining how they are related to representation theory. As is explained in [1], E-groups play the same rôle with respect to the description of certain projective representations (corresponding always to linear covering groups) and they can be treated at the same time without difficulty. (Indeed it is apparent from Definitions 4.3 and 4.6 that E-groups are in certain respects *less* complicated than L-groups.) In any case we will need to have E-groups available when we discuss endoscopy.

Definition 5.1 ([34], section 2). Suppose $^{\vee}H^{\Gamma}$ and $^{\vee}G^{\Gamma}$ are weak E-groups (Definition 4.3). An *L-homomorphism from* $^{\vee}H^{\Gamma}$ *to* $^{\vee}G^{\Gamma}$ is a morphism $\epsilon : {}^{\vee}H^{\Gamma} \to {}^{\vee}G^{\Gamma}$ of algebraic groups, with the property that the diagram

$$\begin{array}{ccc} {}^{\vee}H^{\Gamma} & \xrightarrow{\epsilon} & {}^{\vee}G^{\Gamma} \\ & \searrow \swarrow & \\ & \Gamma & \end{array}$$

(cf. Definition 4.3) commutes. (Even when $^{\vee}H^{\Gamma}$ and $^{\vee}G^{\Gamma}$ are E-groups, we do *not* require that ϵ should respect the distinguished conjugacy classes $\mathcal{D}$, $\mathcal{E}$, or $\mathcal{S}$ of Definitions 4.6, 4.12, and 4.14.) Two such morphisms ϵ and ϵ' are said to be *equivalent* if they are conjugate by the action of $^{\vee}G$; that is, if there is an element $g \in {}^{\vee}G$ so that $\epsilon' = \text{Ad}(g) \circ \epsilon$.

A central feature of Langlands' (local) philosophy is that an equivalence class of L-homomorphisms should give rise to something like a map ("transfer") from representations of (real forms of) H to representations of G. In particular, any object we construct from an E-group that is intended to correspond to representations of G ought to be covariant with respect to L-homomorphisms. The first object of this kind that we will consider is the set of "Langlands parameters" of Definition 1.5. To describe these more carefully, we begin by recalling the definition of the Weil group for $\mathbb{R}$.

Definition 5.2 (see [34], [10], or [51]). The *Weil group of* $\mathbb{R}$ is the real Lie group $W_\mathbb{R}$ generated by $\mathbb{C}^\times$ and a distinguished element j, subject to the relations

$$j^2 = -1 \in \mathbb{C}^\times, \quad jzj^{-1} = \overline{z}. \qquad (5.2)(*)$$

We define a homomorphism from $W_\mathbb{R}$ to the Galois group Γ of $\mathbb{C}/\mathbb{R}$ by sending $\mathbb{C}^\times$ to the identity and j to complex conjugation. (This gives the exact sequence

$$1 \to \mathbb{C}^\times \to W_\mathbb{R} \to \Gamma \to 1$$

of (1.4)(b).) Again we emphasize that the Weil group is *not* a complex Lie group.

Suppose $^\vee G^\Gamma$ is a weak E-group. A *quasiadmissible homomorphism* (or *Langlands parameter*) ϕ from $W_\mathbb{R}$ to $^\vee G^\Gamma$ is a continuous group homomorphism satisfying

(a) the diagram

$$\begin{array}{ccc} W_\mathbb{R} & \xrightarrow{\phi} & {}^\vee G^\Gamma \\ & \searrow \quad \swarrow & \\ & \Gamma & \end{array}$$

is commutative; and

(b) $\phi(\mathbb{C}^\times)$ consists of semisimple elements of $^\vee G$.

Two such homomorphisms are said to be *equivalent* if they are conjugate by the action of $^\vee G$. The set of Langlands parameters for $^\vee G^\Gamma$ is written

$$P(^\vee G^\Gamma) = \{\, \phi : W_\mathbb{R} \to {}^\vee G^\Gamma \mid \phi \text{ is quasiadmissible}\,\}.$$

We make $^\vee G$ act on $P(^\vee G^\Gamma)$ by conjugation on the range of a homomorphism. The set of equivalence classes of Langlands parameters — that is, the set of orbits of $^\vee G$ on $P(^\vee G^\Gamma)$ — is written

$$\Phi(^\vee G^\Gamma) = \{\, {}^\vee G \text{ orbits on } P(^\vee G^\Gamma)\,\}.$$

Definition 5.3 (see [34], [10]). Suppose G is a connected complex reductive algebraic group endowed with an inner class of real forms. If $(^\vee G^\Gamma, \mathcal{D})$ is any L-group for G (Definition 4.6), then we write

$$\Phi(G/\mathbb{R}) = \Phi(^\vee G^\Gamma).$$

5. Langlands parameters and L-homomorphisms

The omission of $^\vee G^\Gamma$ from the notation is justified by Corollary 4.8: if $((^\vee G^\Gamma)', \mathcal{D}')$ is any other L-group for the same inner class of real forms, then the corollary provides a *canonical* bijection from $\Phi(^\vee G^\Gamma)$ to $\Phi((^\vee G^\Gamma)')$. A little more generally, if $(^\vee G^\Gamma, \mathcal{D})$ is an E-group with second invariant z (Definition 4.6), then we write

$$\Phi^z(G/\mathbb{R}) = \Phi(^\vee G^\Gamma).$$

The set of Langlands parameters behaves well under L-homomorphisms.

Proposition 5.4. *Suppose $^\vee H^\Gamma$ and $^\vee G^\Gamma$ are weak E-groups, and $\epsilon : {}^\vee H^\Gamma \to {}^\vee G^\Gamma$ is an L-homomorphism. Then composition with ϵ defines a map*

$$P(\epsilon) : P(^\vee H^\Gamma) \to P(^\vee G^\Gamma)$$

on Langlands parameters, which descends to a map

$$\Phi(\epsilon) : \Phi(^\vee H^\Gamma) \to \Phi(^\vee G^\Gamma)$$

on equivalence classes.

This is obvious. We begin now the proof of Proposition 1.6.

Lemma 5.5 *Suppose H is a complex Lie group, with Lie algebra $\mathfrak{h}$. Then the set of continuous homomorphisms ϕ from $\mathbb{C}^\times$ into H is in one-to-one correspondence with the set of pairs $(\lambda, \mu) \in \mathfrak{h} \times \mathfrak{h}$, subject to the following conditions:*

a) $[\lambda, \mu] = 0$; *and*
b) $\exp(2\pi i(\lambda - \mu)) = 1$.

We omit the elementary proof; essentially the same result may be found in Lemma 2.8 of [34], or Proposition 2.10 of [1]. We will need a formula for ϕ, however; it is

$$\phi(e^t) = \exp(t\lambda + \bar{t}\mu) \qquad (t \in \mathbb{C}). \qquad (5.5)(a)$$

This is usually written more succinctly (if perhaps a little less clearly) as

$$\phi(z) = z^\lambda \bar{z}^\mu \qquad (z \in \mathbb{C}^\times). \qquad (5.5)(b)$$

We restate Proposition 1.6 in our present slightly more general setting.

Proposition 5.6. Suppose ${}^\vee G^\Gamma$ is a weak E-group (Definition 4.3). The set $P({}^\vee G^\Gamma)$ of quasiadmissible homomorphisms from $W_\mathbb{R}$ into ${}^\vee G^\Gamma$ (Definition 5.2) may be identified with the set of pairs (y, λ) satisfying the following conditions:
a) $y \in {}^\vee G^\Gamma - {}^\vee G$, and $\lambda \in {}^\vee \mathfrak{g}$ is a semisimple element;
b) $y^2 = \exp(2\pi i \lambda)$; and
c) $[\lambda, \mathrm{Ad}(y)\lambda] = 0$.

Proof. Suppose ϕ is a quasiadmissible homomorphism. By Lemma 5.4, the restriction of ϕ to $\mathbb{C}^\times$ determines two commuting elements λ and μ of ${}^\vee \mathfrak{g}$ such that

$$\phi(z) = z^\lambda \bar{z}^\mu \qquad (z \in \mathbb{C}^\times). \qquad (5.7)(a)$$

By Definition 5.2(b), the elements λ and μ are semisimple. Since j acts on $\mathbb{C}^\times$ by complex conjugation, we have

$$\mathrm{Ad}(\phi(j))(\lambda) = \mu. \qquad (5.7)(b)$$

Define

$$y = \exp(\pi i \lambda) \phi(j) \in {}^\vee G^\Gamma - {}^\vee G; \qquad (5.7)(c)$$

the last inclusion is a consequence of Definition 5.2(a). This gives (a) of the Proposition. Since $\mathrm{Ad}(\exp(\pi i \lambda))$ fixes λ, we have

$$\mathrm{Ad}(y)(\lambda) = \mu. \qquad (5.7)(d)$$

Since λ and μ commute, this gives (c) of the proposition. Now

$$\begin{aligned}
y^2 &= \exp(\pi i \lambda) \phi(j) \exp(\pi i \lambda) \phi(j) \\
&= \exp(\pi i \lambda) \phi(j) \exp(\pi i \lambda) \phi(j)^{-1} \phi(j^2) \\
&= \exp(\pi i \lambda) \exp(\pi i \mu) \phi(-1) \qquad \text{(by (5.7)(b) and (5.2)(*))} \\
&= \exp(\pi i (\lambda + \mu)) \exp(\pi i (\lambda - \mu)) \qquad \text{(by (5.5)(a))} \\
&= \exp(2\pi i \lambda),
\end{aligned}$$

as required by (b) of the proposition.

Conversely, suppose we are given y and λ satisfying (a)–(c) of the proposition. Define μ by (5.7)(d); then λ and μ are commuting semisimple elements of ${}^\vee \mathfrak{g}$. By (b) of the proposition,

$$\exp(2\pi i \lambda) = \exp(2\pi i \mathrm{Ad}(y)(\lambda)).$$

5. Langlands parameters and L-homomorphisms

So λ and μ satisfy the conditions of Lemma 5.5, and we can define a homomorphism ϕ from $\mathbb{C}^\times$ into ${}^\vee G$ by (5.7)(a). We extend this to $W_\mathbb{R}$ by setting

$$\phi(j) = \exp(-\pi i \lambda) y. \qquad (5.7)(e)$$

That the relations (5.2)(*) are preserved, and that ϕ is a quasiadmissible homomorphism, can be proved by reversing the arguments given for the first half of the proposition. We leave the details to the reader. Q.E.D.

When y, λ, and ϕ are related as in Proposition 5.6 — that is, when (5.7)(a)–(e) hold — we write

$$\phi = \phi(y, \lambda), \quad y = y(\phi), \quad \lambda = \lambda(\phi) \qquad (5.8)(a)$$

It will also be convenient to write (following (1.7)(c))

$$e(\lambda) = \exp(2\pi i \lambda) \in H \qquad (5.8)(b)$$

whenever H is a complex Lie group and $\lambda \in \mathfrak{h}$.

Corollary 5.9. *Suppose ${}^\vee G^\Gamma$ is a weak E-group, $\phi \in P({}^\vee G^\Gamma)$ is a Langlands parameter, and y, λ are as in Proposition 5.6. Define*

$$\begin{aligned}
{}^\vee G(\lambda) &= \text{centralizer in } {}^\vee G \text{ of } e(\lambda) \\
L(\lambda) &= \{ g \in {}^\vee G \mid \operatorname{Ad}(g)\lambda = \lambda \} \\
L(\lambda, y \cdot \lambda) &= \{ g \in {}^\vee G \mid \operatorname{Ad}(g)\lambda = \lambda, \operatorname{Ad}(g)(y \cdot \lambda) = y \cdot \lambda \} \\
{}^\vee G_\phi &= \text{centralizer in } {}^\vee G \text{ of } \phi(W_\mathbb{R}) \\
K(y) &= \text{centralizer in } {}^\vee G \text{ of } y.
\end{aligned}$$

a) *We have*

$${}^\vee G_\phi \subset L(\lambda, y \cdot \lambda) \subset L(\lambda) \subset {}^\vee G(\lambda).$$

All four groups are complex reductive algebraic subgroups of ${}^\vee G$; the middle two are Levi subgroups of parabolics (and therefore connected).

b) *Conjugation by y defines an involutive automorphism θ_y of ${}^\vee G(\lambda)$ preserving $L(\lambda, y \cdot \lambda)$.*

c) *The group of fixed points of θ_y on ${}^\vee G(\lambda)$ is $K(y)$. We have*

$${}^\vee G_\phi = K(y) \cap L(\lambda, y \cdot \lambda) = K(y) \cap L(\lambda).$$

In particular,

d) $^\vee G_\phi$ *may be described as the set of fixed points of an involutive automorphism of the connected reductive (Levi) subgroup* $L(\lambda, y \cdot \lambda)$ *of* $^\vee G$.

This is immediate from Proposition 5.6.

Recall now from Definition 1.16 the algebraic universal cover $^\vee G^{alg}$ of $^\vee G$. We write

$$1 \to \pi_1(^\vee G)^{alg} \to {}^\vee G^{alg} \to {}^\vee G \to 1. \qquad (5.10)(a)$$

(Recall that $\pi_1(^\vee G)^{alg}$ is the projective limit of the finite quotient groups of $\pi_1(^\vee G)$.) If H is any subgroup of $^\vee G$, then we write H^{alg} for its inverse image in $^\vee G^{alg}$. (When we consider H as a subgroup of several different weak E-groups, this notation is ambiguous; we will write instead $H^{alg, ^\vee G}$.) This gives an exact sequence

$$1 \to \pi_1(^\vee G)^{alg} \to H^{alg} \to H \to 1. \qquad (5.10)(b)$$

We should offer here some justification for the introduction of these covering groups. One way to eliminate them is to change the definition of strong real form (Definition 2.13) to require $\delta^2 = 1$. Then many of our principal results (beginning with Theorem 1.18) hold with no covering groups at all. The problem with this approach is that not every real form is represented by a strong real form with $\delta^2 = 1$. (For example, the real form $SU(2)$ of $SL(2)$ is eliminated.) For this reason we believe that the use of some covering groups is unavoidable. It is still of interest to understand exactly which real forms require which coverings of $^\vee G$, and we will investigate this carefully in Chapter 10 (Theorem 10.11).

We can now introduce the parameters for the complete Langlands classification (Theorem 1.18).

Definition 5.11. Suppose $^\vee G^\Gamma$ is a weak E-group, and $\phi \in P(^\vee G^\Gamma)$. Use the notation of Corollary 5.9. Define

$$A_\phi^{loc} = {}^\vee G_\phi / (^\vee G_\phi)_0,$$

the *Langlands component group for* ϕ. Similarly, set

$$A_\phi^{loc,alg} = {}^\vee G_\phi^{alg} / \left({}^\vee G_\phi^{alg}\right)_0,$$

the *universal component group for* ϕ. (We will occasionally need to write $A_{\phi, ^\vee G}^{loc, alg}$ to avoid ambiguity. When $^\vee G^\Gamma$ is an E-group for G, we simplify

5. Langlands parameters and L-homomorphisms 61

this to $A_{\phi,G}^{loc,alg}$.) It is a consequence of Corollary 5.9(d) that A_ϕ^{loc} is a (finite) product of copies of $\mathbb{Z}/2\mathbb{Z}$, and that $A_\phi^{loc,alg}$ is abelian. This is not entirely obvious; the argument is given at (12.11)(e) below. By (5.10)(b), there is a right exact sequence

$$\pi_1({}^\vee G)^{alg} \to A_\phi^{loc,alg} \to A_\phi^{loc} \to 1.$$

(The kernel of the first map is the intersection of $\pi_1({}^\vee G)^{alg}$ with the identity component of ${}^\vee G_\phi^{alg}$.) A *complete Langlands parameter for* ${}^\vee G^\Gamma$ is a pair (ϕ, τ), with τ an irreducible representation of $A_\phi^{loc,alg}$. Two such parameters are called equivalent if they are conjugate under the obvious action of ${}^\vee G^{alg}$. We write $\Xi({}^\vee G^\Gamma)$ for the set of equivalence classes of complete Langlands parameters. As in Definition 5.3, the case of L-groups merits special attention: if G is a connected complex reductive group endowed with an inner class of real forms, we write

$\Xi(G/\mathbb{R})$ = equivalence classes of complete Langlands parameters for the L-group of G.

(This is not the definition that was given in the introduction, but we will see in Chapter 7 that it is equivalent.) More generally, if $({}^\vee G^\Gamma, \mathcal{D})$ is an E-group for G^Γ with second invariant z (Definition 4.6) then we write $\Xi^z(G/\mathbb{R})$ for $\Xi({}^\vee G^\Gamma)$. If $(({}^\vee G^\Gamma)', \mathcal{D}')$ is another E-group with second invariant z, then Proposition 4.7(b) provides a canonical bijection from $\Xi({}^\vee G^\Gamma)$ onto $\Xi(({}^\vee G^\Gamma)')$. This justifies the omission of the E-group from the notation.

Although we will make no use of it, the following proposition may provide a little motivation for this definition.

Proposition 5.12. *Suppose ${}^\vee G$ is a weak E-group. Then there is a natural one-to-one correspondence between equivalence classes of complete Langlands parameters for ${}^\vee G^\Gamma$ (Definition 5.11), and irreducible ${}^\vee G^{alg}$-equivariant local systems on ${}^\vee G$-orbits on $P({}^\vee G^\Gamma)$ (Definition 5.2).*

This is immediate from the definitions (see also Lemma 7.3 below).

Particularly in connection with endoscopy, we will occasionally need to consider a slightly different situation. Suppose that Q is any pro-finite group, and that we have an extension of pro-algebraic groups

$$1 \to Q \to {}^\vee G^Q \to {}^\vee G \to 1. \qquad (5.13)(a)$$

If $\epsilon : H \to {}^\vee G$ is any morphism of algebraic groups, we can pull back the extension (5.13)(a) to

$$\begin{array}{ccccccccc} 1 & \to & Q & \to & H^Q & \to & H & \to & 1 \\ & & \downarrow & & \downarrow \epsilon^Q & & \downarrow \epsilon & & \\ 1 & \to & Q & \to & {}^\vee G^Q & \to & {}^\vee G & \to & 1 \end{array} \qquad (5.13)(b)$$

(The case of (5.10) has $Q = \pi_1({}^\vee G)^{alg}$, and ϵ the inclusion of a subgroup. In almost all the examples we consider, Q will be a quotient of $\pi_1({}^\vee G)^{alg}$.) If $\phi \in P({}^\vee G^\Gamma)$, we define the *Q-component group for ϕ* by

$$A^{loc,Q}_\phi = {}^\vee G^Q_\phi / ({}^\vee G^Q_\phi)_0. \qquad (5.13)(c)$$

If we need to emphasize the group, we write $A^{loc,Q}_{\phi, {}^\vee G}$. There is a right exact sequence

$$Q \to A^{loc,Q}_\phi \to A^{loc}_\phi \to 1. \qquad (5.13)(d)$$

A *complete Langlands parameter for* ${}^\vee G^\Gamma$ *of type* Q is a pair (ϕ, τ), with ϕ a Langlands parameter and τ an irreducible representation of $A^{loc,Q}_\phi$. Two such parameters are called equivalent if they are conjugate by the action of ${}^\vee G^Q$. We write $\Xi({}^\vee G^\Gamma)^Q$ for the set of equivalence classes of complete Langlands parameters of type Q.

Suppose now that we have an L-homomorphism

$$\epsilon : {}^\vee H^\Gamma \to {}^\vee G^\Gamma, \qquad (5.14)(a)$$

(Definition 5.1), and that we are given compatible pro-finite extensions

$$\begin{array}{ccccccccc} 1 & \to & Q_H & \to & {}^\vee H^{Q_H} & \to & {}^\vee H & \to & 1 \\ & & \downarrow & & \downarrow \epsilon^\bullet & & \downarrow \epsilon & & \\ 1 & \to & Q & \to & {}^\vee G^Q & \to & {}^\vee G & \to & 1 \end{array} \qquad (5.14)(b)$$

(In our examples, ${}^\vee G^Q$ will usually be the algebraic universal covering of ${}^\vee G$, and ${}^\vee H^\Gamma$ will be (roughly) the centralizer in ${}^\vee G^\Gamma$ of a semisimple element. We will take $Q_H = Q$, and ${}^\vee H^Q$ the induced extension (cf. (5.13)(b)). In this case ${}^\vee H^Q$ will be a quotient of the algebraic universal cover of ${}^\vee H$.) Suppose now that $\phi \in P({}^\vee H^\Gamma)$, so that $\epsilon \circ \phi \in P({}^\vee G^\Gamma)$ (Proposition 5.4). Then the map

$$\epsilon^\bullet : {}^\vee H^{Q_H} \to {}^\vee G^Q \qquad (5.14)(c)$$

5. Langlands parameters and L-homomorphisms

carries the centralizer of the image of ϕ to the centralizer of the image of $\epsilon \circ \phi$; so we get an induced homomorphism of component groups

$$A^{loc}(\epsilon) : A^{loc,Q_H}_{\phi, ^\vee H} \to A^{loc,Q}_{\epsilon \circ \phi, ^\vee G}. \qquad (5.14)(d)$$

Notice that this map appears to go in the wrong direction from the point of view of Langlands functoriality principle: a complete Langlands parameter of type Q_H for $^\vee H^\Gamma$ induces a Langlands parameter for $^\vee G^\Gamma$, but not a representation of the Q-component group. (The problem is unrelated to our extensions; it occurs even if $Q = Q_H = \{1\}$.) We will return to this point in Chapter 26, using the ideas of Langlands and Shelstad.

6. Geometric parameters

We continue the analysis of the preceding chapter, turning now to the new geometric parameters defined in the introduction. A calculation illustrating several of the technical difficulties is outlined in Example 6.22. The reader may wish to refer to it while reading this chapter.

Definition 1.7 of the introduction was made as succinctly as possible; we repeat it here, including some useful auxiliary ideas. Suppose H is a complex reductive group, with Lie algebra $\mathfrak{h}$, and $\lambda \in \mathfrak{h}$ is a semisimple element. Set

$$\mathfrak{h}(\lambda)_n = \{\mu \in \mathfrak{h} \mid [\lambda, \mu] = n\mu\} \quad (n \in \mathbb{Z}) \qquad (6.1)(a)$$

$$\mathfrak{h}(\lambda) = \sum_{n \in \mathbb{Z}} \mathfrak{h}(\lambda)_n \qquad (6.1)(b)$$

$$\mathfrak{l}(\lambda) = \mathfrak{h}(\lambda)_0 = \text{ centralizer of } \lambda \text{ in } \mathfrak{h} \qquad (6.1)(c)$$

$$\mathfrak{n}(\lambda) = \sum_{n=1,2,\ldots} \mathfrak{h}(\lambda)_n \qquad (6.1)(d)$$

$$\mathfrak{p}(\lambda) = \mathfrak{l}(\lambda) + \mathfrak{n}(\lambda) \qquad (6.1)(e)$$

Next, define

$$e(\lambda) = \exp(2\pi i \lambda) \in H. \qquad (6.2)(a)$$

$$H(\lambda) = \text{ centralizer in } H \text{ of } e(\lambda) \qquad (6.2)(b)$$

$$L(\lambda) = \text{ centralizer in } H \text{ of } \lambda \qquad (6.2)(c)$$

$N(\lambda)$ = connected unipotent subgroup with Lie algebra $\mathfrak{n}(\lambda)$
$$(6.2)(d)$$

$$P(\lambda) = L(\lambda)N(\lambda) \qquad (6.2)(e)$$

Lemma 6.3. *Suppose H is a complex reductive algebraic group, and $\lambda \in \mathfrak{h}$ is a semisimple element. Use the notation of (6.1) and (6.2).*
a) *$H(\lambda)$ is a reductive subgroup of H with Lie algebra $\mathfrak{h}(\lambda)$.*
b) *$L(\lambda)$ is a reductive subgroup of $H(\lambda)$ with Lie algebra $\mathfrak{l}(\lambda)$. It is connected if H is.*
c) *$N(\lambda)$ is a connected unipotent subgroup of $H(\lambda)$ with Lie algebra $\mathfrak{n}(\lambda)$.*
d) *$P(\lambda)$ is a parabolic subgroup of $H(\lambda)$ with Lie algebra $\mathfrak{p}(\lambda)$, and Levi decomposition $P(\lambda) = L(\lambda)N(\lambda)$. It is connected if H is.*
e) *$\mathrm{Ad}(P(\lambda)) \cdot \lambda = \mathrm{Ad}(N(\lambda)) \cdot \lambda = \lambda + \mathfrak{n}(\lambda)$. More precisely, the adjoint action of $N(\lambda)$ defines an isomorphism of varieties*

$$N(\lambda) \simeq \lambda + \mathfrak{n}(\lambda), \qquad n \mapsto \mathrm{Ad}(n)\lambda.$$

This is elementary. (Perhaps the most subtle part is (e). This goes back at least to [19], Lemma 8; the hypotheses there are a little more special, but the proof extends without change. A convenient reference is [62], equation (4) on page 55.) The affine space in (e) is just the canonical flat of Definition 1.7:

$$\mathcal{F}(\lambda) = \mathrm{Ad}(P(\lambda)) \cdot \lambda = \mathrm{Ad}(N(\lambda)) \cdot \lambda = \lambda + \mathfrak{n}(\lambda). \qquad (6.4)$$

Proposition 6.5. *Suppose H is a complex reductive algebraic group, and $\lambda \in \mathfrak{h}$ is a semisimple element. Use the notation of (6.1)–(6.4), and suppose $\lambda' \in \mathcal{F}(\lambda)$. Then*
a) *λ' is a semisimple element of $\mathfrak{h}$, conjugate to λ by $\mathrm{Ad}(N(\lambda))$;*
b) *$e(\lambda) = e(\lambda')$, $H(\lambda) = H(\lambda')$, $N(\lambda) = N(\lambda')$, $P(\lambda) = P(\lambda')$, $\mathcal{F}(\lambda) = \mathcal{F}(\lambda')$; and*
c) *the stabilizer in H of the set $\mathcal{F}(\lambda)$ is $P(\lambda)$. More precisely, if $\mathrm{Ad}(h) \cdot \mathcal{F}(\lambda)$ has a non-trivial intersection with $\mathcal{F}(\lambda)$, then $h \in P(\lambda)$.*

Proof. Part (a) is immediate from (6.4). Write $\lambda' = \mathrm{Ad}(n) \cdot \lambda$, for some $n \in N(\lambda) \subset H(\lambda)$. Obviously $\mathrm{Ad}(n)$ carries objects defined in terms of λ to those defined in terms of λ'. Since $e(\lambda)$ is central in $H(\lambda)$, the first claim of (b) follows. For the next three, use the fact that $N(\lambda)$ normalizes $H(\lambda)$, $N(\lambda)$, and $P(\lambda)$. For the last claim of (b), use the fact that $\mathcal{F}(\lambda)$ is already a homogeneous space for $N(\lambda)$. For (c), suppose $h \in H$ carries λ to another element λ' of $\mathcal{F}(\lambda)$. By (a), there is an element $n \in N(\lambda)$ with $\lambda' = \mathrm{Ad}(n) \cdot \lambda$. Since n and h carry λ to the same element, the element $l = n^{-1}h$ belongs to the stabilizer $L(\lambda)$ of λ. Hence $h = nl$ belongs to $P(\lambda)$, as we wished to show. Q.E.D.

Suppose now that Λ is a canonical flat in $\mathfrak{h}$. We define

$$e(\Lambda) = e(\lambda), \quad H(\Lambda) = H(\lambda), \quad N(\Lambda) = N(\lambda), \quad P(\Lambda) = P(\lambda) \quad (6.6)$$

for any $\lambda \in \Lambda$; this is well-defined by Proposition 6.5(b). We write $\mathfrak{h}(\Lambda)$, etc., for the Lie algebras.

Proposition 6.7. *Suppose H is a complex reductive algebraic group, and $\Lambda \subset \mathfrak{h}$ is a canonical flat. Use the notation of (6.6).*

a) *Λ is an affine space for the vector space $\mathfrak{n}(\Lambda)$, and a homogeneous space for $P(\Lambda)$. These structures are compatible (via the adjoint action of $P(\Lambda)$ on $\mathfrak{n}(\Lambda)$). We can therefore form the induced bundle*

$$H \times_{P(\Lambda)} \Lambda \to H/P(\Lambda),$$

which is an affine bundle for the vector bundle

$$H \times_{P(\Lambda)} \mathfrak{n}(\Lambda) \to H/P(\Lambda).$$

b) *The inclusion of Λ in $\mathfrak{h}$ induces a map*

$$H \times_{P(\Lambda)} \Lambda \to \mathfrak{h}, \quad (h, \lambda) \mapsto \mathrm{Ad}(h) \cdot \lambda.$$

This map is an isomorphism onto the conjugacy class containing Λ. The fibers of the bundle structure map to the canonical flats in the conjugacy class.

Proof. (A brief discussion of induced bundles can be found in the appendix of [61]; this result has some overlap with those of section 3 in that paper. Recall that a point of the induced bundle is an equivalence class of pairs (h, λ), with $h \in H$ and $\lambda \in \Lambda$; the equivalence relation is $(hp, \lambda) \sim (h, p \cdot \lambda)$ for $p \in P(\Lambda)$.) Part (a) and the existence of the map in (b) are formal (cf. Proposition A.2(b) in [61], for example). That the map is an isomorphism follows from Proposition A.2(d) of [61], or a direct calculation: the fiber over λ consists of the equivalence classes of pairs (h, λ') with $\mathrm{Ad}(h) \cdot \lambda' = \lambda$. By (c) of Proposition 6.5, it follows that $h \in P(\Lambda)$, and thus that (h, λ') is equivalent to (e, λ). This shows that the map is bijective. Since the domain and range are homogeneous spaces, it must be an isomorphism. For the last claim, obviously the fiber of the bundle over the identity coset $eP(\Lambda)$ maps to the canonical flat Λ. The general statement then follows from the H-equivariance of the map. Q.E.D.

Example 6.8. Most semisimple elements (a dense open set) of a reductive Lie algebra have no integral eigenvalues except zero; and the zero eigenspace is the unique Cartan subalgebra to which they belong. For such an element λ, we have $P(\lambda) = L(\lambda) = T$, a Cartan subgroup of H; $\mathcal{F}(\lambda) = \{\lambda\}$; the unipotent group $N(\lambda)$ is trivial; the bundles of Proposition 6.7 are trivial on H/T; and Proposition 6.7(b) simply identifies the conjugacy class of λ with H/T.

At the other extreme, suppose λ has all of its eigenvalues integral. Then $H(\lambda)$ contains the identity component of H, and $P(\lambda)$ is a parabolic subgroup of H. The partition of the $\mathrm{Ad}(H)$ orbit of λ into canonical flats can be viewed as a Lagrangian foliation of a symplectic structure on the orbit; but since this is an adjoint rather than a coadjoint orbit, the symplectic structure is not canonical and this is not a particularly good point of view.

Suppose for example that $H = SL(2, \mathbb{C})$. Write $\{h, e, f\}$ for the standard basis of the Lie algebra, and choose

$$\lambda = 1/2 h = \begin{pmatrix} 1/2 & 0 \\ 0 & -1/2 \end{pmatrix} \qquad (6.8)(a)$$

Then

$$\mathfrak{h}(\lambda)_{-1} = \mathbb{C}f, \quad \mathfrak{h}(\lambda)_0 = \mathbb{C}h, \quad \mathfrak{h}(\lambda)_1 = \mathbb{C}e \qquad (6.8)(b)$$

$$\mathfrak{h}(\lambda) = \mathfrak{h}, \quad \mathfrak{l}(\lambda) = \mathbb{C}h, \quad \mathfrak{n}(\lambda) = \mathbb{C}e, \quad \mathfrak{p}(\lambda) = \mathbb{C}h + \mathbb{C}e \qquad (6.8)(c)$$

$$e(\lambda) = -I \in SL(2, \mathbb{C}), \quad H(\lambda) = SL(2, \mathbb{C}) \qquad (6.8)(d)$$

$$\mathcal{F}(\lambda) = \left\{ \begin{pmatrix} 1/2 & t \\ 0 & -1/2 \end{pmatrix} \mid t \in \mathbb{C} \right\} \qquad (6.8)(e)$$

The orbit of λ is the quadric

$$\mathrm{Ad}(H) \cdot \lambda = \left\{ \begin{pmatrix} a & b \\ c & -a \end{pmatrix} \mid a^2 + bc = 1/4 \right\}; \qquad (6.8)(f)$$

its partition into the conjugates of the line (6.8)(e) is just one of the two standard rulings of a quadric surface ([20], Exercise I.2.15).

Definition 6.9. Suppose $^\vee G^\Gamma$ is an E-group. A *geometric parameter* for $^\vee G^\Gamma$ is a pair (y, Λ) satisfying

(a) $y \in {}^\vee G^\Gamma - {}^\vee G$;
(b) $\Lambda \subset {}^\vee \mathfrak{g}$ is a canonical flat (Definition 1.7); and
(c) $y^2 = e(\Lambda)$ (notation (6.6)).

The set of geometric parameters for $^\vee G^\Gamma$ is written $X(^\vee G^\Gamma)$. We make $^\vee G$ act on $X(^\vee G^\Gamma)$ by conjugation. Two geometric parameters are called *equivalent* if they are conjugate by this action. (We do not introduce a separate notation for the set of equivalence classes, since this will turn out to be naturally identified with $\Phi(^\vee G^\Gamma)$.)

If $\mathcal{O}$ is a $^\vee G$ orbit of semisimple elements in $^\vee \mathfrak{g}$, then as in Lemma 1.9 we define

$$X(\mathcal{O}, {}^\vee G^\Gamma) = \{\, (y, \Lambda) \in X(^\vee G^\Gamma) \mid \Lambda \subset \mathcal{O} \,\}.$$

We begin now our analysis of the spaces $X(\mathcal{O}, {}^\vee G^\Gamma)$. Fix a $^\vee G$ orbit of semisimple elements

$$\mathcal{O} \subset {}^\vee \mathfrak{g}, \tag{6.10}(a)$$

and write $\mathcal{F}(\mathcal{O})$ for the set of canonical flats in $\mathcal{O}$. If we fix such a flat Λ, then Proposition 6.5(c) provides an isomorphism

$$\mathcal{F}(\mathcal{O}) \simeq {}^\vee G/P(\Lambda). \tag{6.10}(b)$$

We also need to consider the conjugacy class of $e(\Lambda)$ in $^\vee G$. This is

$$\mathcal{C}(\mathcal{O}) = \{\, g e(\Lambda) g^{-1} \mid g \in {}^\vee G \,\}. \tag{6.10}(c)$$

By (6.2)(b), there is an isomorphism

$$\mathcal{C}(\mathcal{O}) \simeq {}^\vee G/{}^\vee G(\Lambda). \tag{6.10}(d)$$

Because $e(\Lambda)$ is a semisimple element, the conjugacy class $\mathcal{C}(\mathcal{O})$ is closed in $^\vee G$. From the form of Definition 1.8, it is clear that we also need to consider the subvariety

$$\mathcal{I}(\mathcal{O}) = \{\, y \in {}^\vee G^\Gamma - {}^\vee G \mid y^2 \in \mathcal{C}(\mathcal{O}) \,\} \tag{6.10}(e)$$

6. Geometric parameters

It will be convenient also to consider

$$\mathcal{I}(\Lambda) = \{\, y \in {}^\vee G^\Gamma - {}^\vee G \mid y^2 = e(\Lambda)\,\}. \qquad (6.10)(f)$$

Finally, for $y \in \mathcal{I}(\mathcal{O})$, we write

$$K(y) = \text{centralizer of } y \text{ in } {}^\vee G. \qquad (6.10)(g)$$

Now the definition of the map e (cf. (6.2)(a) and (6.6)) on $\mathcal{F}(\mathcal{O})$ involves the exponential map, and so is not algebraic in nature. Nevertheless, we have

Lemma 6.11. *In the setting of (6.10), the map*

$$e : \mathcal{F}(\mathcal{O}) \to \mathcal{C}(\mathcal{O}), \quad \Lambda \mapsto e(\Lambda)$$

is a smooth projective algebraic morphism.

Proof. By (6.10)(b) and (d), the map is just the natural quotient map of algebraic homogeneous spaces

$$ {}^\vee G / P(\Lambda) \to {}^\vee G / {}^\vee G(\Lambda).$$

Since $P(\Lambda)$ is parabolic in ${}^\vee G(\Lambda)$, the map is projective. Q.E.D.

The next lemma will be the key to the finiteness claim of Lemma 1.9.

Lemma 6.12. *Suppose H is a complex reductive algebraic group (possibly disconnected), and $z \in H$ centralizes the identity component H_0. Consider the set*

$$I = \{\, y \in H \mid y^2 = z\,\}.$$

Then I is the union of finitely many orbits under the conjugation action of H_0.

Proof. The conjugation action of $y \in I$ on H_0 defines an involutive automorphism θ_y. Fix a connected component C of H, and consider the subset $I \cap C$ of I. Since H has finitely many components, it suffices to show that $I \cap C$ has finitely many H_0 orbits. The involutions in the set

$$\Theta(C) = \{\, \theta_y \mid y \in I \cap C\,\}$$

all differ by inner automorphisms (since C is a coset of H_0). In particular, they have a common restriction θ_Z to the center $Z(H_0)$, so they are distinguished by their restrictions to the semisimple commutator subgroup of H_0. A complex semisimple algebraic group admits only finitely many conjugacy classes of involutive automorphisms; indeed Cartan showed that these classes correspond bijectively to the equivalence classes of real forms. It follows that $\Theta(C)$ is a finite union of H_0 orbits. To complete the proof, we must study the fibers of the map

$$I \cap C \to \Theta(C), \qquad y \mapsto \theta_y.$$

It is enough to show that each fiber is a finite union of orbits under the conjugation action of $Z(H_0)$. Now the fiber over θ_y is clearly

$$\{\, yw \mid w \in Z(H_0), w\theta_Z(w) = e \,\} = yZ_1.$$

On the other hand, the $Z(H_0)$ conjugacy class of y is

$$\{\, yw \mid w = v\theta_Z(v)^{-1}, \text{ some } v \in Z(H_0) \,\} = yZ_2.$$

The two subgroups of $Z(H_0)$ appearing here are algebraic, and $Z_2 \subset Z_1$. Since they have the same Lie algebra (namely the -1 eigenspace of the differential of θ_Z), it follows that Z_2 has finite index in Z_1, as we wished to show. Q.E.D.

Proposition 6.13. *In the setting (6.10), the orbits of the conjugation action of $^\vee G$ on $\mathcal{I}(\mathcal{O})$ are in one-to-one correspondence (by intersection with $\mathcal{I}(\Lambda)$) with $^\vee G(\Lambda)$ orbits on $\mathcal{I}(\Lambda)$. All of these orbits are closed, and there are finitely many of them.*

Fix a single orbit $\mathcal{I}_1(\mathcal{O})$ of $^\vee G$ on $\mathcal{I}(\mathcal{O})$. Then the natural map

$$\mathcal{I}_1(\mathcal{O}) \to \mathcal{C}(\mathcal{O}), \qquad y \mapsto y^2$$

is a smooth algebraic morphism.

Proof. The bijection is formal. That the orbits are closed follows from the fact that they are finite unions of connected components of semisimple conjugacy classes in the (disconnected) reductive group $^\vee G^\Gamma$. For the finiteness, we consider the reductive group

$$H = \text{centralizer of } e(\Lambda) \text{ in } {}^\vee G^\Gamma.$$

Obviously any element y of $\mathcal{I}(\Lambda)$ belongs to H (since y always commutes with y^2). Furthermore $z = e(\Lambda)$ is central in H. It follows that the set I of Lemma 6.12 contains $\mathcal{I}(\Lambda)$. Since $^\vee G(\Lambda)$ contains the identity component of H, the finiteness we need follows from Lemma 6.12.

For the last claim, fix an element $y_1 \in \mathcal{I}_1(\mathcal{O})$ with $y_1^2 = e(\Lambda)$. Then

$$\mathcal{I}_1(\mathcal{O}) \simeq {^\vee G}/K(y_1), \qquad (6.14)(a)$$

and the morphism in question is just the quotient map

$$^\vee G/K(y_1) \to {^\vee G}/{^\vee G(\Lambda)}. \qquad (6.14)(b)$$

Of course this map is smooth. Q.E.D.

Here is the last general ingredient we need to complete our description of the geometric parameter space.

Lemma 6.15. *Suppose G is an algebraic group, H is a closed subgroup of G, and A and B are closed subgroups of H. Then G/A and G/B both carry natural maps to G/H, so we can form the fiber product*

$$X = G/A \times_{G/H} G/B \to G/H.$$

On the other hand, H acts on H/A and on H/B, so it acts on the product $H/A \times H/B$. We can therefore form the induced bundle

$$Y = G \times_H (H/A \times H/B) \to G/H.$$

Finally, A acts on the homogeneous space H/B, so we can form the induced bundle

$$Z = G \times_A H/B \to G/A \to G/H$$

(and similarly with A and B exchanged).

a) *X, Y, and Z are smooth bundles over G/H.*
b) *Y is isomorphic to X by the map sending the equivalence class of $(g, (h_1 A, h_2 B)) \in Y$ to $(gh_1 A, gh_2 B) \in X$. This map is a G-equivariant isomorphism of fiber bundles from Y to X.*
c) *Z is isomorphic to X by the map sending the equivalence class of (g, hB) to (gA, ghB). This map is a G-equivariant isomorphism of fiber bundles from Y to X.*
d) *The isomorphism in (b) induces a natural bijection from the orbits of H on $(H/A \times H/B)$ onto the orbits of G on X.*
e) *The isomorphism in (c) induces a natural bijection from the orbits of A on H/B onto the orbits of G on X.*

f) *The following four sets are in natural one-to-one correspondence: the orbits of H on $(H/A \times H/B)$; the (A,B) double cosets in H; the orbits of A on H/B; and the orbits of B on H/A.*

Proof. An induced bundle is smooth if and only if the inducing space is smooth; so the smoothness of Y (respectively Z) is a consequence of the smoothness of $(H/A \times H/B)$ (respectively H/B). As for X, the spaces G/A and G/B are both smooth over G/H, so their fiber product is smooth over G/H as well ([20], Proposition III.10.1(d)). For (b), the G-equivariant maps from an induced bundle to a G-space are determined by the H-equivariant maps from the inducing space. The map in (b) corresponds in this way to the natural inclusion of $(H/A \times H/B)$ in the fiber product X. That this map is a bijection is formal. That it is an isomorphism can be deduced by computing its differential (because of the smoothness established in (a)). Part (c) is similar; or one can show directly that Z is isomorphic to Y. For (d) and (e), the G-orbits on an induced bundle are in natural bijection with the orbits on the inducing space. Part (f) is elementary group theory. Q.E.D.

Proposition 6.16. *Suppose $^\vee G^\Gamma$ is a weak E-group. Fix an orbit $\mathcal{O}$ of $^\vee G$ on the semisimple elements of $^\vee \mathfrak{g}$, and use the notation of (6.6) and (6.10). In particular, we have smooth algebraic maps*

$$\mathcal{F}(\mathcal{O}) \to \mathcal{C}(\mathcal{O}), \qquad \mathcal{I}(\mathcal{O}) \to \mathcal{C}(\mathcal{O}).$$

The variety $X(\mathcal{O}, {}^\vee G^\Gamma)$ of Definition 6.9 is the fiber product

$$X(\mathcal{O}, {}^\vee G^\Gamma) = \mathcal{F}(\mathcal{O}) \times_{\mathcal{C}(\mathcal{O})} \mathcal{I}(\mathcal{O}).$$

A little more explicitly, it has the following structure. List the orbits of $^\vee G$ on $\mathcal{I}(\mathcal{O})$ as $\mathcal{I}_1(\mathcal{O}), \ldots, \mathcal{I}_r(\mathcal{O})$ (Proposition 6.13). Fix a canonical flat $\Lambda \in \mathcal{F}(\mathcal{O})$. This gives a reductive group $^\vee G(\Lambda)$ and a parabolic subgroup $P(\Lambda)$ (cf. (6.6) and Lemma 6.3). For each i, choose a point

$$y_i \in \mathcal{I}_i(\mathcal{O}), \qquad y_i^2 = e(\Lambda).$$

Then conjugation by y_i defines an involutive automorphism θ_i of $^\vee G(\Lambda)$ with fixed point set $K_i = K(y_i)$ (cf. (6.10)(g)). The variety $X(\mathcal{O}, {}^\vee G^\Gamma)$ is the disjoint union of r closed subvarieties

$$X_i(\mathcal{O}, {}^\vee G^\Gamma) = \mathcal{F}(\mathcal{O}) \times_{\mathcal{C}(\mathcal{O})} \mathcal{I}_i(\mathcal{O}).$$

The ith subvariety looks like

$$X_i(\mathcal{O}, {}^\vee G^\Gamma) \simeq {}^\vee G \times_{{}^\vee G(\Lambda)} ({}^\vee G(\Lambda)/K_i \times {}^\vee G(\Lambda)/P(\Lambda))$$
$$\simeq {}^\vee G \times_{K_i} ({}^\vee G(\Lambda)/P(\Lambda)).$$

In particular, the orbits of ${}^\vee G$ on $X_i(\mathcal{O}, {}^\vee G^\Gamma)$ are in one-to-one correspondence with the orbits of K_i on the partial flag variety ${}^\vee G(\Lambda)/P(\Lambda)$. These orbits are finite in number. The isotropy group of the action of ${}^\vee G$ at the point $x = (y, \Lambda)$ is

$${}^\vee G_x = K(y) \cap P(\Lambda).$$

We will see later (Proposition 7.14) that the correspondence between K_i orbits on ${}^\vee G(\Lambda)/P(\Lambda)$ and ${}^\vee G$ orbits on $X_i(\mathcal{O})$ not only is a bijection of sets, but also preserves the closure relations and the nature of the singularities of closures.

Proof. Everything but the last finiteness assertion is immediate from Lemma 6.15 and Proposition 6.13. (The description of $X(\mathcal{O}, {}^\vee G^\Gamma)$ as a fiber product is just a restatement of Definition 1.8.) For the finiteness, we must show that if H is a reductive group, Z is a partial flag variety for H (that is, a homogeneous space for which an isotropy group contains a Borel subgroup of H_0) and $K \subset H$ is the group of fixed points of an involution, then K has finitely many orbits on Z. After straightforward reductions, we may assume that H is connected and semisimple, and that Z is the variety of Borel subgroups. In that case the result is well-known. (A complete description of the orbits may be found in [42]; the finiteness result is even older.) Q.E.D.

With this description of the geometric parameter space, we can consider now its relationship to the Langlands parameters of Chapter 5.

Proposition 6.17. *Suppose ${}^\vee G^\Gamma$ is a weak E-group. Then there is a natural ${}^\vee G$-equivariant map p from the set of Langlands parameters for ${}^\vee G^\Gamma$ (Definition 5.2) onto the set of geometric parameters for ${}^\vee G^\Gamma$ (Definition 6.9). In terms of the description of Langlands parameters in Proposition 5.6, this map is*

$$p: P({}^\vee G^\Gamma) \to X({}^\vee G^\Gamma), \quad p(\phi(y, \lambda)) = (y, \mathcal{F}(\lambda))$$

(notation (5.8), (6.4)). The map p induces a bijection from equivalence classes of Langlands parameters (that is, ${}^\vee G$ orbits on $P({}^\vee G^\Gamma)$) onto equivalence classes of geometric parameters (that is, ${}^\vee G$ orbits on

$X({}^\vee G^\Gamma))$. If $x = p(\phi)$, then the isotropy group ${}^\vee G_\phi$ (Corollary 5.9) is a Levi subgroup of the isotropy group ${}^\vee G_x$ (Proposition 6.16). In particular, the fiber $p^{-1}(x)$ is a principal homogeneous space for the unipotent radical of ${}^\vee G_x$.

Before embarking on the proof, we need some general facts about parabolic subgroups and involutions.

Lemma 6.18. *Suppose G is a connected complex reductive algebraic group, $P \subset G$ is a parabolic subgroup, and θ is an involutive automorphism of G. (We do not assume that θ preserves P.) Write $K \subset G$ for the group of fixed points of θ.*

a) There is a maximal torus $T \subset P$ such that $\theta T = T$.

From now on we fix such a torus. There is a unique Levi decomposition $P = LN$ with $T \subset L$.

b) The group $L_\theta = L \cap \theta L$ is a connected θ-stable reductive subgroup of G containing T.

c) The group N_θ generated by $N \cap \theta P$ and $P \cap \theta N$ is a connected θ-stable unipotent subgroup of G, normalized by L_θ.

d) The group $P_\theta = P \cap \theta P$ is a connected algebraic subgroup of G, with θ-stable Levi decomposition $P_\theta = L_\theta N_\theta$.

e) We have

$$L \cap K = L_\theta \cap K, \quad P \cap K = P_\theta \cap K = (L_\theta \cap K)(N_\theta \cap K).$$

The last formula is a Levi decomposition of the algebraic group $P \cap K$.

Proof. For (a), we may as well replace P by a Borel subgroup contained in it. In that case the existence of T is established in [42]. Now write $R = R(G,T)$ for the set of roots. Because θ preserves T, θ acts on R as an automorphism of order 2. Write

$$R(P,T) = R(L,T) \cup R(N,T)$$

for the corresponding sets of roots. Fix an element $\lambda \in \mathfrak{t}$ with the property that

$$\alpha(\lambda) = 0 \quad (\alpha \in R(L,T)), \qquad \alpha(\lambda) > 0 \quad (\alpha \in R(N,T)).$$

Then $\theta\lambda$ also belongs to $\mathfrak{t}$, and has the corresponding properties for θP. For (b), L is the centralizer in G of λ, so

$$L \cap \theta L = \{g \in G \mid \mathrm{Ad}(g)\lambda = \lambda,\, \mathrm{Ad}(g)(\theta\lambda) = \theta\lambda\}.$$

6. Geometric parameters

Since λ and $\theta\lambda$ are commuting semisimple elements of $\mathfrak{g}$, it follows that $L \cap \theta L$ is connected and reductive. The rest of (b) is clear.

For (c) and (d), one computes easily that the roots of T in L_θ and P_θ are

$$R(L_\theta, T) = \{\, \alpha \in R \mid \alpha(\lambda) = 0 \text{ and } \alpha(\theta\lambda) = 0 \,\}$$

$$R(P_\theta, T) = \{\, \alpha \in R \mid \alpha(\lambda) \geq 0 \text{ and } \alpha(\theta\lambda) \geq 0 \,\}.$$

The groups $N \cap \theta P$ and $P \cap \theta N$ are obviously unipotent algebraic, and therefore connected. The corresponding roots are

$$R(N \cap \theta P, T) = \{\, \alpha \in R \mid \alpha(\lambda) > 0 \text{ and } \alpha(\theta\lambda) \geq 0 \,\}$$

$$R(P \cap \theta N, T) = \{\, \alpha \in R \mid \alpha(\lambda) \geq 0 \text{ and } \alpha(\theta\lambda) > 0 \,\}.$$

Now set

$$R_\theta^+ = R(N \cap \theta P, T) \cup R(P \cap \theta N, T).$$

Because this set is easily seen to be closed under addition, it is the set of roots of T in the group N_θ. To prove (c), we must show that N_θ is unipotent. This is a consequence of the fact that all the roots in R_θ^+ lie on one side of the hyperplane defined by $\lambda + \theta\lambda$.

For (d), observe first that

$$R(P_\theta, T) = R(L_\theta, T) \cup R(N_\theta, T),$$

a disjoint union. It follows from (b) and (c) that $L_\theta N_\theta$ is a Levi decomposition of the identity component of P_θ; it remains to show that P_θ is connected. So suppose $p \in P_\theta$; we want to show that p is in the identity component. Now pTp^{-1} is a maximal torus in P_θ; so by the conjugacy of maximal tori in connected algebraic groups, we can find an element p_0 in the identity component of P_θ conjugating pTp^{-1} to T. After replacing p by $p_0 p$, we may therefore assume that p normalizes T. Now the normalizer of T in P is contained in L; so $p \in L$. Similarly, the normalizer of T in θP is contained in θL, so $p \in \theta L$. Therefore $p \in L_\theta$, which we already know is a connected subgroup of P_θ.

Part (e) is clear: since θ respects the Levi decomposition in P_θ, an element of $P_\theta \cap K$ must have both its Levi components in K. Q.E.D.

Proof of Proposition 6.17. That p is a well-defined $^\vee G$-equivariant map is clear from Proposition 5.6 and Definition 6.9. The first problem is to show that p is surjective. So fix $x = (y, \Lambda) \in X(^\vee G^\Gamma)$. Obviously

$$p^{-1}(x) = \{\,\phi(y,\lambda) \mid \lambda \in \Lambda, [\lambda, y\cdot\lambda] = 0\,\}. \qquad (6.19)(a)$$

Use the notation of (6.6); then $P(\Lambda)$ is a parabolic subgroup of $^\vee G(\Lambda)$ (Lemma 6.3), and the conjugation action of y defines an involutive automorphism θ of $^\vee G(\Lambda)$, with fixed points $K(y)$. Apply Lemma 6.18 to this situation. We conclude first of all that $P(\Lambda)$ contains a θ-stable maximal torus T. The semisimple $P(\Lambda)$-conjugacy class $\Lambda \subset \mathfrak{p}(\Lambda)$ must meet the Lie algebra $\mathfrak{t}$ in some element λ. Then λ and $y\cdot\lambda$ both belong to $\mathfrak{t}$, so they commute; so $\phi(y,\lambda) \in p^{-1}(x)$ by (6.19)(a).

To prove that p is a bijection on the level of orbits, we must show that the isotropy group

$$G_x = P(\Lambda) \cap K(y) \qquad (6.19)(b)$$

(cf. Proposition 6.16) acts transitively on the fiber $p^{-1}(x)$. Now any element of G_x is fixed by θ, so it must also belong to $\theta P(\Lambda)$. Consequently

$$G_x = P_\theta(\Lambda) \cap K(y) = (L_\theta(\lambda) \cap K(y))(N_\theta(\Lambda) \cap K(y)). \qquad (6.19)(c)$$

So suppose $\phi(y, \lambda')$ is another point in the fiber. By Lemma 6.3(e),

$$\lambda' = \mathrm{Ad}(n)\lambda \qquad (6.20)(a)$$

for a unique $n \in N(\Lambda)$. By (6.19)(a), we have

$$[n\cdot\lambda, \theta(n\cdot\lambda)] = 0. \qquad (6.20)(b)$$

Now the centralizer of λ in $^\vee\mathfrak{g}$ is $\mathfrak{l}(\lambda) \subset \mathfrak{p}(\Lambda)$. It follows that the centralizer of $\theta(n\cdot\lambda)$ is

$$\theta(n\cdot\mathfrak{l}(\lambda)) \subset \theta(n\cdot\mathfrak{p}(\Lambda)) = \theta\mathfrak{p}(\Lambda).$$

From this and (6.20)(b), we deduce that

$$n\cdot\lambda \in \mathfrak{p}(\Lambda) \cap \theta\mathfrak{p}(\Lambda).$$

The latter Lie algebra is described in Lemma 6.18. In conjunction with (6.3)(e), this allows us to conclude (using notation from Lemma 6.18)

6. Geometric parameters

that

$$n \cdot \lambda \in (\lambda + \mathfrak{n}(\Lambda)) \cap (\mathfrak{l}_\theta(\lambda) + \mathfrak{n}_\theta(\Lambda)) = \lambda + (\mathfrak{n}(\Lambda) \cap \mathfrak{n}_\theta(\Lambda)).$$

Now the isomorphism in Lemma 6.3(e) respects T-stable subgroups of $N(\Lambda)$; so the conclusion we draw (using the uniqueness in Lemma 6.3(e)) is

$$n \in N(\Lambda) \cap N_\theta(\Lambda).$$

We now rewrite (6.20)(b) by applying $\theta(n^{-1})$ to each term in the bracket. The conclusion is

$$[(\theta(n^{-1})n) \cdot \lambda, \theta\lambda] = 0. \qquad (6.20)(c)$$

Since $N_\theta(\Lambda)$ is preserved by θ, the element $m = \theta(n^{-1})n$ belongs to $N_\theta(\Lambda)$. The discussion of roots in Lemma 6.18 leads to a factorization

$$N_\theta(\Lambda) = (N(\Lambda) \cap \theta L(\lambda))(N(\Lambda) \cap \theta N(\Lambda))(L(\lambda) \cap \theta N(\Lambda));$$

write $m = m_1 m_2 m_3$ accordingly. Then m_3 fixes λ, and m_1 fixes $\theta\lambda$, so (6.20)(c) leads to

$$[m_2 \cdot \lambda, \theta\lambda] = 0.$$

But Lemma 6.3(e) tells us that the first term here is of the form $\lambda + X_2$, with $X_2 \in \mathfrak{n}(\Lambda) \cap \theta\mathfrak{n}(\Lambda)$. Such an element commutes with $\theta\lambda$ only if $X_2 = 0$, which implies that $m_2 = e$. On the other hand, the definition of m shows that $\theta m = m^{-1}$, from which we conclude that $m_1 = \theta(m_3^{-1})$. Define $n' = nm_3^{-1}$; then $\theta((n')^{-1})n' = e$. Assembling all of this, we find

$$n' \in N_\theta(\Lambda) \cap K(y), \qquad n' \cdot \lambda = \lambda'. \qquad (6.20)(d)$$

Clearly $n' \cdot \phi(y, \lambda) = \phi(y, \lambda')$. Since n' belongs to the unipotent radical of G_x (cf. (6.19)(c)), this completes the proof of Proposition 6.17. Q.E.D.

Corollary 6.21. *Suppose $^\vee H^\Gamma$ and $^\vee G^\Gamma$ are weak E-groups, and $\epsilon: {^\vee H^\Gamma} \to {^\vee G^\Gamma}$ is an L-homomorphism (Definition 5.1). Then there is a natural map*

$$X(\epsilon): X(^\vee H^\Gamma) \to X(^\vee G^\Gamma)$$

on geometric parameters, compatible with the maps $P(\epsilon)$, $\Phi(\epsilon)$ of Proposition 5.4 via the maps p of Proposition 6.17. Explicitly,

$$X(\epsilon)(y, \Lambda) = (\epsilon(y), \epsilon(\Lambda)).$$

(Here $\epsilon(\Lambda)$ denotes the unique canonical flat containing $d\epsilon(\Lambda)$.)

Fix a $^\vee H$ orbit $\mathcal{O} \subset {}^\vee\mathfrak{h}$ of semisimple elements, and define $\epsilon(\mathcal{O})$ to be the unique $^\vee G$ orbit containing $d\epsilon(\mathcal{O})$. Then $X(\epsilon)$ restricts to a morphism of algebraic varieties

$$X(\mathcal{O}, \epsilon) : X(\mathcal{O}, {}^\vee H^\Gamma) \to X(\epsilon(\mathcal{O}), {}^\vee G^\Gamma).$$

If ϵ is injective, then $X(\mathcal{O}, \epsilon)$ is a closed immersion.

Proof. The first point that is not quite obvious is that the image under $d\epsilon$ of a canonical flat Λ is contained in a single canonical flat for $^\vee\mathfrak{g}$. To see this, fix $\lambda \in \Lambda$, and use the notation of (6.1)–(6.4). Write $\lambda' = d\epsilon(\lambda)$, $\Lambda' = \mathcal{F}(\lambda')$. Clearly

$$^\vee\mathfrak{h}(\lambda)_n = (d\epsilon)^{-1} {}^\vee\mathfrak{g}(\lambda')_n$$

for every n. From this it follows that

$$\Lambda = \lambda + \sum_{n>0} {}^\vee\mathfrak{h}(\lambda)_n \subset (d\epsilon)^{-1}(\lambda') + \sum_{n>0} (d\epsilon)^{-1} {}^\vee\mathfrak{g}(\lambda')_n = (d\epsilon)^{-1}(\Lambda').$$

It follows that $d\epsilon(\Lambda) \subset \Lambda'$, as we wished to show. Notice also that if $d\epsilon$ is injective, this containment becomes an equality:

$$\Lambda = (d\epsilon)^{-1}(\Lambda').$$

For the last assertion, we first verify that $X(\epsilon)$ is injective if ϵ is. So suppose

$$X(\epsilon)(y_1, \Lambda_1) = X(\epsilon)(y_2, \Lambda_2) = (y', \Lambda').$$

This means first of all that $\epsilon(y_i) = y$, and therefore that $y_1 = y_2$. Next, the last observation in the preceding paragraph shows that $\Lambda_i = (d\epsilon)^{-1}(\Lambda')$, and therefore that $\Lambda_1 = \Lambda_2$. This proves that $X(\epsilon)$ is injective. Essentially the same argument proves that $X(\mathcal{O}, \epsilon)$ is a closed immersion; we leave the remaining details to the reader. Q.E.D.

6. Geometric parameters

Example 6.22. Suppose $G = GL(3, \mathbb{C})$, endowed with the inner class of real forms including $U(2,1)$ and $U(3)$. The based root datum of G (Definition 2.10) is the quadruple

$$(\mathbb{Z}^3, \{(1, -1, 0), (0, 1, -1)\}, \mathbb{Z}^3, \{(1, -1, 0), (0, 1, -1)\}),$$

with the standard pairing between the two copies of $\mathbb{Z}^3$. The automorphism t of Proposition 2.12 associated to the inner class of real forms acts on $\mathbb{Z}^3$ by $t(a, b, c) = (-c, -b, -a)$. As a dual group for G we can choose ${}^\vee G = GL(3, \mathbb{C})$. Define an automorphism θ of ${}^\vee G$ by

$$\theta g = \operatorname{Ad} \begin{pmatrix} 0 & 0 & 1 \\ 0 & 1 & 0 \\ 1 & 0 & 0 \end{pmatrix} ({}^t g^{-1}).$$

Then θ is an involutive automorphism preserving the standard Cartan and Borel subgroups ${}^d T \subset {}^d B$; the induced automorphism of the based root datum is t. We can therefore construct a weak E-group ${}^\vee G^\Gamma$ generated by ${}^\vee G$ and an element ${}^\vee \delta$ of order two, subject to the relations

$$({}^\vee \delta) g ({}^\vee \delta^{-1}) = \theta g \qquad (g \in {}^\vee G).$$

(If we take as the distinguished set $\mathcal{S}$ of Definition 4.14 the conjugacy class of $({}^\vee \delta, {}^d B)$, then ${}^\vee G^\Gamma$ is an L-group for G.)

Consider the elements

$$\lambda = \begin{pmatrix} 1 & 0 & 0 \\ 0 & 1 & 0 \\ 0 & 0 & 0 \end{pmatrix}, \quad y = {}^\vee \delta, \quad y \cdot \lambda = \begin{pmatrix} 0 & 0 & 0 \\ 0 & -1 & 0 \\ 0 & 0 & -1 \end{pmatrix}.$$

Notice that $e(\lambda) = e = y^2$, and λ commutes with $y \cdot \lambda$; so there is a Langlands parameter $\phi = \phi(y, \lambda)$. Since $e(\lambda) = e$, ${}^\vee G(\lambda) = {}^\vee G$. The corresponding parabolic $P(\lambda)$ is the standard parabolic in $GL(3)$ with Levi subgroup $GL(2) \times GL(1)$, and the canonical flat is

$$\Lambda = \left\{ \begin{pmatrix} 1 & 0 & a \\ 0 & 1 & b \\ 0 & 0 & 0 \end{pmatrix} \;\middle|\; a, b \in \mathbb{C} \right\}.$$

Of course $x = (y, \Lambda)$ is a geometric parameter. The Levi subgroup $L(\lambda, y \cdot \lambda)$ is the diagonal subgroup; its intersection with $K(y)$ (which is a form of $O(3)$) is

$$^\vee G_\phi = \left\{ \begin{pmatrix} z & 0 & 0 \\ 0 & \pm 1 & 0 \\ 0 & 0 & z^{-1} \end{pmatrix} \mid z \in \mathbb{C}^\times \right\}.$$

The Langlands component group A_ϕ (Definition 5.11) therefore has order 2. The group $P_\theta(\Lambda)$ of Lemma 6.18 is just the standard Borel subgroup. The isotropy group of the geometric parameter x has unipotent radical equal to the θ-invariant upper triangular unipotent matrices; this group is

$$K(y) \cap N_\theta(\Lambda) = \left\{ \begin{pmatrix} 1 & a & -a^2/2 \\ 0 & 1 & -a \\ 0 & 0 & 1 \end{pmatrix} \mid a \in \mathbb{C} \right\}.$$

The full stabilizer of x in $^\vee G$ is the semidirect product of $^\vee G_\phi$ with this unipotent group, namely

$$^\vee G_x = \left\{ \begin{pmatrix} z & \pm a & -z^{-1}a^2/2 \\ 0 & \pm 1 & -z^{-1}a \\ 0 & 0 & z^{-1} \end{pmatrix} \mid a \in \mathbb{C}, z \in \mathbb{C}^\times \right\}.$$

To conclude this chapter, we offer a variant of Proposition 6.16. Although the version presented first seems to be the most natural one, it has the technical disadvantage that the group $^\vee G(\Lambda)$ need not be connected. The only subtlety in the variant is the definition

$$K(y)^0 = K(y) \cap {}^\vee G(\Lambda)_0 \qquad (6.23)(a)$$

for $y \in \mathcal{I}(\Lambda)$ (cf. (6.10)). In the notation of Lemma 6.12, this is the group of fixed points of the involution θ_y on the connected group $^\vee G(\Lambda)_0$. We have

$$K(y)_0 \subset K(y)^0 \subset K(y), \qquad (6.23)(b)$$

and both of these inclusions may be proper.

Proposition 6.24. *Suppose $^\vee G^\Gamma$ is a weak E-group. Fix an orbit $\mathcal{O}$ of $^\vee G$ on the semisimple elements of $^\vee \mathfrak{g}$, and use the notation of (6.6), (6.10), and (6.23). Fix a canonical flat $\Lambda \subset \mathcal{O}$, and write $\mathcal{P}^0(\Lambda)$ for the $^\vee G(\Lambda)_0$ orbit of Λ in the set of canonical flats. List the orbits of $^\vee G(\Lambda)_0$ on $\mathcal{I}(\Lambda)$ as $\mathcal{I}_1^0(\Lambda), \ldots, \mathcal{I}_s^0(\Lambda)$ (notation (6.10)(f) and Lemma 6.12). For*

6. Geometric parameters

each j, choose a point $y_j \in \mathcal{I}_j^0(\Lambda)$. Then conjugation by y_j defines an involutive automorphism θ_j of $^\vee G(\Lambda)_0$ with fixed point set $K_j^0 = K(y_j)^0$ (cf. (6.23)). From the chosen base points Λ and y_j we get isomorphisms

$$\mathcal{P}^0(\Lambda) \simeq {}^\vee G(\Lambda)_0/P(\Lambda), \qquad \mathcal{I}_j^0(\Lambda) \simeq {}^\vee G(\Lambda)_0/K_j^0.$$

The variety $X(\mathcal{O}, {}^\vee G^\Gamma)$ is the disjoint union of s closed connected smooth subvarieties $X_j(\mathcal{O}, {}^\vee G^\Gamma)$. The jth subvariety looks like

$$\begin{aligned}X_j(\mathcal{O}, {}^\vee G^\Gamma) &\simeq {}^\vee G \times_{{}^\vee G(\Lambda)_0} (\mathcal{I}_j^0(\Lambda) \times \mathcal{P}^0(\Lambda)) \\ &\simeq {}^\vee G \times_{{}^\vee G(\Lambda)_0} ({}^\vee G(\Lambda)_0/K_j^0 \times {}^\vee G(\Lambda)_0/P(\Lambda)) \\ &\simeq {}^\vee G \times_{K_j^0} ({}^\vee G(\Lambda)_0/P(\Lambda)).\end{aligned}$$

In particular, the orbits of $^\vee G$ on $X_j(\mathcal{O}, {}^\vee G^\Gamma)$ are in one-to-one correspondence with the orbits of K_j^0 on the partial flag variety $^\vee G(\Lambda)_0/P(\Lambda)$. These orbits are finite in number.

The proof is identical to that given for Proposition 6.16. (It is helpful to keep in mind that $P(\Lambda)$ is connected (Lemma 6.3(d)) and therefore contained in $^\vee G(\Lambda)_0$.) As with Proposition 6.16, we will eventually want the stronger information about the orbit correspondence contained in Proposition 7.14(c).

7. Complete geometric parameters and perverse sheaves

In this chapter we continue our analysis of the geometric parameter spaces described in the preceding chapter. To begin, we recall some terminology from Definition 1.22 in the introduction. This will be applied first of all in the setting of Definition 6.9, to the action of ${}^\vee G^{alg}$ on $X(\mathcal{O}, {}^\vee G^\Gamma)$.

Definition 7.1. Suppose Y is a complex algebraic variety on which the pro-algebraic group H acts with finitely many orbits. A *geometric parameter for H acting on Y* is a (closed) point of Y. Two such parameters are called *equivalent* if they differ by the action of H. The set of equivalence classes of geometric parameters — that is, the set of orbits of H on Y — is written $\Phi(Y, H)$.

Suppose $y \in Y$; write H_y for the isotropy group of the action at y. The *(local) equivariant fundamental group at y* is

$$A_y^{loc} = H_y/(H_y)_0, \qquad (7.1)(a)$$

the (pro-finite) group of connected components of H_y. If y' is equivalent to y (that is, if it belongs to the same H orbit) then we can find $h \in H$ with $h \cdot y = y'$. It follows that conjugation by h carries H_y isomorphically onto $H_{y'}$, so we get an isomorphism

$$A_y^{loc} \simeq A_{y'}^{loc}. \qquad (7.1)(b)$$

The coset hH_y is uniquely determined, so this isomorphism is unique up to inner automorphisms.

Suppose $S \in \Phi(Y, H)$ is an orbit of H on Y. The *equivariant fundamental group of S* is the pro-finite group

$$A_S^{loc} = A_y^{loc} \qquad (y \in S). \qquad (7.1)(c)$$

The isomorphisms in (7.1)(b) show that A_S^{loc} is well-defined up to inner automorphism. This means that we can safely discuss conjugacy classes in A_S^{loc}, or equivalence classes of representations, but it is dangerous to speak of particular elements as well-defined entities.

7. Complete geometric parameters and perverse sheaves

A *(local) complete geometric parameter for H acting on Y* is a pair (y, τ) with $y \in Y$ and $\tau \in \widehat{A_y^{loc}}$ an irreducible representation. Two such parameters (y, τ) and (y', τ') are called *equivalent* if there is an element of H carrying y to y' so that the induced isomorphism (7.1)(b) carries τ to τ'. The set of equivalence classes of (local) complete geometric parameters is written $\Xi(Y, H)$.

A *complete geometric parameter for H acting on Y* is a pair $(S, \mathcal{V})$, with $S \in \Phi(Y, H)$ an orbit of H on Y and $\mathcal{V}$ an irreducible H-equivariant local system on S. (By a "local system" we will understand a vector bundle with a flat connection; perhaps this is not the most reasonable definition, but we are considering only smooth varieties and complex coefficients).

Of course the terminology "geometric parameter" for a point is a little ridiculous, but we include it to maintain consistency with the case of Definition 6.9.

We should make a few remarks about the "pro-algebraic" condition. This means that H is the limit of an inverse system $\{H_i \mid i \in I\}$ of algebraic groups indexed by a directed set I. It is harmless and convenient to assume that all the morphisms

$$h_{ji} : H_i \to H_j \qquad (i \geq j) \qquad (7.2)(a)$$

in the inverse system are surjective. (In our examples, the H_i will all be certain finite coverings of a fixed group H_1.) Then the limit morphisms

$$h_j : H \to H_j \qquad (7.2)(b)$$

are surjective as well. The identity component of H is by definition the inverse limit of the identity components of the H_i. To say that H acts on Y means that for a cofinal set $I_Y \subset I$, the various $\{H_i \mid i \in I_Y\}$ act compatibly on Y; we may as well assume (by shrinking I) that they all do (as will be the case in our examples). Each H_i will have an algebraic isotropy subgroup $H_{i,y}$ at a point y. These groups again form an inverse system of algebraic groups, whose limit is what we called H_y. It is easy to check that A_y^{loc} (defined as a quotient of inverse limits) is naturally isomorphic to the inverse limit of the quotients:

$$A_y^{loc} \simeq \varprojlim H_{i,y}/(H_{i,y})_0. \qquad (7.2)(c)$$

Again we have surjective maps

$$a_i : A_y^{loc} \to H_{i,y}/(H_{i,y})_0. \qquad (7.2)(d)$$

By a (finite-dimensional) representation of A_y^{loc} we understand a representation that factors through one of the maps a_i (and therefore automatically through all a_j with $j \geq i$). For our purposes this can be taken as a definition; but if A_y^{loc} is given the inverse limit topology, then it is a theorem that any continuous irreducible representation has this property.

Here are some standard and elementary facts.

Lemma 7.3. *In the setting of Definition 7.1, suppose $y \in Y$. Write $S = H \cdot y$ for the orbit of y.*

a) *If H is algebraic, connected, and simply connected, then the local equivariant fundamental group A_y^{loc} is canonically isomorphic to the topological fundamental group $\pi_1(S; y)$ of homotopy classes of loops in S based at y.*

b) *More generally, there is always a natural map*

$$\pi_1(S; y) \to A_y^{loc}.$$

This map is surjective whenever H is connected algebraic, and injective whenever H is simply connected algebraic. In particular, any representation of A_y^{loc} gives rise to a representation of $\pi_1(S; y)$.

c) *Suppose (τ, V) is a representation of A_y^{loc}. Regard τ as a representation of H_y trivial on the identity component. Then the induced bundle*

$$\mathcal{V} = H \times_{H_y} V \to S$$

carries an H-invariant flat connection.

d) *Suppose $\mathcal{V}$ is an H-equivariant vector bundle on S carrying an H-invariant flat connection. Write V for the fiber of $\mathcal{V}$ at y, and τ for the isotropy action of H_y on V. Then τ is trivial on $(H_y)_0$, and so factors to a representation of A_y^{loc}.*

e) *The constructions of (c) and (d) establish a natural bijection between the equivalence classes of local complete geometric parameters for H acting on Y, and the complete geometric parameters.*

We will use Lemma 7.3(e) to identify the set $\Xi(Y, H)$ of equivalence classes of local complete geometric parameters. Thus if $\xi \in \Xi(Y, H)$, we will write

$$\xi = (S, \mathcal{V}) = (S_\xi, \mathcal{V}_\xi), \tag{7.4}$$

and say that a pair (y, τ) in the equivalence class "represents ξ." The

7. Complete geometric parameters and perverse sheaves

assertions about fundamental groups are included to explain how H-equivariant local systems are related to arbitrary local systems.

Proof. We begin by defining the map in (b). Suppose

$$\gamma : [0,1] \to S, \quad \gamma(0) = \gamma(1) = y$$

is a loop. Identify S with the quotient H/H_y; then γ can be lifted to a continuous map

$$\tilde{\gamma} : [0,1] \to H, \quad \tilde{\gamma}(0) = e.$$

The element $h = \tilde{\gamma}(1)$ must map to y, so $h \in H_y$. Any two liftings $\tilde{\gamma}$ differ by a continuous map from $[0,1]$ into H_y starting at e. Such a map must take values in $(H_y)_0$, so the class of $\overline{h} \in A_y^{loc}$ is independent of the lifting. Similarly, a homotopy of the loop γ does not affect $\overline{h}$; so we get the map in (b). To see that the map is surjective when H is connected algebraic, fix $h \in H_y$. A path connecting e to h descends to a loop in H/H_y that is the required preimage of $\overline{h}$. For the injectivity when H is simply connected algebraic, suppose γ is a loop at y mapping to the class of e. This means that the lifting $\tilde{\gamma}$ satisfies $\tilde{\gamma}(1) = h \in (H_y)_0$. Choose a path connecting h to the identity in $(H_y)_0$; modifying $\tilde{\gamma}$ by this path gives a new lifting $\tilde{\gamma}'$ satisfying $\tilde{\gamma}'(1) = e$. This new lifting is therefore a loop. Since H is simply connected, $\tilde{\gamma}'$ is homotopic to a trivial map. This homotopy descends to a homotopy from γ to a point in H/H_y. This proves (b), and (a) follows.

We omit the elementary arguments for (c) and (d), but a few remarks on the definitions in the pro-algebraic case may be helpful. For part (c), recall that τ is really a representation of some component group $H_{i,y}/(H_{i,y})_0$ (cf. (7.2)). Then the algebraic vector bundle

$$\mathcal{V} = H_i \times_{H_{i,y}} V \to S$$

carries an action of H_i, and thus compatible actions of all H_j with $j \geq i$. Therefore it is an "H-equivariant vector bundle." Conversely, to say that $\mathcal{V}$ is H-equivariant means that it carries an action of some H_i. Consequently the isotropy action on V factors to $H_{i,y}$. Part (e) is immediate from (c) and (d). Q.E.D.

Lemma 7.5. *Suppose $^\vee G^\Gamma$ is a weak E-group, $\phi \in P(^\vee G^\Gamma)$ is a Langlands parameter, and $x = p(\phi) \in X(^\vee G^\Gamma)$ is a geometric parameter (cf. Proposition 6.17). Then the natural inclusion of isotropy groups*

$$^\vee G_\phi \hookrightarrow {}^\vee G_x$$

induces an isomorphism of component groups

$$^\vee G_\phi/(^\vee G_\phi)_0 \simeq {^\vee G_x}/(^\vee G_x)_0.$$

If we consider isotropy groups in the algebraic universal cover $^\vee G^{alg}$ (Definition 1.16 and (5.10)), we get

$$^\vee G_\phi^{alg}/(^\vee G_\phi^{alg})_0 \simeq {^\vee G_x^{alg}}/(^\vee G_x^{alg})_0.$$

Similarly, in the setting (5.13),

$$^\vee G_\phi^Q/(^\vee G_\phi^Q)_0 \simeq {^\vee G_x^Q}/(^\vee G_x^Q)_0.$$

Proof. The inclusion of isotropy groups comes from the equivariance of the map p. By Proposition 6.17, $^\vee G_x$ is the semidirect product of $^\vee G_\phi$ and a connected unipotent group U_x. It follows that $(^\vee G_x)_0$ must be the semidirect product of $(^\vee G_\phi)_0$ and U_x. The first claim follows. The the second is a special case of the third, so we consider the third. The identity component $(U_x^Q)_0$ is a covering group of U_x. Now a unipotent algebraic group is simply connected, so $(U_x^Q)_0$ must map isomorphically onto U_x, and we identify them henceforth. It follows easily that $^\vee G_x^Q$ must be the semidirect product of $^\vee G_\phi^Q$ and U_x. Now we can argue as for the first claim to get the isomorphism of component groups. Q.E.D.

Definition 7.6. Suppose $^\vee G^\Gamma$ is a weak E-group, and $X(^\vee G^\Gamma)$ is the corresponding set of geometric parameters (Definition 6.9). In order to place ourselves precisely in the setting of Definition 7.1, we can implicitly restrict attention to a single variety $X(\mathcal{O}, {^\vee G^\Gamma})$ (see Proposition 6.16). For $x \in X(^\vee G^\Gamma)$, we consider the (local) equivariant fundamental groups for the actions of $^\vee G$ and $^\vee G^{alg}$,

$$A_x^{loc} = {^\vee G_x}/(^\vee G_x)_0, \qquad A_x^{loc,alg} = {^\vee G_x^{alg}}/(^\vee G_x^{alg})_0.$$

By Lemma 7.5, there are natural isomorphisms

$$A_x^{loc} \simeq A_\phi^{loc}, \qquad A_x^{loc,alg} \simeq A_\phi^{loc,alg}$$

for any Langlands parameter $\phi \in p^{-1}(x)$ (cf. Definition 5.11). We therefore refer to these groups as the *Langlands component group for x* and the *universal component group for x*. If S is an orbit on $X(^\vee G^\Gamma)$, we write following Definition 7.1

$$A_S^{loc} = A_x^{loc}, \qquad A_S^{loc,alg} = A_x^{loc,alg} \qquad (x \in S);$$

7. Complete geometric parameters and perverse sheaves

again these groups are defined up to inner automorphism.

A *(local) complete geometric parameter for* $^\vee G^\Gamma$ is a pair (x,τ) with $x \in X(^\vee G^\Gamma)$ and $\tau \in \widehat{A_x^{loc,alg}}$ an irreducible representation. Two such parameters are called *equivalent* if they are conjugate by $^\vee G^{alg}$. Lemma 7.5 provides a natural map from complete Langlands parameters for $^\vee G^\Gamma$ onto (local) complete geometric parameters, inducing a bijection on equivalence classes. We may therefore write $\Xi(^\vee G^\Gamma)$ for the set of equivalence classes (cf. Definition 5.11).

As in (7.4), we identify $\Xi(^\vee G^\Gamma)$ with the set of pairs $(S, \mathcal{V})$, with S an orbit on $X(^\vee G^\Gamma)$ and $\mathcal{V}$ an irreducible $^\vee G^{alg}$-equivariant local system on S; we call such a pair a *complete geometric parameter for* $X(^\vee G^\Gamma)$. If $\xi \in \Xi(^\vee G^\Gamma)$, we write

$$\xi = (S_\xi, \mathcal{V}_\xi)$$

as in (7.4). A (local) complete geometric parameter (x,τ) in the corresponding equivalence class will sometimes be called a *representative of* ξ.

All of this discussion can be extended to the setting (5.13). We write

$$A_x^{loc,Q} = {^\vee G_x^Q}/({^\vee G_x^Q})_0,$$

the *Q-component group for* x. The set of complete geometric parameters for $^\vee G^Q$ acting on $X(^\vee G^\Gamma)$ is written $\Xi(^\vee G^\Gamma)^Q$; we call these *complete geometric parameters of type* Q.

We return now to the general context of Definition 7.1, and expand on the discussion in Definition 1.22.

Definition 7.7. Suppose Y is a smooth complex algebraic variety on which the pro-algebraic group H acts with finitely many orbits. Write $\mathcal{D}_Y$ for the sheaf of algebraic differential operators on Y ([12], VI.1). Define

$\mathcal{C}(Y, H)$ = category of H-equivariant constructible sheaves
  of complex vector spaces on Y.

A definition of constructible sheaves may be found in [11], V.3. Next, set

$\mathcal{P}(Y, H)$ = category of H-equivariant perverse sheaves
  of complex vector spaces on Y.

The definition of perverse sheaves is discussed in [9] (see especially the introduction.) Again it is most convenient to take the analytic topology on Y, but the equivariance condition implies that one can work algebraically. (See [9], section 6. The main point is that all the local systems involved, which are described by Lemma 7.3, arise from representations of π_1 factoring through finite quotients.) This will be crucial for our eventual application of the results of [40].

Finally, we will consider

$\mathcal{D}(Y, H)$ = category of H-equivariant regular holonomic sheaves of $\mathcal{D}_Y$-modules on Y.

In this case it is most convenient to work in the Zariski topology; then [12] is a convenient reference. For an (algebraic) definition of regular holonomic, we refer to [12], section VII.11. (By [12], Theorem VII.12.11, any H-equivariant coherent sheaf of $\mathcal{D}_Y$-modules on Y is automatically regular holonomic.)

Each of these three categories is abelian, and every object has finite length. The corresponding Grothendieck groups are written

$$K\mathcal{C}(Y, H), \quad K\mathcal{P}(Y, H), \quad K\mathcal{D}(Y, H).$$

Each is a free abelian group with a basis corresponding to the (finite) set of irreducible objects in the corresponding category.

The next two results recall the the most important relationships among these three categories. The first is essentially obvious from the definitions. (It is the definitions that are deep.)

Lemma 7.8 ([9].) *In the setting of Definition 7.7, suppose P is an H-equivariant perverse sheaf on Y. Then the cohomology sheaves $H^i P$ are H-equivariant constructible sheaves on Y. A short exact sequence of perverse sheaves gives rise to a long exact sequence of cohomology sheaves, so there is an additive map*

$$\chi : \mathrm{Ob}\,\mathcal{P}(Y, H) \to K\mathcal{C}(Y, H), \quad \chi(P) = \sum (-1)^i H^i P.$$

The map χ induces an isomorphism

$$\chi : K\mathcal{P}(Y, H) \to K\mathcal{C}(Y, H)$$

of Grothendieck groups.

Theorem 7.9 (the Riemann-Hilbert correspondence — see [12], Theorem VIII.14.4). *In the setting of Definition 7.7, there is an equivalence of categories (the "de Rham functor")*

$$DR : \mathcal{D}(Y, H) \to \mathcal{P}(Y, H).$$

7. Complete geometric parameters and perverse sheaves

This induces an isomorphism

$$DR : K\mathcal{D}(Y, H) \to K\mathcal{P}(Y, H)$$

of Grothendieck groups.

We will use the isomorphisms of Lemma 7.8 and Theorem 7.9 to identify the three Grothendieck groups, writing simply

$$K(Y, H) \simeq K\mathcal{C}(Y, H) \simeq K\mathcal{P}(Y, H) \simeq K\mathcal{D}(Y, H) \qquad (7.10)(a)$$

We will also need notation for the irreducible objects in the three categories. These irreducible objects are in each case parametrized by the set $\Xi(Y, H)$ of complete geometric parameters (Definition 7.1). Fix $\xi = (S, \mathcal{V}) \in \Xi$. We need a little notation from [9], section 1.4. Write

$$j = j_S : S \to \overline{S}, \qquad i = i_S : \overline{S} \to Y \qquad (7.10)(b)$$

for the inclusion of S in its closure, and the inclusion of the closure in Y. Write $d = d(S) = d(\xi)$ for the dimension of S. Regard the local system $\mathcal{V}$ as a constructible sheaf on S. Applying to this first the functor $j_!$ of extension by zero, and then the direct image i_*, we get a complex

$$\mu(\xi) = i_* j_! \mathcal{V} \in \mathrm{Ob}\,\mathcal{C}(Y, H), \qquad (7.10)(c)$$

the *extension of ξ by zero*. Its irreducibility, and the fact that every irreducible has this form, are consequences of the adjunction formulas relating i_* and $j_!$ with the restriction functors i^* and j^* ([9], 1.4.1). Next, consider the complex $\mathcal{V}[-d]$ consisting of the single sheaf $\mathcal{V}$ in degree $-d$. This is an H-equivariant perverse sheaf on S. Applying to it the "intermediate extension functor" $j_{!*}$ ([9], Definition 1.4.22) followed by the direct image i_*, we get a perverse sheaf on Y:

$$P(\xi) = i_* j_{!*} \mathcal{V}[-d] \in \mathrm{Ob}\,\mathcal{P}(Y, H), \qquad (7.10)(d)$$

the *perverse extension of ξ*. Its irreducibility, and the fact that every irreducible has this form, are consequences of Théorème 4.3.1 of [9]. Finally, we define

$$D(\xi) = DR^{-1}(P(\xi)) \in \mathrm{Ob}\,\mathcal{D}(Y, H). \qquad (7.10)(e)$$

It is a consequence of Theorem 7.9 that every irreducible H-equivariant $\mathcal{D}_Y$-module on Y is of this form. (Of course one can define $D(\xi)$ somewhat more directly as a $\mathcal{D}_Y$-module. The starting point is the sheaf

$\mathcal{M}$ of germs of sections of $\mathcal{V}$ on S. Since $\mathcal{V}$ has an H-equivariant flat connection, $\mathcal{M}$ has the structure of a an H-equivariant $\mathcal{D}_S$-module ([12], section VI.1.6). One now applies the (subtle) notion of direct image for $\mathcal{D}$-modules to get a $\mathcal{D}_Y$-module. The first complication is that the inclusion j need not be affine, so there are higher direct images to consider. The second is that (even if j is affine) the direct image will not be irreducible; one has to pass to an appropriate subquotient. None of this is particularly difficult, given the machinery of [12], but it is easier simply to quote Theorem 7.9.)

The sets $\{\mu(\xi) \mid \xi \in \Xi\}$, $\{P(\xi) \mid \xi \in \Xi\}$, and $\{D(\xi) \mid \xi \in \Xi\}$ are all bases of the Grothendieck group $K(Y, H)$. By Theorem 7.9 and the definitions of (7.10), the last two bases are exactly the same. However, the first two are *not* identified by the isomorphism of Lemma 7.8. As in Definition 1.22, we can write in $K(Y, H)$

$$\mu(\xi) = (-1)^{d(\xi)} \sum_{\gamma \in \Xi(Y,H)} m_g(\gamma, \xi) P(\gamma) \qquad (7.11)(a)$$

with $m_g(\gamma, \xi)$ an integer. The matrix m_g is called the *geometric multiplicity matrix*. It follows easily from the definitions that

$$m_g(\xi, \xi) = 1, \quad m_g(\gamma, \xi) \neq 0 \text{ only if } S_\gamma \subset (\overline{S_\xi} - S_\xi) \quad (\gamma \neq \xi).$$
$$(7.11)(b)$$

It is also useful to consider the inverse matrix $c_g = m_g^{-1}$, the *geometric character matrix*. By its definition, c_g satisfies

$$P(\gamma) = \sum_{\xi \in \Xi(Y,H)} (-1)^{d(\xi)} c_g(\xi, \gamma) \mu(\xi). \qquad (7.11)(c)$$

Because of the form of the isomorphism in Lemma 7.8, we have

$$c_g(\xi, \gamma) = (-1)^{d(\xi)} \sum_i (-1)^i (\text{multiplicity of } \mathcal{V}_\xi \text{ in } H^i P(\gamma)|_{S_\xi}).$$
$$(7.11)(d)$$

This says that the entries of c_g are essentially local Euler characteristics for intersection cohomology. As a consequence of (7.11)(d) (or of (7.11)(b)),

$$c_g(\gamma, \gamma) = 1, \quad c_g(\xi, \gamma) \neq 0 \text{ only if } S_\xi \subset (\overline{S_\gamma} - S_\gamma) \quad (\xi \neq \gamma).$$
$$(7.11)(e)$$

Because of (7.11)(b), we can define the *Bruhat order* on $\Xi(Y, H)$ as the

7. Complete geometric parameters and perverse sheaves

smallest partial order with the property that

$$m_g(\gamma, \xi) \neq 0 \text{ only if } \gamma \leq \xi. \qquad (7.11)(f)$$

This refines the partial preorder induced by the closure relation on the underlying orbits. Because the inverse of an upper triangular matrix is upper triangular, we get the same order using c_g in place of m_g.

To understand why we call m_g a multiplicity matrix, suppose for a moment that the orbit S_ξ is affinely embedded in its closure. (For $Y = X(\mathcal{O}, {}^\vee G^\Gamma)$ as in Definition 6.9, this turns out to be automatic if the orbit $\mathcal{O}$ consists of regular elements.) Regarded as an element of the derived category concentrated in degree $-d$, the constructible sheaf $\mu(\xi)[-d]$ is then perverse ([9], Corollary 4.1.3), and the map of Lemma 7.8 is

$$\chi(\mu(\xi)[-d]) = (-1)^d \mu(\xi).$$

On the other hand, as a perverse sheaf $\mu(\xi)[-d]$ must have a finite composition series. The integer $m_g(\gamma, \xi)$ is just the multiplicity of the irreducible perverse sheaf $P(\gamma)$ in this composition series. In particular, it is non-negative. One can give a parallel discussion in terms of $\mathcal{D}$-modules.

We will be forced occasionally to refer to certain derived categories (which are of course central to the basic theory developed in [9] and [11]). In general an object in one of these derived categories will be a complex $\mathcal{S}^\bullet$ of sheaves on Y; we will write $H^i(\mathcal{S}^\bullet)$ for the ith cohomology sheaf. The notion of H-equivariance in such categories is quite subtle, and we will simply not consider it (although a complete treatment of some of the "obvious" assertions we make requires confronting this issue). Write

$$D^b(\mathcal{C}(Y)) \qquad (7.12)(a)$$

for the bounded derived category of sheaves of complex vector spaces on Y (analytic topology) having cohomology sheaves constructible with respect to an algebraic stratification of Y (cf. [9], section 2.2, and [11], section 6.3). Perverse sheaves on Y live in this derived category. Finally, write

$$D^b_{rh}(\mathcal{D}(Y)) \qquad (7.12)(b)$$

for the bounded derived category of sheaves of quasicoherent $\mathcal{D}_Y$-modules having regular holonomic cohomology sheaves (cf. [12], Corollary VII.12.8).

Definition 7.13. Suppose $^\vee G^\Gamma$ is a weak E-group, and $X(^\vee G^\Gamma)$ is the corresponding set of geometric parameters (Definition 6.9). Define

$$\mathcal{C}(X(^\vee G^\Gamma), {}^\vee G^{alg})$$

to be the direct sum over semisimple orbits $\mathcal{O} \subset {}^\vee\mathfrak{g}$ of the categories $\mathcal{C}(X(\mathcal{O}, {}^\vee G^\Gamma), {}^\vee G^{alg})$ of Definition 7.7. We call the objects of this category $^\vee G^{alg}$-*equivariant constructible sheaves on* $X(^\vee G^\Gamma)$. Similarly we define

$$\mathcal{P}(X(^\vee G^\Gamma), {}^\vee G^{alg}), \qquad \mathcal{D}(X(^\vee G^\Gamma), {}^\vee G^{alg}),$$

the $^\vee G^{alg}$-*equivariant perverse sheaves* and the $^\vee G^{alg}$-*equivariant coherent $\mathcal{D}$-modules* on $X(^\vee G^\Gamma)$. These are still abelian categories in which every object has finite length, and we have analogues of Lemma 7.8 and Theorem 7.9. In particular, we can write $KX(^\vee G^\Gamma)$ for the common Grothendieck group of the three categories. The irreducible objects in each category are parametrized by the set $\Xi(^\vee G^\Gamma)$ of complete geometric parameters (Definition 7.6); we write $\mu(\xi)$, $P(\xi)$, and $D(\xi)$ as in (7.10). We get matrices m_g and c_g satisfying (7.11), the *geometric multiplicity matrix* and the *geometric character matrix* for $^\vee G^\Gamma$.

In the setting (5.13), we introduce a superscript Q in this notation. Thus for example $KX(^\vee G^\Gamma)^Q$ is the Grothendieck group of $^\vee G^Q$-equivariant perverse sheaves (or constructible sheaves, or coherent $\mathcal{D}$-modules) on $X(^\vee G^\Gamma)$.

We want to use structure theorems like Proposition 6.16 to study the geometric multiplicity matrix. To do that, we must first understand the behavior of the categories of Definition 7.7 under some simple constructions.

Proposition 7.14. *Suppose Y is a smooth complex algebraic variety on which the pro-algebraic group H acts with finitely many orbits. Suppose G is a pro-algebraic group containing H. Consider the induced bundle*

$$X = G \times_H Y.$$

a) The inclusion

$$i: Y \to X, \qquad i(y) = \text{equivalence class of } (e, y)$$

7. Complete geometric parameters and perverse sheaves

induces a bijection

$$\Phi(Y,H) \to \Phi(X,G), \qquad S \mapsto G \cdot i(S)$$

from H-orbits on Y to G-orbits on X.
b) *If $y \in Y$, then the isotropy group of the G action on X at $i(y)$ is H_y. In particular, the local equivariant fundamental groups satisfy*

$$A_y^{loc} = A_{i(y)}^{loc}$$

(notation as in Definition 7.1). Consequently there is an isomorphism

$$A_S^{loc} \simeq A_{\Phi(i)(S)}^{loc},$$

canonically defined up to inner automorphisms.
c) *The maps of (a) and (b) combine to give a natural bijection*

$$i : \Xi(Y,H) \to \Xi(X,G).$$

In the language of (7.4), the H-equivariant local system $\mathcal{V}$ on S corresponds to

$$i(\mathcal{V}) = G \times_H \mathcal{V}$$

on $G \cdot i(S)$.
d) *There are natural equivalences of categories*

$$\mathcal{C}(Y,H) \simeq \mathcal{C}(X,G), \quad \mathcal{P}(Y,H) \simeq \mathcal{P}(X,G), \quad \mathcal{D}(Y,H) \simeq \mathcal{D}(X,G).$$

These equivalences are compatible with the parametrizations of irreducibles by $\Xi(Y,H) \simeq \Xi(X,H)$ ((7.10) and (c) above) and with the isomorphisms of Grothendieck groups coming from Lemma 7.8 and Theorem 7.9.
e) *The geometric multiplicity and character matrices of (7.11) for (Y,H) and (X,G) coincide (under the bijection of (c) above for the index sets).*

Proof. Parts (a), (b), and (c) are formal; in fact similar ideas appeared already in Lemma 6.15. The equivalences of categories are in some sense "obvious" once the categories are understood, and we will just offer some hints. We discuss only the perverse case; the constructible case is similar but more elementary, and the $\mathcal{D}$-module case is handled by Theorem 7.9. It is helpful to introduce the intermediate space

$$Z = G \times Y,$$

on which the group $G \times H$ acts by

$$(g, h) \cdot (g', y) = (gg'h^{-1}, h \cdot y).$$

It is easy to see that $\Xi(Z, G \times H)$ is in one-to-one correspondence with $\Xi(Y, H)$. The groups G and H each act freely on Z (although the product need not). Dividing by the G action gives a smooth quotient morphism

$$f_G : Z \to Y$$

(projection on the second factor) that respects the action of H. Dividing by the H action gives a smooth quotient morphism

$$f_H : Z \to X$$

(the definition of the induced bundle) that respects the action of G. We can therefore apply Proposition 4.2.5 of [9] to conclude that $f_G^*[\dim G]$ (respectively $f_H^*[\dim H]$) is a fully faithful exact functor from $\mathcal{P}(Y, H)$ (respectively $\mathcal{P}(X, G)$) into $\mathcal{P}(Z, G \times H)$. That the irreducible objects behave properly under these functors is contained in the discussion at the bottom of page 110 in [9]. We leave the rest of the argument to the reader. Part (e) is a formal consequence of (d). Q.E.D.

Proposition 7.15. *Suppose X and Y are smooth algebraic varieties on which the pro-algebraic group H acts with finitely many orbits, and*

$$f : X \to Y$$

is a surjective smooth equivariant morphism having connected fibers of dimension d.
a) *There is a natural inclusion*

$$f^* : \Xi(Y, H) \hookrightarrow \Xi(X, H)$$

from complete geometric parameters for H acting on Y to complete geometric parameters for H acting on X.
b) *The functor $f^*[d]$ is a fully faithful exact functor from $\mathcal{P}(Y, H)$ into $\mathcal{P}(X, H)$, carrying the irreducible $P(\xi)$ to $P(f^*\xi)$ (notation (7.10)(d)).*

7. Complete geometric parameters and perverse sheaves

c) *The inclusion of (a) respects the geometric character matrix of (7.11):*

$$c_g(f^*\xi, f^*\gamma) = c_g(\xi, \gamma) \qquad (\xi, \gamma \in \Xi(Y, H)).$$

Proof. We begin by describing the map f^* on geometric parameters. If S is an orbit of H on Y, of dimension $d(S)$, then $f^{-1}S$ is a smooth H-invariant subset of X, of dimension $d(S)+d$. Since H has only finitely many orbits on X, there is an open orbit f^*S of H on $f^{-1}S$. Since f has connected fibers, f^*S is unique, and is dense in $f^{-1}S$. This defines an injection on the sets of orbits

$$f^* : \Phi(Y, H) \to \Phi(X, H). \qquad (7.16)(a)$$

The image of f^* consists of orbits $S' \subset X$ satisfying any of the three equivalent properties

$$\begin{aligned}&S' \text{ is open in } f^{-1}(f(S')), \\ &\dim f(S') = \dim S' - d, \\ &\dim H_{f(x)} = \dim H_x + d \qquad (x \in S'). \end{aligned} \qquad (7.16)(b)$$

Suppose S' satisfies these conditions. A typical fiber of the map $S' \to f(S')$ is dense in the corresponding fiber of f, and is therefore connected. Consequently

$$H_{f(x)}/H_x \text{ is connected} \qquad (x \in S' \in f^*(\Phi(Y, H))). \qquad (7.16)(c)$$

For the natural map on component groups, it follows that

$$A_x^{loc} \to A_{f(x)}^{loc} \text{ is surjective} \qquad (x \in S' \in f^*(\Phi(Y, H))). \qquad (7.16)(d)$$

Consequently every irreducible representation of $A_{f(x)}^{loc}$ defines by composition an irreducible representation of A_x^{loc}. This construction clearly provides the inclusion of (a) on the level of equivalence classes of local complete geometric parameters. The image of f^* consists of equivalence classes of pairs (x, τ) for which x and its orbit S' satisfy (7.16)(b), and the representation τ is trivial on the kernel of the map (7.16)(d). On the level of local systems, the map f^* sends $(S, \mathcal{V})$ to $(S', \mathcal{V}')$, with

$$S' \text{ open in } f^{-1}S, \mathcal{V}' = (f^*\mathcal{V})|_{S'}. \qquad (7.16)(e)$$

For the rest of the argument, we refer to [9], Proposition 4.2.5. Q.E.D.

It is *not* true that the inclusion f^* respects the geometric multiplicity matrix in general. The matrix c_g for Y can be regarded as an upper left corner block of c_g for X, but such a relationship between two matrices is not preserved by inverting them. (One might hope for help from the fact, implicit in (7.11)(e), that c_g is upper triangular. Unfortunately, the ordering of the basis needed to get upper triangular matrices is incompatible with the ordering needed to make the matrix for Y a corner of the one for X.)

We conclude this chapter with a more general (and correspondingly weaker) functorial relationship between different geometric parameters. We begin with two pairs (X, G) and (Y, H) as in Definition 7.7. That is, X and Y are smooth algebraic varieties, on which the pro-algebraic groups G and H act with finitely many orbits. We suppose given in addition a morphism

$$\epsilon : (Y, H) \to (X, G). \qquad (7.17)(a)$$

This means that $\epsilon : Y \to X$ is a morphism of varieties, $\epsilon : H \to G$ is a morphism of pro-algebraic groups, and

$$\epsilon(h) \cdot \epsilon(y) = \epsilon(h \cdot y) \qquad (h \in H, y \in Y). \qquad (7.17)(b)$$

It follows immediately that ϵ induces a map from the set of orbits of H on Y to the set of orbits of G on X:

$$\Phi(\epsilon) : \Phi(Y, H) \to \Phi(X, G), \qquad \Phi(\epsilon)(H \cdot y) = G \cdot \epsilon(y) \qquad (7.17)(c)$$

(cf. Definition 7.1). Evidently ϵ carries the isotropy group H_y into $G_{\epsilon(y)}$, so we get a homomorphism on the level of equivariant fundamental groups

$$A^{loc}(\epsilon) : A^{loc}_y \to A^{loc}_{\epsilon(y)} \qquad (7.17)(d)$$

(Definition 7.1).

Proposition 7.18. *In the setting of (7.17), there is an exact functor*

$$\epsilon^* : \mathcal{C}(X, G) \to \mathcal{C}(Y, H).$$

The stalk $(\epsilon^* C)_y$ at a point $y \in Y$ is naturally identified with $C_{\epsilon(y)}$. In particular, the representation of A^{loc}_y on this stalk is obtained from the representation of $A^{loc}_{\epsilon(y)}$ on $C_{\epsilon(y)}$ by composition with the homomorphism $A^{loc}(\epsilon)$ of (7.17)(d).

7. Complete geometric parameters and perverse sheaves   97

The functor ϵ^* induces a homomorphism of Grothendieck groups

$$\epsilon^* : K(X, G) \to K(Y, H)$$

(notation (7.10)(a)).

This is obvious: the functor ϵ^* is just the usual inverse image for constructible sheaves. This functor behaves very well on the derived category $D^b(\mathcal{C}(X))$, but not on the abelian subcategory of perverse sheaves. If P is a G-equivariant perverse sheaf on X, then $\epsilon^* P$ will be only an H-equivariant constructible complex on Y; its ordinary cohomology sheaves and its perverse cohomology sheaves will both live in several different degrees. Proposition 7.18 computes the homomorphism ϵ^* in the basis of irreducible constructible sheaves. We will make this explicit in Proposition 23.7.

Here is an example. Suppose we are in the setting of (5.14). That is, we have an L-homomorphism $\epsilon : {}^\vee H^\Gamma \to {}^\vee G^\Gamma$, and compatible pro-finite extensions ${}^\vee H^{Q_H}$ and ${}^\vee G^Q$. The maps $X(\epsilon)$ of Corollary 6.21 and $\epsilon_\bullet$ of (5.14)(c) give

$$\epsilon = (X(\epsilon), \epsilon_\bullet) : (X({}^\vee H^\Gamma), {}^\vee H^{Q_H})) \to (X({}^\vee G^\Gamma), {}^\vee G^Q) \qquad (7.19)(a)$$

as in (7.17). In particular, we have the correspondence of orbits

$$\Phi(\epsilon) : \Phi({}^\vee H^\Gamma) \to \Phi({}^\vee G^\Gamma) \qquad (7.19)(b)$$

and the homomorphisms of equivariant fundamental groups

$$A^{loc}(\epsilon) : A^{loc,Q_H}_y \to A^{loc,Q}_{\epsilon(y)} \qquad (7.19)(c)$$

given by (7.17). These are clearly identified (via Proposition 6.17 and Lemma 7.5) with the ones constructed in Proposition 5.4 and (5.14)(d). Finally, we have from Proposition 7.18 a natural homomorphism

$$\epsilon^* : KX({}^\vee G^\Gamma)^Q \to KX({}^\vee H^\Gamma)^{Q_H}. \qquad (7.19)(d)$$

We will see in Chapter 26 that this last homomorphism provides the formal part of the "Langlands functoriality" associated to the L-homomorphism ϵ.

8. Perverse sheaves on the geometric parameter space

In this chapter we begin to study in detail the categories of Definition 7.13. Specifically, we want to be able to compute the geometric multiplicity and character matrices of (7.11). After the formalities of Propositions 7.14 and 7.15, our main tools will be the results of [8], [14], and [40]. We begin by recalling the first of these.

Suppose Y is a smooth complex algebraic variety. Recall from Definition 7.7 the sheaf $\mathcal{D}_Y$ of algebraic differential operators on Y. We write

$$D_Y = \Gamma \mathcal{D}_Y \qquad (8.1)(a)$$

for the algebra of global sections of $\mathcal{D}_Y$, the *algebra of global differential operators on* Y. If G is an algebraic group acting on Y, then every element of the Lie algebra of G defines a global vector field on Y; that is, a first order differential operator (without constant term). This mapping extends to an algebra homomorphism

$$\psi_Y : U(\mathfrak{g}) \to D_Y \qquad (8.1)(b)$$

called the *operator representation of* $U(\mathfrak{g})$ (see for example [13], section 3.1). The kernel of ψ_Y is a two-sided ideal

$$I_Y = \ker \psi_Y \subset U(\mathfrak{g}). \qquad (8.1)(c)$$

If $\mathcal{M}$ is any sheaf of $\mathcal{D}_Y$-modules, then the vector space $M = \Gamma \mathcal{M}$ is in a natural way a D_Y-module, and therefore (via ψ_Y) a module for $U(\mathfrak{g})/I_Y$. The functor sending $\mathcal{M}$ to the $U(\mathfrak{g})/I_Y$-module M is called the *global sections functor*. In the other direction, if M is any module for $U(\mathfrak{g})/I_Y$, then we may form the tensor product

$$\mathcal{M} = \mathcal{D}_Y \otimes_{\psi_Y(U(\mathfrak{g})/I_Y)} M. \qquad (8.1)(d)$$

This is a sheaf of $\mathcal{D}_Y$-modules on Y. The functor sending M to $\mathcal{M}$ is called *localization*. For formal reasons, there is an adjoint relationship

$$\operatorname{Hom}_{\mathcal{D}_Y}(\mathcal{D}_Y \otimes M, \mathcal{N}) \simeq \operatorname{Hom}_{U(\mathfrak{g})/I_Y}(M, \Gamma \mathcal{N}). \qquad (8.1)(e)$$

8. Perverse sheaves on the geometric parameter space

We need to recall a little algebraic representation theory at this point.

Definition 8.2. A *compatible pair* $(\mathfrak{g}, H)$ consists of a Lie algebra $\mathfrak{g}$ and a pro-algebraic group H, endowed with an algebraic action

$$\mathrm{Ad} : H \to \mathrm{Aut}(\mathfrak{g}),$$

and an injective Lie algebra homomorphism

$$i : \mathfrak{h} \to \mathfrak{g}$$

compatible with the differential of Ad. A *compatible* $(\mathfrak{g}, H)$-*module* is a $\mathfrak{g}$-module M endowed with an algebraic representation π of H, satisfying the compatibility conditions

$$\pi(h)(X \cdot m) = (\mathrm{Ad}(h)(X)) \cdot \pi(h) \cdot m$$

and

$$d\pi(Z) \cdot m = i(Z) \cdot m$$

for $h \in H$, $X \in \mathfrak{g}$, $m \in M$, and $Z \in \mathfrak{h}$. We allow M to be infinite-dimensional; to say that the representation of H is algebraic means that each vector belongs to a finite-dimensional H-invariant subspace on which H acts algebraically in the obvious sense. Fix an ideal $I \subset U(\mathfrak{g})$. The category of finite length compatible $(\mathfrak{g}, H)$-modules annihilated by I is denoted

$$\mathcal{F}(\mathfrak{g}, H, I).$$

This idea is essentially due to Harish-Chandra; the definition was formalized by Lepowsky.

Theorem 8.3 (Beilinson-Bernstein localization theorem — see [8], [13], Theorem 3.8, and [14], Theorem 1.9). *Suppose G is a complex connected algebraic group, and Y is a complete homogeneous space for G.*
  a) *The operator representation $\psi_Y : U(\mathfrak{g}) \to \mathcal{D}_Y$ of (8.1)(b) is surjective.*
  b) *The global section and localization functors provide an equivalence of categories between quasicoherent sheaves of $\mathcal{D}_Y$-modules on Y and modules for $U(\mathfrak{g})/I_Y$.*

100   The Langlands Classification and Irreducible Characters

c) *Suppose $(\mathfrak{g}, H)$ is a compatible pair (Definition 8.2), and that H acts (compatibly) with finitely many orbits on Y. Then the functors in (b) provide an equivalence of categories between $\mathcal{D}(Y, H)$ (Definition 7.7) and $\mathcal{F}(\mathfrak{g}, H, I_Y)$ (Definition 8.2).*

To understand this result, one should keep in mind that the complete homogeneous spaces for an algebraic group G are precisely the quotients G/P, with P a parabolic subgroup. The unipotent radical of G must act trivially on such a homogeneous space, so there is essentially no loss of generality in assuming G to be reductive. In part (b), what is being established is an analogue of the relationship between modules and quasicoherent sheaves on affine algebraic varieties ([20], Corollary II.5.5). The case of projective space (with G equal to $GL(n)$) is treated in [12], VII.9. Once (b) is proved, part (c) is nearly formal; the only point is to show that an H-equivariant coherent $\mathcal{D}_Y$-module is automatically regular holonomic ([12], Theorem VII.12.11).

For the rest of this chapter, we will fix

$$^\vee G^\Gamma, \qquad \mathcal{O} \subset {}^\vee\mathfrak{g} \qquad (8.4)(a)$$

a weak E-group and a semisimple orbit in its Lie algebra. Fix a canonical flat $\Lambda \subset \mathcal{O}$, and define

$$e(\Lambda) \in {}^\vee G, \qquad P(\Lambda) \subset {}^\vee G(\Lambda) \qquad (8.4)(b)$$

as in (6.6). Recall (Lemma 6.3) that $P(\Lambda)$ is a connected parabolic subgroup of the reductive group $^\vee G(\Lambda)$. As in Proposition 6.24, write

$$\mathcal{P}^0(\Lambda) = {}^\vee G(\Lambda)_0 \cdot \Lambda \subset \mathcal{O} \qquad (8.4)(c)$$

for the indicated orbit of Λ in the variety of all canonical flats in $\mathcal{O}$. Proposition 6.5 identifies $\mathcal{P}^0(\Lambda)$ naturally with the variety of conjugates of $P(\Lambda)$ in $^\vee G(\Lambda)_0$ (by sending each flat in $\mathcal{P}^0(\Lambda)$ to its stabilizer). In particular,

$$\mathcal{P}^0(\Lambda) \simeq {}^\vee G(\Lambda)_0 / P(\Lambda). \qquad (8.4)(d)$$

This space is a complete homogeneous space for $^\vee G(\Lambda)_0$, and we will apply Theorem 8.3 to it. Recall the orbit decomposition

$$\mathcal{I}(\Lambda) = \mathcal{I}_1^0(\Lambda) \cup \ldots \cup \mathcal{I}_s^0(\Lambda) \qquad (8.4)(e)$$

of Proposition 6.24, and choose a representative $y_j \in \mathcal{I}_j^0(\Lambda)$ for each orbit. Define K_j^0 as in Proposition 6.24, so that

$$\mathcal{I}_j^0(\Lambda) \simeq {}^\vee G(\Lambda)_0 / K_j^0. \qquad (8.4)(f)$$

Recall from (5.10) that we write $K_j^{0,alg}$ for the preimage of K_j^0 in ${}^\vee G^{alg}$

Theorem 8.5. *Suppose ${}^\vee G^\Gamma$ is a weak E-group and $\mathcal{O} \subset {}^\vee \mathfrak{g}$ a semisimple orbit in its Lie algebra. Use the notation of (8.4). Then the category of ${}^\vee G^{alg}$-equivariant perverse sheaves on $X(\mathcal{O}, {}^\vee G^{alg})$ (Definitions 6.9, 7.7, and 7.13) is equivalent to the direct sum of the categories $\mathcal{F}({}^\vee \mathfrak{g}(\Lambda), K_j^{0,alg}, I_{\mathcal{P}^0(\Lambda)})$ (Definition 8.2) of compatible modules of finite length annihilated by the kernel $I_{\mathcal{P}^0(\Lambda)}$ of the operator representation on $\mathcal{P}^0(\Lambda)$ (see (8.1)).*

This is an immediate consequence of Proposition 6.24, Proposition 7.14, Theorem 7.9, and Theorem 8.3.

Next, we want to reduce (that is, to reduce the calculation of the geometric multiplicity and character matrices) to the case of regular Λ. For that we need a kind of "translation functor" for these categories.

Definition 8.6. In the setting of (8.5), suppose $\mathcal{O}'$ is another orbit of semisimple elements. A *translation datum from $\mathcal{O}$ to $\mathcal{O}'$* is a ${}^\vee G$-conjugacy class $\mathcal{T}$ of pairs (Λ_1, Λ_1'), subject to the following conditions.
(a) $\Lambda_1 \subset \mathcal{O}$ and $\Lambda_1' \subset \mathcal{O}'$ are canonical flats.
(b) $P(\Lambda_1') \subset P(\Lambda_1)$, and $e(\Lambda_1) = e(\Lambda_1')$.

There are several other ways of specifying the same data. If we fix a canonical flat $\Lambda \subset \mathcal{O}$, then it is equivalent to specify a single canonical flat $\Lambda' \subset \mathcal{O}'$, subject to the conditions
(c) $P(\Lambda') \subset P(\Lambda)$, and $e(\Lambda) = e(\Lambda')$.
(Necessarily Λ' will be contained in $\mathfrak{p}(\Lambda)$.) In this case two canonical flats define the same translation datum if and only if they are conjugate by $P(\Lambda)$.

If we fix a point $\lambda \in \mathcal{O}$, then it is equivalent to specify another point $\lambda' \in \mathcal{O}'$, subject to
(d) $P(\lambda') \subset P(\lambda)$, $L(\lambda') \subset L(\lambda)$, and $e(\lambda) = e(\lambda')$.
(Necessarily λ' will belong to $\mathfrak{l}(\lambda)$.) In this case two points λ_1', λ_2' define the same translation family if and only if they are conjugate by $L(\lambda)$.

Finally, we could fix a pair (λ, T), with T a maximal torus and $\lambda \in \mathcal{O} \cap \mathfrak{t}$. Then the translation datum is determined by the choice of another point $\lambda' \in \mathcal{O}' \cap \mathfrak{t}$, subject to

(e) For every root α of T in $^\vee G$,

$$\alpha(\lambda) \in \mathbb{N} - \{0\} \Rightarrow \alpha(\lambda') \in \mathbb{N} - \{0\},$$

and

$$\lambda - \lambda' \in X_*(T).$$

(Of course the lattice on the right here is the kernel of the normalized exponential mapping e on $\mathfrak{t}$.) Two such points λ_1', λ_2' determine the same translation datum if and only if they are conjugate by the Weyl group of T in $L(\lambda)$.)

Lemma 8.7. *Suppose $^\vee G^\Gamma$ is a weak E-group and $\mathcal{O} \subset {^\vee\mathfrak{g}}$ a semisimple orbit in its Lie algebra. Then there is a regular semisimple orbit $\mathcal{O}' \subset {^\vee\mathfrak{g}}$ and a translation datum $\mathcal{T}$ from $\mathcal{O}$ to $\mathcal{O}'$ (Definition 8.6)*

Proof. We can use the last description (e) of translation data in Definition 8.6. With that notation, choose a set R^+ of positive roots for T in $^\vee G$ containing the roots of T in $N(\lambda)$ (notation (6.1) and (6.2)). Let μ be any element of $X_*(T)$ on which the positive roots take strictly positive values; for example, the sum of the positive coroots. Then $\lambda' = \lambda + \mu$ is a regular semisimple element satisfying the conditions of Definition 8.6(e), so the corresponding conjugacy class

$$\mathcal{T} = {^\vee G} \cdot (\mathcal{F}(\lambda), \mathcal{F}(\lambda'))$$

is the translation datum we need. Q.E.D.

Proposition 8.8. *Suppose $^\vee G^\Gamma$ is a weak E-group, $\mathcal{O}$, $\mathcal{O}'$ are semisimple orbits in its Lie algebra, and $\mathcal{T}$ is a translation datum from $\mathcal{O}$ to $\mathcal{O}'$.*

a) The set $\mathcal{T}$ is the graph of a $^\vee G$-equivariant smooth proper morphism

$$f_\mathcal{T} : \mathcal{F}(\mathcal{O}') \to \mathcal{F}(\mathcal{O})$$

(notation (6.10)).

b) The morphism in (a) induces an equivariant smooth proper morphism

$$f_\mathcal{T} : X(\mathcal{O}', {^\vee G^\Gamma}) \to X(\mathcal{O}, {^\vee G^\Gamma}).$$

Consequently (Proposition 7.15) the category of $^\vee G^{alg}$-equivariant perverse sheaves on $X(\mathcal{O}, {^\vee G^\Gamma})$ is equivalent to a full subcategory of

8. Perverse sheaves on the geometric parameter space

$^\vee G^{alg}$-equivariant perverse sheaves on $X(\mathcal{O}', {}^\vee G^\Gamma)$. *This inclusion respects the geometric character matrix.*

Fix a pair $(\Lambda, \Lambda') \in \mathcal{T}$.

c) The objects of (8.4) for Λ and Λ' are related by

$$e(\Lambda) = e(\Lambda'), \quad {}^\vee G(\Lambda)_0 = {}^\vee G(\Lambda')_0, \quad \mathcal{I}(\Lambda) = \mathcal{I}(\Lambda').$$

The morphism of (a) restricts to a smooth ${}^\vee G(\Lambda)_0$*-equivariant proper morphism*

$$f_\mathcal{T} : \mathcal{P}^0(\Lambda') \to \mathcal{P}^0(\Lambda).$$

d) The kernel $I_{\mathcal{P}^0(\Lambda)}$ of the operator representation of $U({}^\vee\mathfrak{g}(\Lambda))$ contains the kernel $I_{\mathcal{P}^0(\Lambda')}$. Consequently (for any $y \in \mathcal{I}(\Lambda)$) the category $\mathcal{F}({}^\vee\mathfrak{g}(\Lambda), K^{0,alg}(y), I_{\mathcal{P}^0(\Lambda)})$ (Definition 8.2) is a full subcategory of $\mathcal{F}({}^\vee\mathfrak{g}(\Lambda'), K^{0,alg}(y), I_{\mathcal{P}^0(\Lambda')})$.

e) The inclusions of categories in (b) and (d) correspond under the equivalences of Theorem 8.5.

Proof. Part (a) is a consequence of Definition 8.6(b) and (6.10)(b). (That the morphism is projective follows from the fact that $P(\Lambda')$ is parabolic in ${}^\vee G(\Lambda)$ (Lemma 6.3), and therefore also in $P(\Lambda) \subset {}^\vee G(\Lambda)$.) The second condition in Definition 8.6(b) implies that $\mathcal{C}(\mathcal{O}) = \mathcal{C}(\mathcal{O}')$ (notation 6.10(c)). Now (b) follows from the description of the spaces as fiber products in Proposition 6.16. Part (c) is immediate from the definitions; the last statement is just (a) applied to the group ${}^\vee G(\Lambda)_0$. Part (d) is just the obvious statement that the kernel of the operator representation on a homogeneous space grows as the isotropy group grows. (It is instructive to find a description of the kernel in terms of the isotropy subgroup. Halfway through this exercise you will no longer believe that the preceding assertion is obvious, but in the end you will see that it is after all.) Notice that the inclusion of categories in (d) is really that: an object of the smaller category is by definition an object of the larger one as well. Part (e) is an immediate consequence of the dense thicket of definitions from [9] and [12] that it conceals; we will not attempt even to sketch an argument. Q.E.D.

Lemma 8.7 and Proposition 8.8 reduce the calculation of the geometric character matrix — and therefore also its inverse, the multiplicity matrix — to the case of a regular orbit. We recall now very briefly how that case is treated in [40]. Suppose we are in the setting (8.4). Write $\Xi(\mathcal{O}, {}^\vee G^\Gamma)$ for the corresponding set of complete geometric parameters

(Definition 7.6). By Propositions 6.24 and 7.14, there is a natural one-to-one correspondence

$$\Xi(\mathcal{O}, {}^\vee G^\Gamma) = \coprod_j \Xi(\mathcal{P}^0(\Lambda), K_j^{0,alg}). \qquad (8.9)(a)$$

This decomposition respects the geometric multiplicity and character matrices. (In particular, if ξ and ξ' correspond to different values of j on the right, then $m_g(\xi, \xi') = c_g(\xi, \xi') = 0$.)

Suppose now that $\mathcal{O}$ is regular. Then the groups $L(\lambda)$ of Lemma 6.3 are all maximal tori, so $P(\Lambda)$ is a Borel subgroup of ${}^\vee G(\Lambda)_0$, and the variety $\mathcal{P}^0(\Lambda)$ may be identified with the variety $\mathcal{B}$ of Borel subgroups of ${}^\vee G(\Lambda)_0$. The parameter set corresponding to j on the right in (8.9)(a) is now

$$\mathcal{D}_j = \Xi(\mathcal{B}, K_j^{0,alg}). \qquad (8.9)(b)$$

This is almost precisely the setting of Definition 1.1 in [40]. (The only difference, aside from a harmless change of ground field, is that our group $K_j^{0,alg}$ is a covering of the fixed point group of an involution, rather than the fixed point group itself. The arguments of [LV] carry over to this setting unchanged.) For every pair (ξ, ξ') of elements of $\mathcal{D}_j$, Theorem 1.11 of [40] provides a polynomial

$$P_g(\xi, \xi') \in \mathbb{Z}[u], \qquad (8.9)(c)$$

the *Kazhdan-Lusztig polynomial*. The algorithm of [40] for computing this polynomial involves an action of the Hecke algebra associated to the Weyl group of ${}^\vee G(\Lambda)_0$ on the free $\mathbb{Z}[u, u^{-1}]$-module with basis $\mathcal{D}_j$. Theorem 1.12 of [40] asserts that these polynomials compute the geometric character matrix:

$$c_g(\xi, \xi') = (-1)^{d(\xi)-d(\xi')} P_g(\xi, \xi')(1). \qquad (8.9)(d)$$

To be a little more precise: (7.11)(d) says that $c_g(\xi, \xi')$ is an Euler characteristic. The theorem in [40] says that the cohomology groups in the Euler characteristic vanish in every other degree, and that the dimensions of the remaining ones are given by the coefficients of $P_g(\xi, \xi')$. We will recall some additional details in Chapter 16, in connection with the proof of Theorem 1.24.

9. The Langlands classification for tori

In this chapter we establish Theorem 1.18 (the bijection between irreducible representations of strong real forms and complete geometric parameters on an L-group) in the special case of a torus. We will treat the general case of Theorem 1.18 by reduction to this special case. (Of course such an approach seems unlikely to succeed for groups over other local fields, where not all representations are constructed from characters of tori.)

So suppose that T is an algebraic torus; that is, a connected complex reductive abelian algebraic group. As in Definition 2.10, we write

$$X_*(T) = \text{Hom}_{alg}(\mathbb{C}^\times, T), \qquad (9.1)(a)$$

the *lattice of (rational) one-parameter subgroups of T*. We can use the exponential map to identify the Lie algebra of $\mathbb{C}^\times$ with $\mathbb{C}$. A one-parameter subgroup is determined by its differential, which is a complex-linear map from $\mathbb{C}$ to $\mathfrak{t}$. Such a map is in turn determined by its value at $1 \in \mathbb{C}$; so $X_*(T)$ is identified with a lattice in $\mathfrak{t}$. As is well-known ([1], Proposition 2.2)

$$X_*(T) \simeq (1/2\pi i)\ker(\exp) = \ker(e) \subset \mathfrak{t}. \qquad (9.1)(b)$$

Here exp refers to the exponential mapping in T, and e is as in (6.2)(a). We can recover T from $X_*(T)$ by the natural isomorphism

$$X_*(T) \otimes_\mathbb{Z} \mathbb{C}^\times \simeq T, \qquad \phi \otimes z \mapsto \phi(z). \qquad (9.1)(c)$$

On the level of Lie algebras, this gives

$$X_*(T) \otimes_\mathbb{Z} \mathbb{C} \simeq \mathfrak{t}, \qquad \phi \otimes z \mapsto d\phi(z) \qquad (9.1)(d)$$

$$\mathfrak{t}^* \simeq \text{Hom}_\mathbb{Z}(X_*(T), \mathbb{C}), \qquad \lambda \mapsto (\phi \mapsto \lambda(d\phi(1))). \qquad (9.1)(e)$$

Dually, write

$$X^*(T) = \text{Hom}_{alg}(T, \mathbb{C}^\times), \qquad (9.2)(a)$$

the *lattice of (rational) characters of T*. A character is determined by its differential, which is a linear functional on $\mathfrak{t}^*$; the resulting lattice may be identified as

$$X^*(T) = \{\lambda \in \mathfrak{t}^* \mid \lambda(X_*(T)) \subset \mathbb{Z}\}. \qquad (9.2)(b)$$

This exhibits the natural pairing between $X_*(T)$ and $X^*(T)$ as the restriction of the pairing between $\mathfrak{t}$ and $\mathfrak{t}^*$. The analogue of (9.1)(c) is

$$T \simeq \mathrm{Hom}_{\mathbb{Z}}(X^*(T), \mathbb{C}^\times), \quad t \mapsto (\tau \mapsto \tau(t)). \qquad (9.2)(c)$$

Similarly,

$$\mathfrak{t} \simeq \mathrm{Hom}_{\mathbb{Z}}(X^*(T), \mathbb{C}), \quad X \mapsto (\tau \mapsto d\tau(X)) \qquad (9.2)(d)$$

$$X^*(T) \otimes_{\mathbb{Z}} \mathbb{C} \simeq \mathfrak{t}^*, \quad \tau \otimes z \mapsto (X \mapsto z d\tau(X)). \qquad (9.2)(e)$$

By the remark after Definition 4.2, a dual group ${}^\vee T$ for T is determined up to unique isomorphism. Explicitly, the formulas in (9.1)(c) and (9.2)(c) show that we may choose

$$^\vee T = X^*(T) \otimes_{\mathbb{Z}} \mathbb{C}^\times \quad \text{or} \quad {}^\vee T = \mathrm{Hom}_{\mathbb{Z}}(X_*(T), \mathbb{C}^\times). \qquad (9.3)(a)$$

In any case Definition 4.2, (9.1)(d)–(e), and (9.2)(d)–(e) provide natural isomorphisms

$$^\vee \mathfrak{t} \simeq \mathfrak{t}^*, \quad {}^\vee \mathfrak{t}^* \simeq \mathfrak{t}. \qquad (9.3)(b)$$

Proposition 9.4 (Langlands — see [34], Lemma 2.8, or [1], Theorem 4.4.) *Suppose T is an algebraic torus defined over $\mathbb{R}$. Fix an L-group $(^\vee T^\Gamma, \mathcal{D})$ for T (Definition 4.6). Then there is a natural bijection*

$$\phi \to \pi(\phi)$$

from the set $\Phi(T/\mathbb{R}) = \Phi(^\vee T^\Gamma)$ of equivalence classes of Langlands parameters (Definitions 5.2 and 5.3) onto the set $\Pi(T(\mathbb{R}))$ of equivalence classes of irreducible representations (that is, continuous homomorphisms into $\mathbb{C}^\times$) of $T(\mathbb{R})$. In this bijection, the differential $d\pi(\phi)$ (an element of $\mathfrak{t}^*$) is identified by (9.3)(b) with the parameter $\lambda(\phi) \in {}^\vee \mathfrak{t}$ (cf. (5.8)(a)). The trivial character of $T(\mathbb{R})$ is attached to the parameters $\phi(^\vee\delta, 0)$ with $^\vee\delta \in \mathcal{D}$ (notation (5.8)(a)); these constitute a single equivalence class.

9. The Langlands Classification for tori

Notice that the last assertion shows clearly the dependence of the classification on the L-group structure (and not just the weak E-group).

This beautiful result is *not* as it stands a special case of Theorem 1.18, for it mentions neither strong real forms of $T(\mathbb{R})$ nor complete geometric parameters. We will analyze each of these ideas in turn and compare them to get that result. Our main tool is the following elementary version of Pontriagin duality.

Lemma 9.5. *Suppose T is an algebraic torus; use the notation of (9.2). There is a natural inclusion-reversing bijection between algebraic subgroups of T and sublattices (by which we mean simply subgroups) of $X^*(T)$. To an algebraic subgroup $S \subset T$ corresponds the sublattice*

$$L = \{\, \tau \in X^*(T) \mid \tau(s) = 1, \text{ all } s \in S \,\}. \tag{a}$$

Dually, to a sublattice L corresponds the algebraic subgroup

$$S = \{\, s \in T \mid \tau(s) = 1, \text{ all } \tau \in L \,\}. \tag{b}$$

Suppose the subgroups $S_1 \subset S_2$ correspond to sublattices $L_1 \supset L_2$. Then restriction of characters from T to S_1 defines a natural isomorphism

$$\mathrm{Hom}_{alg}(S_1/S_2, \mathbb{C}^\times) \simeq L_2/L_1. \tag{c}$$

Suppose the subgroup S corresponds to the sublattice L. Then the identity component $S_0 \subset S$ corresponds to the lattice

$$L^0 = \{\, \tau \in X^*(T) \mid n\tau \in L \text{ for some positive integer } n \,\} \supset L. \tag{d}$$

Suppose θ is an algebraic automorphism of T, and a is the transpose automorphism of $X^*(T)$ (so that $\tau(\theta t) = (a\tau)(t)$ for all $\tau \in X^*(T)$ and $t \in T$). Then the algebraic subgroup T^θ of fixed points of θ corresponds to the sublattice

$$(1-a)X^*(T) = \{\, \tau - a\tau \mid \tau \in X^*(T) \,\}. \tag{e}$$

Dually, the sublattice $X^*(T)^a$ of fixed points of a corresponds to the (connected) subgroup

$$(1-\theta)T = \{\, t(\theta t)^{-1} \mid t \in T \,\}. \tag{f}$$

This is well-known and elementary; we omit the proof.

We turn now to the description of complete geometric parameters for tori.

Proposition 9.6. *Suppose $^\vee T^\Gamma$ is a weak E-group for the algebraic torus T. Write a for the corresponding automorphism of the based root datum (Proposition 4.4) and θ for the automorphism of $^\vee T$ defined by any element of $^\vee T^\Gamma - {}^\vee T$. Suppose $x \in X(^\vee T^\Gamma)$ is a geometric parameter (Definition 6.9).*

a) The stabilizer $^\vee T_x$ of x for the action of $^\vee T$ on geometric parameters is the group $^\vee T^\theta$ of fixed points of θ.

b) The Langlands component group A_x^{loc} is

$$^\vee T^\theta / {}^\vee T_0^\theta = {}^\vee T^\theta / (1+\theta){}^\vee T$$

c) The group of characters of the Langlands component group is given by a natural isomorphism

$$\widehat{A}_x^{loc} \simeq X_*(T)^{-a}/(1-a)X_*(T).$$

Since T is a torus, geometric parameters are in one-to-one correspondence with Langlands parameters (even before passing to equivalence classes) by Proposition 6.17.

Proof. Part (a) is immediate from Proposition 6.16 (or from Corollary 5.9, or from the definitions directly). Part (b) follows from (a). Part (c) follows from (b) using Lemma 9.5(c), (e) and (f). Q.E.D.

We now extend this result to algebraic coverings. Suppose $\widetilde{^\vee T}$ is a finite covering of an algebraic torus $^\vee T$. The lattice of rational characters of $\widetilde{^\vee T}$ must contain $X^*(^\vee T)$ as a sublattice of finite index:

$$X^*(^\vee T) \subset X^*(\widetilde{^\vee T}). \qquad (9.7)(a)$$

We may therefore regard $X^*(\widetilde{^\vee T})$ as a lattice in the rational vector space generated by $X^*(^\vee T)$:

$$X^*(\widetilde{^\vee T}) \subset X^*(^\vee T) \otimes_{\mathbb{Z}} \mathbb{Q}. \qquad (9.7)(b)$$

Conversely, any lattice in this vector space containing $X^*(^\vee T)$ as a subgroup of finite index is the character lattice of a finite covering of $^\vee T$. It follows that the group on the right (which is no longer finitely generated,

9. The Langlands Classification for tori

and so not really a "lattice") is the group of rational characters of the pro-algebraic group $^\vee T^{alg}$. Extending the notation of (9.2), we write

$$X^*(^\vee T^{alg}) = X^*(^\vee T) \otimes_{\mathbb{Z}} \mathbb{Q}. \qquad (9.7)(c)$$

Equivalently,

$$X^*(^\vee T^{alg}) = X_*(T) \otimes_{\mathbb{Z}} \mathbb{Q}. \qquad (9.7)(d)$$

Proposition 9.8. *Suppose we are in the setting of Proposition 9.6; use the notation there and in (9.7). Write as usual $^\vee T_x^{alg}$ for the preimage of $^\vee T_x$ in $^\vee T^{alg}$, and $(^\vee T_x^{alg})_0$ for its identity component.*

a) *The group of rational characters of $^\vee T^{alg}$ trivial on $^\vee T_x^{alg}$ is identified via (9.7)(d) with $(1-a)X_*(T)$.*
b) *The group of rational characters of $^\vee T^{alg}$ trivial on $(^\vee T_x^{alg})_0$ is identified via (9.7)(d) with $(X_*(T) \otimes_{\mathbb{Z}} \mathbb{Q})^{-a}$.*
c) *The group of characters of the universal component group (Definition 7.6) is given by a natural isomorphism*

$$\widehat{A}_x^{loc,alg} \simeq (X_*(T) \otimes_{\mathbb{Z}} \mathbb{Q})^{-a}/(1-a)X_*(T).$$

The proof is parallel to that of Proposition 9.6, so we omit it. (For (c) we are using the fact that any rational character of a pro-algebraic subgroup of a reductive abelian pro-algebraic group extends to the whole group. This is a formal consequence of the corresponding fact for algebraic groups.)

Next, we consider the notion of strong real form in the case of tori. We make use of the following elementary result, which is in some sense dual to Lemma 9.5.

Lemma 9.9. *Suppose T is an algebraic torus; use the notation (9.1). Write T^{fin} for the subgroup of T consisting of elements of finite order. Define*

$$\mathfrak{t}_{\mathbb{Q}} = X_*(T) \otimes_{\mathbb{Z}} \mathbb{Q} \subset \mathfrak{t}$$

(cf. (9.1)(d)).

a) *The normalized exponential mapping e, given by $e(\tau) = \exp(2\pi i \tau)$ (cf. (6.2)(a)) defines an isomorphism*

$$\mathfrak{t}/X_*(T) \simeq T.$$

b) *The preimage of T^{fin} under the isomorphism in (a) is precisely $t_\mathbb{Q}$, so we have a natural isomorphism*

$$t_\mathbb{Q}/X_*(T) \simeq T^{fin}.$$

c) *Suppose σ is an antiholomorphic automorphism of T. Then the isomorphism of (a) carries $-d\sigma$ to σ:*

$$e(-d\sigma(\tau)) = \sigma(e(\tau)).$$

d) *In the setting of (c), write a for the automorphism of $X_*(T)$ induced by σ (Proposition 2.12). Write a also for the corresponding ($\mathbb{Q}$-linear) automorphism of $t_\mathbb{Q}$. Then the isomorphism of (b) carries $-a$ to σ:*

$$e(-a(\tau)) = \sigma(e(\tau)).$$

In particular, $a = d\sigma$ on $t_\mathbb{Q}$.

Proof. Part (a) is immediate from (9.1)(b). Part (b) follows from (a). Part (c) is clear. For (d), one must examine the definition of a in the proof of Proposition 2.12, and use the fact that if $z \in \mathbb{C}^\times$ has finite order, then $\bar{z} = z^{-1}$. Q.E.D.

Proposition 9.10. *Suppose T^Γ is a weak extended group for T (Definition 2.13). Write σ for the antiholomorphic involution on T defined by conjugation by any element of $T^\Gamma - T$. Consider the three subgroups*

$$T^{-\sigma,fin} = \{\, t \in T \mid t\sigma(t) \text{ has finite order}\,\}$$
$$T^{-\sigma} = \{\, t \in T \mid t\sigma(t) = 1\,\}$$
$$T_0^{-\sigma} = \text{identity component of } T^{-\sigma} = \{\, s\sigma(s^{-1}) \mid s \in T\,\}.$$

a) *Left multiplication defines a natural simply transitive action of $T^{-\sigma,fin}$ on the set of strong real forms of T^Γ (Definition 2.13).*
b) *Two strong real forms are equivalent precisely when they lie in the same coset of $T_0^{-\sigma}$. Consequently there is a natural simply transitive action of the quotient $T^{-\sigma,fin}/T_0^{-\sigma}$ on the set of equivalence classes of strong real forms.*
c) *The mapping $\tau \mapsto e(\tau/2)$ (cf. Lemma 9.9) maps the -1 eigenspace $t_\mathbb{Q}^{-a}$ into $T^{-\sigma,fin}$. The preimage of the subgroups $T^{-\sigma}$ is the lattice $X_*(T)^{-a}$. The preimages of the subgroups $T_0^{-\sigma} \subset T^{-\sigma}$ are the*

9. The Langlands Classification for tori

lattices

$$(1-a)X_*(T) \subset X_*(T)^{-a}.$$

d) *After composition with the quotient, the map of (c) becomes a surjection onto $T^{-\sigma,fin}/T_0^{-\sigma}$.*

e) *There are natural isomorphisms*

$$\mathfrak{t}_{\mathbb{Q}}^{-a}/(1-a)X_*(T) \simeq T^{-\sigma,fin}/T_0^{-\sigma}$$

$$X_*(T)^{-a}/(1-a)X_*(T) \simeq T^{-\sigma}/T_0^{-\sigma}.$$

Proof. The first thing to prove is the equivalence of the two definitions of $T_0^{-\sigma}$. The Lie algebra of $\{s\sigma(s^{-1})\}$ is $\{\tau - d\sigma(\tau)\}$. Since σ has order two, this is equal to the -1-eigenspace of $d\sigma$, which in turn is the Lie algebra of $T^{-\sigma}$. Since $\{s\sigma(s^{-1})\}$ is clearly connected, the claim follows.

Now consider (a). Suppose $t \in T$, and δ is a strong real form. Then

$$(t\delta)^2 = t(\delta t \delta^{-1})\delta^2 = t\sigma(t)\delta^2.$$

Since δ^2 is assumed to have finite order, it follows that $t\delta$ is a strong real form if and only if $t \in T^{-\sigma,fin}$. A similar calculation shows that

$$s(t\delta)s^{-1} = t(s\sigma(s^{-1}))\delta,$$

from which (b) follows.

For (c), suppose that $\tau \in \mathfrak{t}_{\mathbb{Q}}^{-a}$. Then

$$e(\tau/2)\sigma(e(\tau/2)) = e(\tau/2)e(-a\tau/2) = e(\tau/2)e(\tau/2) = e(\tau).$$

Here we use successively Lemma 9.9(d) and the assumption on τ. Now the first two claims of (c) are clear.

For the rest of the argument, we will need to use the fact that

$$T = T_0^{-\sigma}T_0^{\sigma}. \tag{9.11}$$

(The two factors may have a non-trivial intersection; the claim is simply that every element of T is a product.) This follows from the corresponding assertion on the Lie algebra, which in turn follows from the fact that σ has order 2.

For the third part of (c), suppose that $\tau \in \mathfrak{t}_\mathbb{Q}^{-a}$ and that $e(\tau/2) = s\sigma(s^{-1})$. Because of (9.11), we may as well assume that $s \in T_0^{-\sigma}$. By Lemma 9.9(c), $s = e(\beta)$, with β in the $+1$-eigenspace of $d\sigma$. The assumption on s says (in light of Lemma 9.9(a) and (c)) that for some $\gamma \in X_*(T)$,

$$\tau/2 = \beta/2 + d\sigma(\beta/2) + \gamma = \beta + \gamma.$$

The first conclusion is that $\beta \in \mathfrak{t}_\mathbb{Q}$. By Lemma 9.9(c), β is in the $+1$-eigenspace of a. Consequently $\tau/2$ is the projection of γ on the -1-eigenspace of a. That is,

$$\tau/2 = (\gamma - a\gamma)/2,$$

which gives the third part of (c).

For (d), (9.11) and Lemma 9.9(c) imply that every coset in $T^{-\sigma,fin}/T_0^{-\sigma}$ has a representative $t = e(\tau/2)$ with τ in the -1-eigenspace of $d\sigma$. As in (a) we compute $t\sigma(t) = e(\tau)$; so τ must actually belong to the -1-eigenspace of $d\sigma$ on $\mathfrak{t}_\mathbb{Q}$. By Lemma 9.9(d), τ belongs to $\mathfrak{t}_\mathbb{Q}^{-a}$, as we wished to show.

Part (e) is an immediate consequence of (c) and (d). Q.E.D.

Corollary 9.12. *Suppose $^\vee T^\Gamma$ is a weak E-group for the algebraic torus T (Definition 4.3), and T^Γ is a weak extended group for the corresponding real form (Definition 2.13). Suppose $x \in X(^\vee T^\Gamma)$ is a geometric parameter (Definition 6.9). Then there is a natural simply transitive action of the group $\widehat{A}^{loc,alg}$ (of characters of the universal component group) on the set of equivalence classes of strong real forms of T^Γ.*

Proof. Combine Proposition 9.8(c) with Proposition 9.10(d). Q.E.D.

Proof of Theorem 1.18 for tori. The extra datum $\mathcal{W}$ needed to make an extended group for a torus from a weak extended group is precisely an equivalence class of strong real forms (cf. Definition 1.12). The action of Corollary 9.12 therefore gives a natural bijection between strong real forms and characters of the universal component group (sending the distinguished class of strong real forms to the trivial character). In conjunction with Proposition 9.4, this gives Theorem 1.18 for tori.
Q.E.D.

10. Covering groups and projective representations

In order to deduce Theorem 1.18 from the special case established in the last chapter, we will need to exploit a relationship between characters of tori in G and representations of G. This relationship is most natural when it is formulated in terms of certain coverings of the tori related to "ρ-shifts" for G (see for example Theorem 1.37 or Theorem 6.8 in [58]). We will therefore need for tori a version of Theorem 1.18 that describes representations of such coverings. It is just as easy to treat coverings of general groups; in any case this will be necessary when we discuss endoscopy.

Definition 10.1. Suppose G^Γ is a weak extended group for G (Definition 2.13). A connected finite covering group

$$1 \to F \to \widetilde{G} \to G \to 1$$

is said to be *distinguished* if the following two conditions are satisfied.

(a) For every $x \in G^\Gamma - G$, the conjugation action σ_x of x on G lifts to an automorphism $\widetilde{\sigma}_x$ of $\widetilde{G}$.

It is equivalent to require this condition for a single x. As x varies, $\widetilde{\sigma}_x$ changes by inner automorphisms; so its restriction $\widetilde{\sigma}_Z$ to $Z(\widetilde{G})$ is independent of the choice of x. Necessarily $\widetilde{\sigma}_Z$ preserves the subgroup F of $Z(\widetilde{G})$. We can now formulate the second condition on $\widetilde{G}$.

(b) The automorphism $\widetilde{\sigma}_Z$ sends every element of F to its inverse.

Write G^{sc} for the simply-connected covering group of G, and $\pi_1(G)$ for the kernel of the covering map. This covering automatically satisfies the condition analogous to (a) above, so that we can define an involutive automorphism σ_Z^{sc} of $Z(G^{sc})$, which preserves the subgroup $\pi_1(G)$. We may regard F as a quotient of $\pi_1(G)$:

$$1 \to K \to \pi_1(G) \to F \to 1.$$

Then condition (a) amounts to the requirement that K should be preserved by σ_Z^{sc}, and condition (b) to

$$K \supset \{ y \cdot \sigma_Z^{sc}(y) \mid y \in \pi_1(G) \}.$$

The *canonical covering* G^{can} is the projective limit of all the distinguished coverings of G. This is a pro-algebraic group, of which each finite-dimensional representation factors to some distinguished finite cover of G. We write (in analogy with (5.10))

$$1 \to \pi_1(G)^{can} \to G^{can} \to G \to 1.$$

The group $\pi_1(G)^{can}$ depends on the inner class of real forms under consideration. It is a pro-finite abelian group, the inverse limit of certain finite quotients of $\pi_1(G)$.

Lemma 10.2. *Suppose G is a connected complex reductive algebraic group, and $^\vee G$ is a dual group (Definition 4.2).*

a) *The group of complex characters of $\pi_1(G)$ is naturally isomorphic to the center of $^\vee G$:*

$$Z(^\vee G) \simeq \mathrm{Hom}(\pi_1(G), \mathbb{C}^\times).$$

Write this isomorphism as $z \mapsto \chi_z$.

Suppose G is endowed with an inner class of real forms. Write σ_Z^{sc} for the induced action of any of these real forms on $\pi_1(G)$. Dually, write a for the automorphism of the based root datum of G defined by the inner class of real forms (Definition 2.12), and θ_Z for the automorphism of $Z(^\vee G)$ induced by any automorphism of $^\vee G$ corresponding to a (cf. Proposition 4.4).

b) *The automorphisms σ_Z^{sc} and θ_Z composed with inversion are transposes of each other with respect to the isomorphism in (a). That is, if $z \in Z(^\vee G)$ and $p \in \pi_1(G)$, then*

$$\chi_z(p) = \chi_{\theta_Z(z)^{-1}}(\sigma_Z^{sc}(p)).$$

c) *The character χ_z is trivial on the subgroup $\{p\sigma_Z^{sc}(p)\}$ of $\pi_1(G)$ if and only if $\theta_Z(z) = z$.*

d) *The group of continuous characters of $\pi_1(G)^{can}$ (Definition 10.1) is naturally isomorphic to the group of elements of finite order in $Z(^\vee G)^{\theta_Z}$.*

Proof. Part (a) is well-known and elementary (see for example [1], Proposition 10.1). Part (b) is immediate from the definitions, and (c) follows from (b). Part (d) is a consequence of (c) and the description of $\pi_1(G)^{can}$ in Definition 10.1. (Recall that a continuous character of a profinite group factors through any large enough finite quotient.) Q.E.D.

10. Covering groups and projective representations

Definition 10.3 (cf. Definition 2.13). Suppose G^Γ is a weak extended group for G; use the notation of Definition 10.1 and Proposition 10.2. If δ is a strong real form of G^Γ (Definition 2.13), write $G(\mathbb{R}, \delta)^{can}$ for the preimage of $G(\mathbb{R}, \delta)$ in G^{can}. There is a short exact sequence

$$1 \to \pi_1(G)^{can} \to G(\mathbb{R}, \delta)^{can} \to G(\mathbb{R}, \delta) \to 1.$$

A *canonical projective representation of a strong real form of* G^Γ is a pair (π, δ), subject to

(a) δ is a strong real form of G^Γ; and
(b) π is an admissible representation of $G(\mathbb{R}, \delta)^{can}$.

Equivalence is defined as in Definition 2.13. Suppose $z \in Z({}^\vee G)^{\theta_z}$. We say that (π, δ) is *of type* z if the restriction of π to $\pi_1(G)^{can}$ is a multiple of χ_z (Proposition 10.2(a)). (Notice that every irreducible canonical projective representation has a unique type.) Finally, define

$$\Pi^z(G^\Gamma) = \Pi^z(G/\mathbb{R})$$

to be the set of infinitesimal equivalence classes of irreducible canonical projective representations of type z.

With this definition, we can formulate a mild generalization of Theorem 1.18. (It is important to remember that the coverings appearing in Definition 10.3 are all linear. Extending the Langlands classification to non-linear coverings is a much more difficult and interesting problem.)

Theorem 10.4. *Suppose* $(G^\Gamma, \mathcal{W})$ *is an extended group for* G *(Definition 1.12), and* $({}^\vee G^\Gamma, \mathcal{D})$ *is an E-group for the corresponding inner class of real forms (Definition 4.6). Write* z *for the second invariant of the E-group (Definition 4.6). Then there is a natural bijection between the set* $\Pi^z(G/\mathbb{R})$ *of equivalence classes of irreducible canonical projective representations of strong real forms of* G *of type* z *(Definition 10.3), and the set* $\Xi^z(G/\mathbb{R})$ *of complete geometric parameters for* ${}^\vee G^\Gamma$ *(Definition 7.6 and Definition 5.11).*

We will give the proof in the next four chapters. For the moment, we record only the corresponding generalization of Proposition 9.4. It is helpful first to describe the rational characters of the pro-algebraic group T^{can} when T is an algebraic torus defined over $\mathbb{R}$. A calculation analogous to (9.7) gives

$$X^*(T^{can}) = \{\, \lambda \in X_*({}^\vee T) \otimes_{\mathbb{Z}} \mathbb{Q} \mid \lambda - \theta\lambda \in X_*({}^\vee T) \,\}. \qquad (10.5)(a)$$

Following Lemma 9.9, we write

$$^\vee t_\mathbb{Q} = X_*(^\vee T) \otimes_\mathbb{Z} \mathbb{Q}. \qquad (10.5)(b)$$

Now Lemma 9.9(b) says that the normalized exponential mapping e provides an isomorphism

$$^\vee t_\mathbb{Q}/X_*(^\vee T) \simeq {}^\vee T^{fin}. \qquad (10.5)(c)$$

Clearly $X^*(T^{can})$ is precisely the inverse image of the θ-fixed elements $^\vee T^{fin}$ under this map; so

$$X^*(T^{can})/X^*(T) \simeq (^\vee T^{fin})^\theta. \qquad (10.5)(d)$$

Of course the group on the left may be identified with the characters of $\pi_1(T)^{can}$, and we recover the isomorphism in Lemma 10.2(d).

Proposition 10.6 ([1], Theorem 5.11.) *Suppose T is an algebraic torus defined over $\mathbb{R}$. Fix an E-group $(^\vee T^\Gamma, \mathcal{W})$ for T with second invariant z (Definition 4.6). Then there is a natural bijection*

$$\phi \to \pi(\phi)$$

from the set $\Phi(^\vee T^\Gamma)$ of equivalence classes of Langlands parameters (Definition 5.2) onto the set $\Pi^z(T(\mathbb{R}))$ of equivalence classes of irreducible canonical projective representations of $T(\mathbb{R})$ of type z (Definition 10.3). In this bijection, the differential $d\pi(\phi)$ (an element of $\mathfrak{t}^*$) is identified by (9.3)(b) with the parameter $\lambda(\phi) \in {}^\vee \mathfrak{t}$ (cf. (5.8)(a)).

Suppose $\lambda \in {}^\vee t_\mathbb{Q}$ satisfies $e(\lambda) = z$. By (10.5), λ may be identified with a rational character of T^{can} of type z. By restriction to $T(\mathbb{R})^{can}$, λ defines an irreducible canonical projective representation of $T(\mathbb{R})$ of type z. Its Langlands parameters are the various $\phi(^\vee \delta, \lambda)$ (notation (5.8)(a)) with $^\vee \delta \in \mathcal{D}$; these constitute a single equivalence class.

As in Proposition 9.4, the last assertion is included primarily to show how the classification theorem forces the choice of the class $\mathcal{D}$ in addition to the weak E-group structure.

Corollary 10.7. *Theorem 10.4 is true if G is an algebraic torus.*

Just as in the argument at the end of Chapter 9, we need only apply Corollary 9.12 and Proposition 10.6.

We record some notation based on Theorem 10.4, extending (1.19) in the introduction.

10. Covering groups and projective representations

Definition 10.8. In the setting of Theorem 10.4, suppose $\xi \in \Xi^z(G/\mathbb{R})$ is a complete geometric parameter. A representative for the corresponding equivalence class of representation and strong real form will be written as

$$(\pi(\xi), \delta(\xi)) \in \Pi^z(G/\mathbb{R})$$

(Definition 10.3). If we regard ξ as represented by a complete Langlands parameter (ϕ, τ) (Definition 5.11), we may write instead $(\pi(\phi, \tau), \delta(\phi, \tau))$.

Conversely, suppose $(\pi, \delta) \in \Pi^z(G/\mathbb{R})$. The corresponding complete geometric parameter ξ will be written as

$$\xi(\pi, \delta) = (S(\pi, \delta), \mathcal{V}(\pi, \delta)).$$

Here S is an orbit of ${}^\vee G$ on the geometric parameter space, and $\mathcal{V}$ is a ${}^\vee G^{alg}$-equivariant local system on S. A representative for the corresponding equivalence class of complete Langlands parameters is $(\phi(\pi, \delta), \tau(\pi, \delta))$ (Definition 5.11). More generally, we may use invariants previously attached to complete geometric parameters as if they were attached directly to representations. For example, we write $d(\pi)$ or $d(\pi, \delta)$ for the dimension of the orbit $S(\pi, \delta)$ (cf. (7.10)), and $e(\pi)$ or $e(\pi, \delta)$ for the Kottwitz sign attached to $\xi(\pi, \delta)$ (Definition 15.8 below).

Suppose now that $\phi \in \Phi^z(G/\mathbb{R})$ (Definition 5.3), and let $S_\phi \subset X({}^\vee G^\Gamma)$ be the corresponding orbit (Proposition 6.17). The *L-packet attached to* ϕ is the set of irreducible (canonical projective) representations of strong real forms parametrized by complete geometric parameters supported on S:

$$\begin{aligned}\Pi^z(G/\mathbb{R})_\phi &= \{\, (\pi(\xi), \delta(\xi)) \mid \xi = (S_\phi, \mathcal{V}) \,\} \\ &= \{\, (\pi, \delta) \mid \phi(\pi, \delta) \text{ is equivalent to } \phi \,\}.\end{aligned}$$

Perhaps the least satisfactory aspect of Theorem 10.4 is the appearance of the algebraic universal covering ${}^\vee G^{alg}$. (We can avoid coverings of G, at least in the final result, by using the L-group instead of another E-group.) It is therefore of interest to understand exactly what coverings are needed to classify what representations. This information will be important in our treatment of endoscopy in Chapter 26 as well.

Lemma 10.9. *Suppose we are in the setting of Lemma 10.2.*

a) The group of continuous characters of $\pi_1({}^\vee G)^{alg}$ is naturally isomorphic to the group $Z(G)^{fin}$ of elements of finite order in $Z(G)$. Write χ_z for the character corresponding to z.

b) In the setting of Theorem 10.4, write $z_0 \in Z(G)^{fin}$ for the second invariant of the extended group $(G^\Gamma, \mathcal{W})$ (cf. (3.5)). Suppose that the complete geometric parameter ξ corresponds to the irreducible canonical projective representation $(\pi(\xi), \delta(\xi))$. Set $z_1 = \delta^2 \in Z(G)^{fin}$. Then $\pi_1({}^\vee G)^{alg}$ acts on ξ by the character $\chi_{z_0 z_1}$.

Proof. Part (a) is almost identical to Lemma 10.2. For (b), the definitions underlying Theorem 10.4 (see particularly Chapter 13) reduce it to the case of a torus, where it follows by inspection of the proof of Corollary 9.12. We omit the details.

Here is an important case of the situation considered in (5.13).

Definition 10.10. Suppose ${}^\vee G^\Gamma$ is a weak E-group, and Q is any quotient of $\pi_1({}^\vee G)^{alg}$:

$$1 \to K_Q \to \pi_1({}^\vee G)^{alg} \to Q \to 1 \qquad (10.10)(a)$$

(We assume that K_Q is closed; equivalently, that Q is the inverse limit of its finite quotients.) Define

$$ {}^\vee G^Q = {}^\vee G^{alg}/K_Q, \qquad (10.10)(b)$$

a pro-finite covering of ${}^\vee G$:

$$1 \to Q \to {}^\vee G^Q \to {}^\vee G \to 1. \qquad (10.10)(c)$$

(Notice in particular that ${}^\vee G^{\pi_1({}^\vee G)^{alg}}$ is just ${}^\vee G^{alg}$.) Recall from Definition 7.6 the notion of a complete geometric parameter of type Q; this may be regarded as a complete geometric parameter $(S, \mathcal{V})$ with the property that K_Q acts trivially on $\mathcal{V}$, so

$$\Xi({}^\vee G^\Gamma)^Q \subset \Xi({}^\vee G^\Gamma). \qquad (10.10)(c)$$

In the setting of Theorem 10.4, we may write $\Xi^z(G/\mathbb{R})^Q$ for $\Xi({}^\vee G^\Gamma)^Q$.

Suppose $(G^\Gamma, \mathcal{W})$ is an extended group with second invariant z_0 (cf. (3.5)), and $J \subset Z(G)^{fin}$ is any subgroup. A *strong real form of G of type J* is an element $\delta \in G^\Gamma - G$ such that $\delta^2 \in z_0 J$. (The shift by z_0 guarantees that the real forms in $\mathcal{W}$ are of type J for every J.) We write

$$\Pi(G/\mathbb{R})_J = \{\, (\pi, \delta) \in \Pi(G/\mathbb{R}) \mid \delta^2 \in z_0 J \,\}$$

for the set of equivalence classes of irreducible representations of strong real forms of G of type J, and extend other notation analogously.

10. Covering groups and projective representations

Theorem 10.11. *In the setting of Theorem 10.4, suppose Q is the quotient of $\pi_1({}^\vee G)^{alg}$ by a closed subgroup $K_Q \subset \pi_1({}^\vee G)^{alg}$. Define $J \subset Z(G)^{fin}$ to consist of all those elements for which the corresponding character (Lemma 10.9)(a)) is trivial on K_Q:*

$$J = \{\, z \in Z(G)^{fin} \mid \chi_z(k) = 1,\ \text{all}\ k \in K_Q \,\}.$$

Then the bijection of Theorem 10.4 restricts to a bijection

$$\Xi^z(G/\mathbb{R})^Q \leftrightarrow \Pi^z(G/\mathbb{R})_J.$$

This is immediate from Theorem 10.4 and Lemma 10.9. Notice that Lemma 10.9(a) provides a natural identification of J with the group $\widehat{Q}$ of characters of Q, and conversely. The conclusion of Theorem 10.11 may therefore be written as

$$\Xi^z(G/\mathbb{R})^Q \leftrightarrow \Pi^z(G/\mathbb{R})_{\widehat{Q}}, \qquad (10.12)(a)$$

or as

$$\Pi^z(G/\mathbb{R})_J \leftrightarrow \Xi^z(G/\mathbb{R})^{\widehat{J}}. \qquad (10.12)(b)$$

11. The Langlands classification without L-groups

In this chapter we recall the "elementary" version of the Langlands classification of representations, in which L-groups do not appear (Theorem 11.14 below). Because some of the groups we consider (such as $G(\mathbb{R})^{can}$) are not precisely groups of real points of connected algebraic groups, we need to formulate this result in a slightly more general setting. With possible generalizations in mind, we allow even nonlinear groups. The class of groups we consider is essentially the one in section 5 of [50]. (The only difference is that Springer allows G to be disconnected, and imposes an additional technical hypothesis that is empty if G is connected.) We refer the reader to [50] for basic structural facts and further references.

We will be quoting a number of technical results, particularly from [54]. Unfortunately our hypotheses in this chapter are weaker than those of [54], where it is assumed that $G_\mathbb{R}$ is linear. We offer two possible responses to this problem. The first is that we will quote only results that can be extended routinely to the present setting. The second is that all of the groups considered elsewhere in this book *are* linear.

So suppose G is a connected reductive complex algebraic group defined over $\mathbb{R}$; and suppose $G_\mathbb{R}$ is a real Lie group endowed with a homomorphism

$$G_\mathbb{R} \to G(\mathbb{R}) \qquad (11.1)(a)$$

having finite kernel and cokernel. We use the homomorphism to identify the Lie algebra $\mathfrak{g}_\mathbb{R}$ with $\mathfrak{g}(\mathbb{R})$, and its complexification with $\mathfrak{g}$. A *Cartan subgroup* of $G_\mathbb{R}$ is by definition the centralizer in $G_\mathbb{R}$ of a Cartan subalgebra of $\mathfrak{g}_\mathbb{R}$. Such a subgroup $T_\mathbb{R}$ is the preimage of the real points $T(\mathbb{R})$ of a unique maximal torus defined over $\mathbb{R}$:

$$T_\mathbb{R} \to T(\mathbb{R}). \qquad (11.1)(b)$$

(We should remark that $T_\mathbb{R}$ may be non-abelian under these hypotheses, although this does not happen inside our canonical coverings.) Using such homomorphisms, we can pull extensions of $G(\mathbb{R})$ and its subgroups

11. The Langlands classification without L-groups

back to $G_\mathbb{R}$, getting for example a central extension

$$1 \to \pi_1(G)^{can} \to G_\mathbb{R}^{can} \to G_\mathbb{R} \to 1. \tag{11.1}(c)$$

We may therefore speak of a "canonical projective representation of $G_\mathbb{R}$ of type z" as in Definition 10.3, and we write

$\Pi^z(G_\mathbb{R}) = $ infinitesimal equivalence classes of such irreducible representations. $\hfill(11.1)(d)$

(Here z is an element of finite order in $Z(^\vee G)^{\theta_z}$ if we have a dual group $^\vee G$ available; otherwise we can just think of z directly as a character of $\pi_1(G)^{can}$ as in Lemma 10.2(d).) If $H_\mathbb{R}$ is a subgroup of $G_\mathbb{R}$, we write $H_\mathbb{R}^{can}$ or $H_\mathbb{R}^{can,G}$ for its preimage in $G_\mathbb{R}^{can}$. This is a central extension

$$1 \to \pi_1(G)^{can} \to H_\mathbb{R}^{can,G} \to H_\mathbb{R} \to 1. \tag{11.1}(e)$$

Definition 11.2 (cf. [57], Definition 2.6, and [1], Definition 8.18). In the setting of (11.1), suppose $T_\mathbb{R}$ is a Cartan subgroup of $G_\mathbb{R}$. Fix an element $z \in Z(^\vee G)^{\theta_z}$ of finite order. We may choose as a dual group for T a maximal torus $^\vee T \subset {^\vee G}$; then

$$Z(^\vee G)^{\theta_z} \subset {^\vee T^{\theta_z}},$$

and this embedding is independent of all choices. It therefore makes sense to regard z as an element of finite order in $^\vee T^{\theta_z}$. Recall the element $z(\rho) \in {^\vee T^{\theta_z}}$ constructed in Definition 4.9. (Actually $z(\rho)$ even belongs to $Z(^\vee G)^{\theta_z}$, but this fact is a bit of an accident in the present context, and is better ignored.) In any case $zz(\rho) \in {^\vee T^{\theta_z}}$ is an element of finite order, so it makes sense to speak of canonical projective representations of $T_\mathbb{R}$ of type $zz(\rho)$.

A $G_\mathbb{R}$-*limit character of* $T_\mathbb{R}$ *of type* z is a triple $\Lambda = (\Lambda^{can}, R_{i\mathbb{R}}^+, R_\mathbb{R}^+)$ subject to the following conditions. The first term is an irreducible canonical projective representation

$$\Lambda^{can} \in \Pi^{zz(\rho)}(T_\mathbb{R}). \tag{11.2}(a)$$

This means that Λ^{can} is an irreducible representation of $T_\mathbb{R}^{can,T}$ and that the restriction of Λ^{can} to $\pi_1(T)^{can}$ is a multiple of $\chi_{zz(\rho)}^T$. Write

$$\lambda = d\Lambda^{can} \in \mathfrak{t}^*. \tag{11.2}(b)$$

Next,

$$R_{\mathbb{R}}^+, \quad R_{i\mathbb{R}}^+ \quad (11.2)(c)$$

are positive root systems for the real and imaginary roots of T in G. Finally, we assume that

$$\langle \alpha, \lambda \rangle \geq 0, \quad (\alpha \in R_{i\mathbb{R}}^+). \quad (11.2)(d)$$

If these inequalities are strict (for example if λ is regular), the choice of $R_{i\mathbb{R}}^+$ is forced by (11.2)(d), and we may sometimes omit it in writing the limit character.

The limit character is called *G-regular* if $\langle \alpha, \lambda \rangle \neq 0$ for any root α of T in G.

Attached to each limit character Λ there is a *standard limit representation* $M(\Lambda)$, defined by a procedure outlined in section 8 of [1] or section 2 of [57]. We define our standard limit representations as in [34] (using for example real parabolic induction with non-negative continuous parameter), so that the Langlands subquotients appear as *quotients*. Thus we define

$$\pi(\Lambda) = \text{largest completely reducible quotient of } M(\Lambda), \quad (11.2)(e)$$

the *Langlands quotient of* $M(\Lambda)$. Occasionally we will need the standard representation having a Langlands subrepresentation; this is written

$$\tilde{M}(\Lambda) \supset \pi(\Lambda). \quad (11.2)(f)$$

This representation has exactly the same composition factors and multiplicities as $M(\Lambda)$.

The infinitesimal character of $M(\Lambda)$ or $\pi(\Lambda)$ is given in the Harish-Chandra parametrization (see for example [54], Corollary 0.2.10, or [22], p. 130) by the weight $\lambda = d\Lambda^{can} \in \mathfrak{t}^*$. In particular, the infinitesimal character is regular if and only if the limit character is G-regular.

The definition of limit character given here looks somewhat different from the one in [57], so we will explain briefly their relationship. It is convenient to fix a maximal compact subgroup $K_{\mathbb{R}}$ of $G_{\mathbb{R}}$ so that the corresponding Cartan involution preserves $T_{\mathbb{R}}$. This gives rise to a direct product decomposition

$$T_{\mathbb{R}} = (T_{\mathbb{R}} \cap K_{\mathbb{R}})(A_{\mathbb{R}}) \quad (11.3)(a)$$

with the second factor a vector group. This decomposition in turn leads to

$$T_{\mathbb{R}}^{can} = (T_{\mathbb{R}} \cap K_{\mathbb{R}})^{can}(A_{\mathbb{R}}). \qquad (11.3)(b)$$

The first step is to choose a positive root system R^+ for T in G containing the systems of positive real and imaginary roots already fixed. We may assume that the set of non-imaginary roots in R^+ is preserved by complex conjugation; then

$$\rho(R^+) = \rho(R_{i\mathbb{R}}^+) \text{ on the Lie algebra of } T_{\mathbb{R}} \cap K_{\mathbb{R}}. \qquad (11.3)(c)$$

Now it follows from (10.5) that $\rho(R^+)$ may be regarded as a rational character of $T^{can,T}$ of type $z(\rho)$. Consequently

$$\Lambda^{can} \otimes \rho(R^+) \in \Pi^z(T(\mathbb{R})) \qquad (11.3)(d)$$

may be regarded as a character of $T_{\mathbb{R}}^{can,T}$ of type z. Since the type is in $Z({}^{\vee}G)^{\theta z}$, we can now replace $T_{\mathbb{R}}^{can,T}$ by $T_{\mathbb{R}}^{can,G}$. After tensoring with the sum of the negative compact imaginary roots and an appropriate character of the vector group $A_{\mathbb{R}}$, we therefore get a character $\tilde{\Lambda} \in \Pi^z(T(\mathbb{R}))$ satisfying

$$d\tilde{\Lambda} = \lambda + \rho(R_{i\mathbb{R}}^+) - 2\rho(R_{i\mathbb{R},compact}^+) \qquad (11.3)(e)$$

It is not difficult to show that $\tilde{\Lambda}$ is independent of the choice of R^+; the argument is contained in Lemma 11.5 below. The pair $(\tilde{\Lambda}, \lambda)$ is what is called $(\Gamma, \overline{\gamma})$ in Definition 2.4 of [57]. Write $L_{\mathbb{R}}$ for the centralizer of $A_{\mathbb{R}}$ in $G_{\mathbb{R}}$. The standard representation $M(\Lambda)$ may be constructed by parabolic induction from a limit of discrete series representation on $L_{\mathbb{R}}^{can}$ having lowest $L_{\mathbb{R}} \cap K_{\mathbb{R}}$-type of highest weight $\tilde{\Lambda}$.

As is explained in [1], the dependence of $M(\Lambda)$ on $R_{\mathbb{R}}^+$ is very mild. We recall the result. Suppose $(R_{\mathbb{R}}^+)'$ is another set of positive real roots for T in G. Define

$$\mathfrak{n}(R_{\mathbb{R}}^+, (R_{\mathbb{R}}^+)') = \text{span}(X_\alpha \mid \alpha \in R_{\mathbb{R}}^+ - (R_{\mathbb{R}}^+)'). \qquad (11.4)(a)$$

The Cartan subgroup $T_{\mathbb{R}}$ acts on $\mathfrak{n}(R_{\mathbb{R}}^+, (R_{\mathbb{R}}^+)')$, and the determinant of this action is a real-valued character of $T_{\mathbb{R}}$. We may therefore define a character $\tau(R_{\mathbb{R}}^+, (R_{\mathbb{R}}^+)')$ taking values in $\{\pm 1\}$ by

$$\tau(R_{\mathbb{R}}^+, (R_{\mathbb{R}}^+)')(t) = \text{sgn}\left(\det(\text{Ad}(t) \text{ on } \mathfrak{n}(R_{\mathbb{R}}^+, (R_{\mathbb{R}}^+)'))\right). \qquad (11.4)(b)$$

Lemma 11.5 ([1], Lemma 8.24). *Suppose* $\Lambda = (\Lambda^{can}, R^+_{i\mathbb{R}}, R^+_{\mathbb{R}})$ *is a limit character of* $T_{\mathbb{R}}$ *(Definition 11.2) and* $(R^+_{\mathbb{R}})'$ *is another set of positive real roots of* T *in* G. *Choose sets of positive roots* $R^+ \supset R^+_{i\mathbb{R}}, R^+_{\mathbb{R}}$ *and* $(R^+)' \supset R^+_{i\mathbb{R}}, (R^+_{\mathbb{R}})'$ *for* T *in* G *as in (11.3), and use them to construct characters* $\tilde{\Lambda}$ *and* $\tilde{\Lambda}'$.

a) $\rho(R^+) = \rho((R^+)') \otimes \tau(R^+_{\mathbb{R}}, (R^+_{\mathbb{R}})')$ *on* $(T_{\mathbb{R}} \cap K_{\mathbb{R}})^{can,T}$.
b) $\tilde{\Lambda} = \tilde{\Lambda}' \otimes \tau(R^+_{\mathbb{R}}, (R^+_{\mathbb{R}})')$.
c) *Write*

$$\Lambda' = (\Lambda^{can} \otimes \tau(R^+_{\mathbb{R}}, (R^+_{\mathbb{R}})'), R^+_{i\mathbb{R}}, (R^+_{\mathbb{R}})').$$

Then there is an isomorphism of standard limit representations $M(\Lambda) \simeq M(\Lambda')$.

Proof. With notation analogous to (11.4), it is easy to check that (as rational characters of $T^{can,T}$)

$$\rho(R^+) = \rho((R^+)') \otimes (\det(\text{Ad on } \mathfrak{n}(R^+, (R^+)'))).$$

That is,

$$\rho(R^+)(t) = \rho((R^+)')(t) \prod_{\substack{\alpha \in R^+ \\ \alpha \notin (R^+)'}} \alpha(t).$$

Assume now that $t \in (T_{\mathbb{R}} \cap K_{\mathbb{R}})^{can,T}$. The roots (like all characters of $T_{\mathbb{R}}$) have absolute value 1 on the compact group $T_{\mathbb{R}} \cap K_{\mathbb{R}}$; so the real roots contribute exactly $\tau(R^+_{\mathbb{R}}, (R^+_{\mathbb{R}})')(t)$ to the product. Therefore

$$\rho(R^+)(t) = \rho((R^+)')(t)\tau(R^+_{\mathbb{R}}, (R^+_{\mathbb{R}})')(t) \prod_{\substack{\alpha \in R^+ \text{ complex} \\ \alpha \notin (R^+)'}} \alpha(t).$$

The roots in this last product occur in complex conjugate pairs, since the non-imaginary roots in R^+ and $(R^+)'$ were assumed to be stable under complex conjugation. Such a pair contributes $|\alpha(t)|^2 = 1$ to the product. This proves (a). Part (b) is immediate from (a) and the definition of $\tilde{\Lambda}$. For (c), part (b) shows that the two limit characters correspond to exactly the same limit character in the sense of [57], so they define the same representation. Q.E.D.

Definition 11.6. In the setting of (11.1) and Definition 11.2, suppose $(\Lambda^{can}, R^+_{i\mathbb{R}}, R^+_{\mathbb{R}})$ is a limit character of $T_{\mathbb{R}}$, and $((\Lambda^{can})', (R^+_{i\mathbb{R}})', (R^+_{\mathbb{R}})')$ is a limit character of $T'_{\mathbb{R}}$. We say that these

11. The Langlands classification without L-groups   125

limit characters are *equivalent* if there is a $g \in G_\mathbb{R}$ that conjugates $T'_\mathbb{R}$ to $T_\mathbb{R}$ and $(R^+_{i\mathbb{R}})'$ to $R^+_{i\mathbb{R}}$, and has the following additional property. Write $(R^+_\mathbb{R})'' = \mathrm{Ad}(g)(R^+_\mathbb{R})'$ (a set of positive real roots for T in G), and

$$(\Lambda^{can})'' = \mathrm{Ad}(g)(\Lambda^{can})',$$

a canonical projective representation of $T_\mathbb{R}$. Then our final requirement is

$$(\Lambda^{can})'' = \Lambda^{can} \otimes \tau(R^+_\mathbb{R}, (R^+_\mathbb{R})'').$$

Because of Lemma 11.5(c), equivalent limit characters define equivalent standard limit representations.

With this (rather subtle) notion of equivalence in hand, we can formulate an important special case of the Langlands classification.

Theorem 11.7 ([34]) *Suppose $G_\mathbb{R}$ is a real reductive group as in (11.1). Then the infinitesimal equivalence classes of irreducible admissible representations of $G_\mathbb{R}$ with regular infinitesimal character are in one-to-one correspondence with the equivalence classes of G-regular limit characters, as follows.*

a) *Suppose Λ is a G-regular limit character of a Cartan subgroup $T_\mathbb{R}$ (Definition 11.2). Then the corresponding standard representation $M(\Lambda)$ is non-zero, and has a unique irreducible quotient. In particular, the Langlands quotient representation $\pi(\Lambda)$ is irreducible.*

b) *Suppose π is an irreducible canonical projective representation of $G_\mathbb{R}$ of type z (Definition 11.2), having regular infinitesimal character. Then there is a G-regular limit character Λ of type z with the property that π is infinitesimally equivalent to $\pi(\Lambda)$.*

c) *Suppose Λ and Λ' are G-regular limit characters, and that $\pi(\Lambda)$ is infinitesimally equivalent to $\pi(\Lambda')$. Then Λ is equivalent to Λ' (Definition 11.6).*

This formulation of Langlands' result incorporates the work of several other people. We mention in particular Harish-Chandra's result on the irreducibility of tempered induction at regular infinitesimal character, and Miličić' observation that the Langlands quotient (originally defined as the image of an intertwining operator) is simply a unique irreducible quotient.

Of course we need a result for singular infinitesimal character as well. Before beginning the technical preliminaries to its formulation, let us consider what goes wrong with Theorem 11.7 when we drop the regularity hypotheses. First, the standard limit representation $M(\Lambda)$

may be zero. The simplest example occurs with $G_\mathbb{R} = SU(2)$, and $T_\mathbb{R}$ a compact torus. In that case $z(\rho) = 1$ (since half the sum of the positive roots is already a rational character of the algebraic torus T), so all the coverings under consideration are trivial. We can take Λ^{can} to be the trivial character of $T_\mathbb{R}$, and $R_{i\mathbb{R}}^+$ arbitrary. (There are no real roots.) The corresponding standard representation must have infinitesimal character zero by Definition 11.2; but no representation of $SU(2)$ has infinitesimal character zero. (Harish-Chandra's parametrization assigns the infinitesimal character ρ to the trivial representation.)

Next, the standard representation may be non-zero, but it may have several distinct irreducible quotients. The first example occurs with $G_\mathbb{R} = SL(2, \mathbb{R})$, and $T_\mathbb{R} \simeq \mathbb{R}^\times$ the diagonal subgroup. Again all the coverings are trivial. We take Λ^{can} to be the trivial character of $T_\mathbb{R}$, and $R_\mathbb{R}^+$. Because the restriction to $T_\mathbb{R} \cap K_\mathbb{R} = \{\pm 1\}$ of the rational character ρ is non-trivial, we find that the character $\tilde{\Lambda}$ of (11.3) is the signum character of $\mathbb{R}^\times$. The standard representation is induced from $\tilde{\Lambda}$, and is therefore equal to the reducible unitary principal series representation of $SL(2, \mathbb{R})$: a direct sum of two limits of discrete series. The Langlands quotient is therefore also equal to this direct sum.

We will see that part (b) of Theorem 11.7 remains valid for singular infinitesimal characters. This already provides a kind of counterexample to (c): in the preceding example, the two constituents of the Langlands quotient must themselves be (irreducible) Langlands quotients of some other standard limit representations. In fact the situation is even a little worse: there can be two non-zero standard limit representations having isomorphic and irreducible Langlands quotients, with inequivalent limit characters. The simplest example occurs with $G_\mathbb{R} = GL(2, \mathbb{R})$, $T_\mathbb{R} =$ split Cartan subgroup, and $(T_\mathbb{R})' =$ fundamental Cartan subgroup. The standard representation we want is an irreducible unitary principal series whose restriction to $SL(2, \mathbb{R})$ is reducible; this representation is also a limit of (relative) discrete series. We will write down one of the limit characters, just as an example of the coverings involved. A dual group is $GL(2, \mathbb{C})$, and the element $z(\rho)$ is $-I$. The complex torus T consists of diagonal matrices in $GL(2, \mathbb{C})$, so

$$T \simeq \{ (z_1, z_2) \mid z_i \in \mathbb{C}^\times \} \qquad (11.8)(a)$$

The T-canonical covering of T (Definition 10.1) turns out to be the projective limit of all the finite covers, since T is split. A character of type $z(\rho)$ factors through the double cover

$$\tilde{T} = \{ (z_1, z_2, w) \mid z_i, w \in \mathbb{C}^\times, w^2 = z_1 z_2^{-1} \}. \qquad (11.8)(b)$$

(This is the "square root of ρ double cover" considered in Definition 1.33 of [58], or Definition 8.11 of [1].) Its pullback to $T_\mathbb{R}$ is

$$\tilde{T}_\mathbb{R} = \{ (x_1, x_2, w) \mid x_i \in \mathbb{R}^\times, w \in \mathbb{C}^\times, w^2 = x_1 x_2^{-1} \}. \qquad (11.8)(c)$$

We take Λ^{can} to be the character

$$\Lambda^{can}(x_1, x_2, w) = |w|\operatorname{sgn}(x_1)/w, \qquad (11.8)(d)$$

which takes values in $\{\pm 1, \pm i\}$. Let $R_\mathbb{R}^+$ be the positive root corresponding to the upper triangular matrices. Then the rational character ρ on $\tilde{T}$ is $\rho(z_1, z_2, w) = w$. The character $\tilde{\Lambda}$ of (11.3) from which we induce is now easily computed to be

$$\tilde{\Lambda}(x_1, x_2) = \operatorname{sgn}(x_1). \qquad (11.8)(e)$$

Now it is clear that $M(\Lambda)$ is an irreducible unitary principal series whose restriction to $SL(2, \mathbb{R})$ is the reducible principal series of the preceding example. We leave to the reader the construction of the corresponding limit character on $(T_\mathbb{R})'$.

In order to extend Theorem 11.7 to singular infinitesimal character, we need to restrict the class of limit characters in order to avoid such phenomena. It is a remarkable fact (related to the simple nature of the Galois group of $\mathbb{C}/\mathbb{R}$) that nothing essentially more complicated than these examples can happen. The problem of vanishing is particularly simple.

Proposition 11.9. *In the setting of Definition 11.2, suppose $\Lambda = (\Lambda^{can}, R_{i\mathbb{R}}^+, R_\mathbb{R}^+)$ is a limit character. Then $M(\Lambda) = 0$ if and only if there is a simple root $\alpha \in R_{i\mathbb{R}}^+$ such that α is compact, and $\langle \alpha, \lambda \rangle = 0$.*

(Notice that the first condition provides a subgroup of $G_\mathbb{R}$ locally isomorphic to $SU(2)$, and the second says that, along that subgroup, our limit character looks like the one in the first example after Theorem 11.7.)

Proof. The standard limit representation may be constructed by induction from a parabolic subgroup $P_\mathbb{R} = L_\mathbb{R} N_\mathbb{R}$, so it will be zero if and only if the inducing representation is zero. This provides a reduction to the case $L_\mathbb{R} = G_\mathbb{R}$; that is, to the case of (relative) limits of discrete series. That the condition in the proposition implies that the standard limit representation is zero is easy. For example, it is immediate from one of the two character identities established in [45]. Another approach (cf. [54], Proposition 8.4.3) is to construct the standard limit representation using cohomological parabolic induction. Then an induction by stages argument reduces one to the case of $SU(2)$.

The converse is not quite so easy. The argument is a special case of the one we will give for Theorem 11.14 below, and the reader may refer to that for some additional details. We are still assuming that $L_\mathbb{R} = G_\mathbb{R}$; that is, that $R^+_{i\mathbb{R}}$ is a full set of positive roots of T in G. Fix a dominant regular rational character μ of T, and consider the limit character

$$(\Lambda^{can} \otimes \mu, R^+_{i\mathbb{R}}, R^+_\mathbb{R}).$$

We denote this by the symbol $\Lambda + \mu$. The corresponding weight is $\lambda + d\mu$, which is dominant regular by (11.2)(d) and the choice of μ. Then $M(\Lambda + \mu)$ is a (non-zero irreducible) discrete series representation. The standard limit representation $M(\Lambda)$ is obtained from $M(\Lambda + \mu)$ by applying a Jantzen-Zuckerman translation functor (namely the one denoted $\psi^\lambda_{\lambda+\mu}$ in [54]). By [54], Corollary 7.3.23, it follows that $M(\Lambda) = 0$ if and only if there is a root α in $\tau(M(\Lambda + \mu))$ (the Borho-Jantzen-Duflo τ-invariant) satisfying $\langle \alpha, \lambda \rangle = 0$. Now the τ-invariant is a subset of the set of simple roots for the system of λ-integral roots. By (11.2)(d), a simple λ-integral singular root α must be simple in all of R^+. By hypothesis, α must be noncompact. By [54], Corollary 8.4.7, α does not lie in the τ-invariant; so $M(\Lambda) \neq 0$, as we wished to show. Q.E.D.

We want now a condition analogous to the one in Proposition 11.9 that will rule out the other bad examples given after Theorem 11.7. In each example, there was a real root α orthogonal to λ. Assuming that such roots do not exist certainly eliminates these phenomena, but it also eliminates limit characters that are needed for the classification. The prototypical example is the spherical representation π' of $SL(2, \mathbb{R})$ of infinitesimal character zero. In the notation of the second example after Theorem 11.7, π' is the unique Langlands quotient of — in fact is equal to — the standard limit representation $M(\Lambda')$, with $(\Lambda')^{can}$ the signum character of $T_\mathbb{R}$. Furthermore π' occurs in no other standard limit representation. To get a good classification theorem, we need to keep the limit character Λ' but discard Λ. What distinguishes these two limit characters is their behavior on the non-identity component of $T_\mathbb{R}$. The appropriate generalization of this is provided by the "parity condition."

Definition 11.10 ([53], Proposition 4.5, or [54], Definition 8.3.11). Suppose $G_\mathbb{R}$ is as in (11.1), $T_\mathbb{R}$ is a Cartan subgroup, and α is a real root of T in G. Choose a root subgroup homomorphism

$$\phi_\alpha : SL(2) \to G \qquad (11.10)(a)$$

11. The Langlands classification without L-groups

defined over $\mathbb{R}$, so that ϕ_α(diagonal matrices) $\subset T$. Since $SL(2)$ is simply connected, this lifts to

$$\phi_\alpha^{can} : SL(2) \to G^{can}. \tag{11.10}(b)$$

Since $SL(2,\mathbb{R})$ is connected, this map restricts to

$$\phi_\alpha^{can} : SL(2,\mathbb{R}) \to \text{ image of } G_\mathbb{R}^{can} \subset G(\mathbb{R})^{can}. \tag{11.10}(c)$$

The mapping (11.1)(a) therefore induces an extension of $SL(2,\mathbb{R})$. Passing to its identity component, we get

$$\tilde{\phi}_\alpha^{can} : \widetilde{SL}(2,\mathbb{R}) \to G_\mathbb{R}^{can}, \tag{11.10}(d)$$

with

$$\widetilde{SL}(2,\mathbb{R}) \to SL(2,\mathbb{R}) \tag{11.10}(e)$$

a connected cover. Define

$$\tilde{m} = \exp\begin{pmatrix} 0 & \pi \\ -\pi & 0 \end{pmatrix} \in \widetilde{SL}(2,\mathbb{R}), \quad \tilde{m}_\alpha = \tilde{\phi}_\alpha^{can}(\tilde{m}) \in G_\mathbb{R}^{can}. \tag{11.10}(f)$$

(If $G_\mathbb{R}$ is linear, then $\widetilde{SL}(2,\mathbb{R}) = SL(2,\mathbb{R})$, and $\tilde{m} = -I$. In general $\tilde{m}$ is a generator for $Z(\widetilde{SL}(2,\mathbb{R}))$, and it may have any (finite) order.) The homomorphism ϕ_α is not quite uniquely determined by the conditions we have imposed on it. It is easy to check that changing the choice replaces $\tilde{m}_\alpha$ by $\tilde{m}_\alpha^{\pm 1}$.

Suppose $\Lambda = (\Lambda^{can}, R_{i\mathbb{R}}^+, R_\mathbb{R}^+)$ is a limit character of $T_\mathbb{R}$. Choose any set R^+ of positive roots for T in G. As in (11.3), regard $\Lambda^{can} \otimes \rho$ as a representation of $T_\mathbb{R}^{can,G}$. It may then be applied to the element $\tilde{m}_\alpha$ defined above. We say that α *satisfies the parity condition for* Λ if the eigenvalues of the operator $(\Lambda^{can} \otimes \rho)(\tilde{m}_\alpha)$ are contained in the set

$$-\exp\left(\pm i\pi\langle\alpha^\vee, \lambda + \rho - \rho(R_\mathbb{R}^+)\rangle\right). \tag{11.10}(g)$$

(Because of the $\pm$ in the exponent, this condition is unchanged if we replace $\tilde{m}_\alpha$ by $\tilde{m}_\alpha^{-1}$.) Here it is important to understand that $\rho(R_\mathbb{R}^+)$ refers to the set of positive real roots fixed in Λ, but that ρ refers to the arbitrary set R^+ of positive roots. We do not assume that $R_\mathbb{R}^+$ is contained in R^+. If $G_\mathbb{R}$ is linear (so that $\tilde{m}_\alpha$ is a central element of order

2 in $T_\mathbb{R}^{can,G}$) this condition amounts to

$$\langle \alpha^\vee, \lambda + \rho - \rho(R_\mathbb{R}^+)\rangle = n \in \mathbb{Z}, \text{ and } (\Lambda^{can} \otimes \rho)(\tilde{m}_\alpha) = (-1)^{n+1}. \quad (11.10)(h)$$

Since the notion of limit character we are using is a little different from that in [53] and [54], we need to check that this definition of the parity condition agrees with the (more complicated) one there. The first thing to notice is that our condition (11.10)(g) does not depend on the choice of positive root system R^+. The reason is this. Write $\alpha^\vee \in X_*(T)$ for the coroot corresponding to α. By a calculation in $SL(2)$, the image of $\tilde{m}_\alpha$ in G is

$$m_\alpha = \alpha^\vee(-1) \in T. \quad (11.11)(a)$$

It follows that if γ is any rational character of T, then

$$\gamma(m_\alpha) = (-1)^{\langle \alpha^\vee, \gamma \rangle}. \quad (11.11)(b)$$

In particular, if R^+ and $(R^+)'$ are two systems of positive roots,

$$\left(\rho(R^+) \otimes \rho((R^+)')^{-1}\right)(m_\alpha) = (-1)^{\langle \alpha^\vee, \rho(R^+) - \rho((R^+)')\rangle}. \quad (11.11)(c)$$

This equality shows that the condition at (11.10)(g) does not depend on R^+. (A similar argument shows that the parity condition behaves well under equivalence of limit characters.) To complete the comparison with [53] or [54], one can use the prescription in (11.3) for translating limit characters from one form to the other. The argument is not trivial, but it is straightforward and dull, so we omit it.

Here is an analogue of Proposition 11.9.

Proposition 11.12 (The Hecht-Schmid character identity — see [45] and [53], Proposition 4.5.) *In the setting of Definition 11.2, suppose there is a real root α such that α satisfies the parity condition for Λ (Definition 11.10), and $\langle \alpha, \lambda \rangle = 0$. Then the standard limit representation $M(\Lambda)$ is isomorphic to a direct sum of one or two standard limit representations attached to limit characters on a Cartan subgroup $(T_\mathbb{R})'$ having a strictly smaller split part.*

Definition 11.13 ([57], Definition 2.6.) In the setting of Definition 11.2, the limit character $\Lambda = (\Lambda^{can}, R_{i\mathbb{R}}^+, R_\mathbb{R}^+)$ is called *final* if it satisfies the following two conditions.

(a) Suppose α is a simple root in $R_{i\mathbb{R}}^+$, and $\langle \alpha, \lambda \rangle = 0$. Then α is noncompact.

(b) Suppose α is a real root, and $\langle \alpha, \lambda \rangle = 0$. Then α does not satisfy the parity condition for Λ (Definition 11.10).

We write

$$L^z(G_\mathbb{R}) = \{ \text{ equivalence classes of final limit characters of type } z \ \}.$$

The standard limit representations attached to final limit characters are called *final standard limit representations*.

Theorem 11.14 ([29]; see [57], Theorem 2.6). *Suppose $G_\mathbb{R}$ is a real reductive group as in (11.1). Then the set $\Pi^z(G_\mathbb{R})$ of infinitesimal equivalence classes of irreducible admissible representations of $G_\mathbb{R}$ is in one-to-one correspondence with the set $L^z(G_\mathbb{R})$ of equivalence classes of final limit characters, as follows.*

a) *Suppose Λ is a final limit character of a Cartan subgroup $T_\mathbb{R}$ (Definition 11.13). Then the corresponding standard representation $M(\Lambda)$ is non-zero, and has a unique irreducible quotient. In particular, the Langlands quotient representation $\pi(\Lambda)$ is irreducible.*

b) *Suppose π is an irreducible canonical projective representation of $G_\mathbb{R}$ of type z (Definition 11.2). Then there is a final limit character Λ of type z with the property that π is infinitesimally equivalent to $\pi(\Lambda)$.*

c) *Suppose Λ and Λ' are final limit characters, and that $\pi(\Lambda)$ is infinitesimally equivalent to $\pi(\Lambda')$. Then the limit characters are equivalent (Definition 11.6).*

Any standard limit representation of $G_\mathbb{R}$ (Definition 11.2) is isomorphic to a direct sum of final standard limit representations.

Proof. (The result in [29] is for linear groups, and is in any case formulated in a very different way. Since [57] contains no proof, we outline the argument.) The last claim is clear from Propositions 11.9 and 11.12. For the rest, one can use the Jantzen-Zuckerman translation principle just as it is used in the proof of Proposition 11.9. Here is a sketch. (Another development of the translation principle, emphasizing the connection with the geometric version in Chapter 8, may be found in Chapter 16.) Fix a maximal torus T_a of G and a weight

$$\lambda_a \in \mathfrak{t}_a^*. \tag{11.15}(a)$$

We study representations of infinitesimal character λ_a. Fix also a system of positive roots R_a^+ for T_a in G, chosen so that

$$\langle \alpha, \lambda_a \rangle > 0 \Rightarrow \alpha \in R_a^+. \tag{11.15}(b)$$

Fix a regular dominant rational character $\mu_a \in X^*(T_a)$. By (11.15)(b),

$$\lambda'_a = \lambda_a + \mu_a \qquad (11.15)(c)$$

is a G-regular weight, and satisfies the analogue of (11.15)(b). As in the proof of Proposition 11.9, we consider the translation functor

$$\psi = \psi^{\lambda_a}_{\lambda_a + \mu_a} \qquad (11.15)(d)$$

from modules with the (regular) infinitesimal character $\lambda_a + \mu_a$ to modules with infinitesimal character λ_a. (The functor is defined by tensor product with the finite-dimensional representation of lowest weight $-\mu_a$, followed by projection on the infinitesimal character λ_a.) Finally, define

$$\Delta^0_a = \text{ simple roots for } R^+_a \text{ orthogonal to } \lambda_a. \qquad (11.15)(e)$$

Any irreducible representation π' of infinitesimal character λ'_a has a Borho-Jantzen-Duflo τ-invariant, $\tau(\pi')$ which is a subset of the simple roots for the positive λ'_a-integral roots (cf. [54], Definition 7.3.8.) This set of simple roots includes Δ^0_a, so we can define

$$\tau^0(\pi') = \tau(\pi') \cap \Delta^0_a. \qquad (11.15)(f)$$

Here are the general facts we need from the translation principle.

Proposition 11.16 ([54], Corollary 7.3.23; cf. also [61], Proposition 7.7). *Suppose $G_\mathbb{R}$ is a real reductive group as in (11.1), and suppose we are in the setting (11.15). Then the translation functor ψ of (11.15)(d) is an exact functor from canonical projective representations of $G_\mathbb{R}$ of type z and infinitesimal character λ'_a to canonical projective representations of type z and infinitesimal character λ_a. It has the following additional properties.*

a) *Suppose π' is an irreducible representation of infinitesimal character λ'_a. Then $\psi(\pi')$ is irreducible or zero. The first possibility occurs if and only if $\tau^0(\pi')$ (cf. (11.15)(f)) is empty.*

b) *Suppose π is an irreducible representation of infinitesimal character λ_a. Then there is a unique irreducible representation π' with $\psi(\pi') \simeq \pi$.*

In order to use this result to reduce Theorem 11.14 to the special case of Theorem 11.7, we need to know two more specific things: how the translation functor affects standard limit representations, and how to compute $\tau^0(\pi')$ in terms of the classification of Theorem 11.7. Suppose then that $\Lambda' = ((\Lambda')^{can}, R^+_{i\mathbb{R}}, R^+_\mathbb{R})$ is a $G_\mathbb{R}$-limit character of $T_\mathbb{R}$ of

11. The Langlands classification without L-groups

infinitesimal character λ'_a. Write λ' for the differential of Λ'. Since λ' and λ'_a are regular and define the same infinitesimal character, there is a unique isomorphism

$$i(\lambda'_a, \lambda') : T_a \to T \qquad (11.17)(a)$$

induced by an element of G and carrying λ'_a to λ'. We also write $i(\lambda'_a, \lambda')$ for the induced isomorphism from the root system of T_a in G to that of T, and so on. Put

$$R^+ = i(\lambda'_a, \lambda')(R_a^+), \qquad (11.17)(b)$$

a system of positive roots for T in G. Because of (11.15)(b) and (11.2)(d), we have

$$R^+ \supset R_{i\mathbb{R}}^+. \qquad (11.17)(c)$$

Similarly, define

$$\mu = i(\lambda'_a, \lambda')(\mu_a) \in X^*(T). \qquad (11.17)(d)$$

Then we can define

$$\Lambda' - \mu = ((\Lambda')^{can} \otimes \mu^{-1}, R_{i\mathbb{R}}^+, R_{\mathbb{R}}^+). \qquad (11.17)(e)$$

Because of (11.15)(b) and (11.17)(c), this is a $G_\mathbb{R}$-limit character of $T_\mathbb{R}$ of infinitesimal character λ_a.

Proposition 11.18 *Suppose we are in the setting of (11.15), and Λ' is a $G_\mathbb{R}$-limit character of infinitesimal character λ'_a. Use the notation of (11.17).*

a) The translation functor ψ satisfies

$$\psi(M(\Lambda')) \simeq M(\Lambda' - \mu).$$

b) Write Δ^0 for the image of Δ_a^0 under the bijection $i(\lambda'_a, \lambda')$. Then the τ invariant $\tau^0(\pi(\Lambda'))$ corresponds to the subset of roots in Δ^0 satisfying one of the following three conditions:
   i) α is compact imaginary;
   ii) α is complex, and its complex conjugate $\bar{\alpha}$ is positive; or
   iii) α is real and satisfies the parity condition (Definition 11.10).

c) *Suppose $\tau^0(\pi(\Lambda'))$ is empty. Then $\psi(\pi(\Lambda'))$ is the unique irreducible quotient of $M(\Lambda' - \mu)$. That is,*

$$\psi(\pi(\Lambda')) = \pi(\Lambda' - \mu).$$

Proof. Part (a) is proved for linear groups in [54], Proposition 7.4.1; the argument is unchanged in general. For (b), the τ-invariant depends only on the restriction of a representation to the identity component (in fact only on its annihilator in the enveloping algebra); so we may assume $G_{\mathbb{R}}$ is connected. Then the result is proved in [53], Theorem 4.12. (For linear groups, this is [54], Theorem 8.5.18.)

For (c), suppose that π_1 is an irreducible quotient of $M(\Lambda)$. By Proposition 11.16, there is an irreducible representation π_1' of infinitesimal character λ_a' with $\psi(\pi_1') \simeq \pi_1$. The translation functor ψ has a two-sided adjoint

$$\phi = \psi_{\lambda_a}^{\lambda_a + \mu_a} \qquad (11.19)(a)$$

([54], Proposition 4.5.8). Consequently

$$\operatorname{Hom}(M(\Lambda' - \mu), \pi_1) \simeq \operatorname{Hom}(\psi(M(\Lambda')), \psi(\pi_1'))$$
$$\simeq \operatorname{Hom}(M(\Lambda'), \phi\psi(\pi_1')). \qquad (11.19)(b)$$

(The Hom spaces are infinitesimal homomorphisms; that is, $(\mathfrak{g}, K_{\mathbb{R}})$-module maps for an appropriate maximal compact $K_{\mathbb{R}}$.) Now it is a formal consequence of the adjointness of ϕ and ψ that any irreducible subrepresentation π_2' of $\phi\psi(\pi_1')$ must be a preimage of π_1 under ψ. By Proposition 11.16(b),

$$\operatorname{Hom}(\pi_2', \phi\psi(\pi_1')) = \begin{cases} \mathbb{C}, & \text{if } \pi_2' \simeq \pi_1' \\ 0 & \text{otherwise.} \end{cases} \qquad (11.19)(c)$$

On the other hand, (11.19)(b) implies that the unique irreducible quotient π' of $M(\Lambda')$ must be a composition factor of $\phi\psi(\pi_1')$. But the only composition factor of $\phi\psi(\pi_1')$ having τ^0 empty is π_1' itself ([54], Proposition 7.3.2(b) and Corollary 7.3.21). Consequently $\pi_1' \simeq \pi'$, and $\pi_1 \simeq \psi(\pi')$. The last group in (11.19)(b) is therefore

$$\operatorname{Hom}(M(\Lambda'), \phi\psi(\pi')). \qquad (11.19)(d)$$

The image P' of a non-zero map in this space must have π' as its unique irreducible quotient (since P' is a quotient of $M(\Lambda')$) and as its unique irreducible subrepresentation (by (11.19)(c)). But π' occurs only once as a composition factor of $M(\Lambda')$, so it follows that $P' = \pi'$. Consequently

$$\operatorname{Hom}(M(\Lambda'), \phi\psi(\pi')) \simeq \operatorname{Hom}(\pi', \phi\psi(\pi')) \simeq \mathbb{C}. \qquad (11.19)(e)$$

In conjunction with (11.19)(b), this shows that $\psi(\pi')$ occurs exactly once as a quotient of $M(\Lambda)$, as we wished to show. Q.E.D.

Propositions 11.16 and 11.18 explain rather explicitly how to describe representations of infinitesimal character λ_a in terms of representations of (regular) infinitesimal character λ'_a. The proof of Theorem 11.14 now comes down to checking that these constructions are compatible with the parametrization in terms of final limit characters. Here is the result we need.

Proposition 11.20. *In the setting (11.15), there is a bijection from the set*

{equivalence classes of limit characters Λ' of infinitesimal character λ'_a with $\tau^0(\pi(\Lambda'))$ empty } *onto the set*

{equivalence classes of final limit characters Λ of infinitesimal character λ_a}. *In the notation of (11.17), this bijection ends the class of Λ' to the class of $\Lambda' - \mu$.*

Proof. Suppose first that $\tau^0(\pi(\Lambda'))$ is empty; we want to show that $\Lambda = \Lambda' - \mu$ is final (Definition 11.13). First, Proposition 11.18 guarantees that $M(\Lambda) \neq 0$. By Proposition 11.9, it follows that Λ satisfies condition (11.13)(a). To verify condition (11.13)(b), we need to understand a little about the root system

$$R^0 = \{\alpha \in R(G, T) \mid \langle \alpha, \lambda \rangle = 0\}. \qquad (11.21)(a)$$

First, it has $(R^0)^+ = R^+ \cap R^0$ (notation (11.17)(c)) as a set of positive roots, with Δ^0 the corresponding simple roots. Define

$$\Delta^0_{\mathbb{R}} = \Delta^0 \cap R_{\mathbb{R}}, \qquad R^0_{\mathbb{R}} = R^0 \cap R_{\mathbb{R}}. \qquad (11.21)(b)$$

What we need to show is that

$$\Delta^0_{\mathbb{R}} \text{ is a set of simple roots for } R^0_{\mathbb{R}}. \qquad (11.21)(c)$$

Assume for a moment that we have established this. From Definition 11.10 one can check that

if α and β in $R^0_{\mathbb{R}}$ do not satisfy the parity condition, then neither does $s_\alpha(\beta)$. $\qquad (11.21)(d)$

(For linear groups one can say even more: because of (11.11)(a), the set of coroots for roots in $R_{\mathbb{R}}^0$ not satisfying the parity condition is closed under addition (cf. [54], Lemma 8.6.3). In the non-linear case, one can reduce immediately to the case of split simple groups of rank 2. These can be treated by hand; the calculation is very easy except in type B_2.) By Proposition 11.18(b)(iii) and the hypothesis on Λ', the roots in $\Delta_{\mathbb{R}}^0$ do not satisfy the parity condition. By (11.21)(c) and (d), it follows that no root in $R_{\mathbb{R}}^0$ satisfies the parity condition. This is (11.13)(b).

To prove (11.21)(c), observe first that

$$\text{if } \alpha \in (R^0)^+, \text{ then } \langle -\overline{\alpha}, \lambda \rangle \geq 0. \qquad (11.22)(a)$$

For it is enough to prove this for $\alpha \in \Delta^0$. If α is real or imaginary, the inner product is zero. If α is complex, then the hypothesis on Λ' and Proposition 11.18(b)(ii) guarantee that $-\overline{\alpha} \in R^+$; so the inequality follows from (11.15)(b) (and (11.17)). Now define

$$\begin{aligned} R^{00} &= \{\, \alpha \in R^0 \mid -\overline{\alpha} \in R^0 \,\} \\ &= \{\, \alpha \in R \mid \langle \alpha, \lambda \rangle = \langle -\overline{\alpha}, \lambda \rangle = 0 \,\}. \end{aligned} \qquad (11.22)(b)$$

This is a root system; we define

$$(R^{00})^+ = R^{00} \cap R^+, \qquad \Delta^{00} = \Delta^0 \cap R^{00}. \qquad (11.22)(c)$$

Because of (11.22)(a),

$$\Delta^{00} \text{ is a set of simple roots for } (R^{00})^+. \qquad (11.22)(d)$$

Now $\alpha \mapsto -\overline{\alpha}$ is an involutive automorphism of R^{00}, so we can apply to it Lemma 8.6.1 of [54]. Suppose α is a complex simple root for $(R^{00})^+$. By (11.22)(d), $\alpha \in \Delta^0$; so by Proposition 11.18(b)(ii) and the hypothesis on Λ', $-\overline{\alpha} \in (R^{00})^+$. Lemma 8.6.1 of [54] therefore implies that

the set of non-real roots in $(R^{00})^+$ is stable under $\alpha \mapsto -\overline{\alpha}$.
$$(11.22)(e)$$

Now (11.21)(c) is precisely the conclusion of Lemma 8.6.2 of [54].

We have therefore shown that the map in Proposition 11.20 is well-defined. To see that it is surjective, assume that

$$\Lambda = (\Lambda^{can}, R_{i\mathbb{R}}^+, R_{\mathbb{R}}^+) \text{ is a final limit character of } T_{\mathbb{R}} \text{ of infinitesimal character } \lambda_a. \qquad (11.23)(a)$$

Write R for the set of roots of T in G. Define

$$R^0 = \{\, \alpha \in R \mid \langle \alpha, \lambda \rangle = 0 \,\}, \qquad (11.23)(b)$$

$$R^{00} = \{\, \alpha \in R^0 \mid -\overline{\alpha} \in R^0 \,\}. \qquad (11.23)(c)$$

Choose a set of positive roots $(R^{00})^+$ for R^{00} with the property that

the set of non-real roots in $(R^{00})^+$ is stable under $\alpha \mapsto -\overline{\alpha}$.
$$(11.23)(d)$$
(This is certainly possible.) Possibly after modifying this system by reflections in imaginary roots, we may assume that

$$R_{i\mathbb{R}}^+ \cap R^{00} \subset (R^{00})^+. \qquad (11.23)(e)$$

Now define a positive root system for R^0 by

$$(R^0)^+ = (R^{00})^+ \cup \{\, \alpha \in R^0 \mid \langle -\overline{\alpha}, \lambda \rangle > 0 \,\}. \qquad (11.23)(f)$$

Write

$$\Delta^0 = \text{simple roots for } (R^0)^+. \qquad (11.23)(g)$$

Since λ and λ_a define the same infinitesimal character, there is a unique isomorphism

$$i(\lambda_a, \lambda; \Delta_a^0, \Delta^0) : T_a \to T \qquad (11.24)(a)$$

induced by an element of G, carrying λ_a to λ and Δ_a^0 to Δ^0. Just as in (11.17), we now define

$$R^+ = i(\lambda_a, \lambda; \Delta_a^0, \Delta^0)(R_a^+), \qquad (11.24)(b)$$

a system of positive roots for T in G. Because of (11.23)(e) and (f), and (11.2)(d), we have

$$R^+ \supset R_{i\mathbb{R}}^+. \qquad (11.24)(c)$$

Put

$$\mu = i(\lambda_a, \lambda; \Delta_a^0, \Delta^0)(\mu_a) \in X^*(T). \qquad (11.24)(d)$$

Then we can define

$$\Lambda' = (\Lambda^{can} \otimes \mu, R_{i\mathbb{R}}^+, R_{\mathbb{R}}^+). \qquad (11.24)(e)$$

Because of (11.23)(f) and (11.24)(c), this is a $G_\mathbb{R}$-limit character of $T_\mathbb{R}$ of infinitesimal character λ'_a. We are again in the situation (11.17), and $\Lambda = \Lambda' - \mu$. We need to check that $\tau^0(\pi(\Lambda'))$ is empty. So fix a root $\alpha \in \Delta^0$. If α is imaginary, then (11.24)(c) implies that it is simple in $R^+_{i\mathbb{R}}$; so since Λ is final, α is noncompact. If α is complex, then (11.23)(d) and (f) imply that $-\overline{\alpha}$ is positive. If α is real, then since Λ is final, α does not satisfy the parity condition. By Proposition 11.18(b), α is not in the τ-invariant. This proves the surjectivity of the map in Proposition 11.20.

For the injectivity, suppose Λ'_1 and Λ'_2 both map to the equivalence class of Λ. By Proposition 11.18(c),

$$\psi(\pi(\Lambda'_1)) \simeq \pi(\Lambda) \simeq \psi(\pi(\Lambda'_2)).$$

By Proposition 11.16(b), it follows that $\pi(\Lambda'_1) \simeq \pi(\Lambda'_1)$. By Theorem 11.7(c), Λ'_1 is equivalent to Λ'_2. Q.E.D.

Theorem 11.14 is a formal consequence of Proposition 11.16, Proposition 11.18, Proposition 11.20, and Theorem 11.7 (applied to the infinitesimal character λ'_a in the setting (11.15)). Q.E.D.

12. Langlands parameters and Cartan subgroups

In this chapter we show how to reformulate Theorem 10.4 in terms of Cartan subgroups of extended groups and E-groups. We begin by recasting Theorem 11.14 in the language of extended groups.

Definition 12.1. Suppose G^Γ is a weak extended group for G, $^\vee G$ is a dual group for G, and $z \in Z(^\vee G)^{\theta_z}$ (cf. Definition 10.3). A *Cartan subgroup of* G^Γ is a weak extended group $T^\Gamma \subset G^\Gamma$ with identity component a maximal torus $T \subset G$. The conjugation action of any element $\delta \in T^\Gamma - T$ on T is a real form of T, independent of the choice of δ; we write $T(\mathbb{R})$ for the group of real points.

A *G-limit character of T^Γ of type z* is a pair (δ, Λ) subject to the following conditions. First,

$$\delta \in T^\Gamma - T \text{ is a strong real form of } G \qquad (12.1)(a)$$

(Definition 2.13). Thus $T(\mathbb{R})$ is a Cartan subgroup of $G(\mathbb{R}, \delta)$. We can therefore impose the second requirement

$$\Lambda \text{ is a } G(\mathbb{R}, \delta)\text{-limit character of } T(\mathbb{R}) \text{ of type } z \qquad (12.1)(b)$$

(Definition 11.2). We say that (δ, Λ) is *final* if Λ is (Definition 11.13). We say that (δ, Λ) is *equivalent* to the G-limit character (δ', Λ') of $(T')^\Gamma$ if there is an element $g \in G$ that conjugates δ' to δ, so that $g \cdot \Lambda'$ and Λ are equivalent as limit characters of $G(\mathbb{R}, \delta)$ (Definition 11.6). Write

$$L^z(G/\mathbb{R}) = \{ \text{ equivalence classes of final limit characters } (\delta, \Lambda) \text{ of type } z\}. \qquad (12.1)(c)$$

This definition can be formulated a little more directly, by incorporating some of the earlier definitions to which it refers. We can say that a limit character of T^Γ of type z is a quadruple

$$(\delta, \Lambda^{can}, R^+_{i\mathbb{R}}, R^+_{\mathbb{R}}), \qquad (12.2)(a)$$

subject to the conditions below. First,

$$\delta \in T^\Gamma - T, \text{ and } \delta^2 \in Z(G). \qquad (12.2)(b)$$

Write $T^{can,T}$ for the corresponding canonical cover (Definition 10.1), and $T^{can,T}(\mathbb{R})$ for the preimage of $T(\mathbb{R})$ in this cover (Definition 10.3):

$$1 \to \pi_1(T)^{can} \to T(\mathbb{R})^{can,T} \to T(\mathbb{R}) \to 1. \quad (12.2)(c)$$

Then $zz(\rho) \in {}^\vee T^{\theta z}$ (Definition 11.2) defines a character $\chi^T_{zz(\rho)}$ of $\pi_1(T)^{can}$. The second condition is

Λ^{can} is a character of $T(\mathbb{R})^{can,T}$, and $\Lambda^{can}|_{\pi_1(T)^{can}} = \chi^T_{zz(\rho)}$.
$$(12.2)(d)$$

(Notice that this condition does not involve δ.) Write $R_{i\mathbb{R}}$ for the set of roots assuming imaginary values on $\mathfrak{t}(\mathbb{R})$, and $R_\mathbb{R}$ for those assuming real values. Then

$R_{i\mathbb{R}}^+$ is a set of positive roots for $R_{i\mathbb{R}}$, and $R_\mathbb{R}^+$ for $R_\mathbb{R}$. $\quad (12.2)(e)$

Finally, write $\lambda = d\Lambda^{can} \in \mathfrak{t}^*$. Then we require

$$\langle \alpha, \lambda \rangle \geq 0, \qquad (\alpha \in R_{i\mathbb{R}}^+). \quad (12.2)(f)$$

We leave to the reader the straightforward task of formulating similarly explicit descriptions of the notion of final and the equivalence relation on limit characters.

Theorem 12.3. *Suppose G^Γ is a weak extended group for G (Definition 2.13). Then there is a natural bijection between the set $\Pi^z(G/\mathbb{R})$ of equivalence classes of irreducible canonical projective representations of type z of strong real forms of G (Definition 10.3) and the set $L^z(G/\mathbb{R})$ of equivalence classes of final limit characters of type z of Cartan subgroups in G^Γ (Definition 12.1).*

This is just a reformulation of Theorem 11.14.

We would like a parallel description of the Langlands parameters, in terms of tori in the dual group.

Definition 12.4. Suppose ${}^\vee G^\Gamma$ is a weak E-group (Definition 4.3). A *Cartan subgroup of* ${}^\vee G^\Gamma$ is a weak E-group ${}^d T^\Gamma \subset {}^\vee G^\Gamma$ such that the identity component ${}^d T$ is a maximal torus in ${}^\vee G$, and the inclusion of ${}^d T^\Gamma$ in ${}^\vee G^\Gamma$ is an L-homomorphism (Definition 5.1). Conjugation by any element y of ${}^d T^\Gamma - {}^d T$ defines an involutive automorphism θ of ${}^d T$, which is independent of y. Write

$$A({}^d T^\Gamma) = {}^d T^\theta / ({}^d T^\theta)_0 \quad (12.4)(a)$$

12. Langlands parameters and Cartan subgroups

for the group of connected components of the fixed points of θ. We call this group the *Langlands component group for* $^dT^\Gamma$. As in (5.10) we can form such groups as $^dT^{alg,\vee G}$ (which is a quotient of the algebraic universal covering $^dT^{alg,{}^dT}$). Define

$$A(^dT^\Gamma)^{alg,\vee G} = (^dT^\theta)^{alg,\vee G}/((^dT^\theta)^{alg,\vee G})_0, \qquad (12.4)(b)$$

the *universal component group for* $^dT^\Gamma$ *with respect to* $^\vee G^\Gamma$.

Since $^dT^\Gamma$ is a weak E-group in its own right, we can speak of Langlands parameters for $^dT^\Gamma$ (Definition 5.2). A Langlands parameter ϕ is said to be $^\vee G$-*regular* if for every root α of ^dT in $^\vee G$, we have

$$\alpha(\lambda(\phi)) \neq 0 \qquad (12.4)(c)$$

(notation (5.8)). In general the roots that fail to satisfy (12.4)(c) are called $\lambda(\phi)$-*singular* or ϕ-*singular*.

A *complete Langlands parameter for* $^dT^\Gamma$ *with respect to* $^\vee G^\Gamma$ is a pair (ϕ, τ_1) with ϕ a Langlands parameter for $^dT^\Gamma$, and τ_1 an irreducible representation of $A(^dT^\Gamma)^{alg,\vee G}$ (cf. Definition 5.12). Such a parameter is said to be *equivalent* to a complete parameter (ϕ', τ_1') for $(^dT^\Gamma)'$ if the triple $(^dT^\Gamma, \phi, \tau_1)$ is conjugate by $^\vee G$ to $((^dT^\Gamma)', \phi', \tau_1')$.

To formulate a definition analogous to that of "final," we need a little preliminary notation. Suppose we are in the setting of Definition 12.4. Since θ extends (although not uniquely) to an automorphism of $^\vee G$, it must permute the roots and coroots of ^dT in $^\vee G$. We write

$$^dR = R(^\vee G, {}^dT) \qquad (12.5)(a)$$

$$^dR_\mathbb{R} = \{\alpha \in {}^dR \mid \theta\alpha = -\alpha\} \qquad (12.5)(b)$$

$$^dR_{i\mathbb{R}} = \{\alpha \in {}^dR \mid \theta\alpha = \alpha\}, \qquad (12.5)(c)$$

the *real* and *imaginary roots of* $^dT^\Gamma$. Suppose now that α is a real root. The corresponding coroot $\alpha^\vee$ is a homomorphism from $\mathbb{C}^\times$ to ^dT (or even into $^dT^{alg,\vee G}$). Since α is real, it satisfies

$$\theta(\alpha^\vee(z)) = \alpha^\vee(z^{-1}). \qquad (12.6)(a)$$

In particular, we get a distinguished element of order two

$$m_\alpha = \alpha^\vee(-1) \in {}^dT^\theta. \quad (12.6)(b)$$

Similarly, we write

$$m_\alpha^{alg} = \alpha^\vee(-1) \in ({}^dT^\theta)^{alg,\vee G}. \quad (12.6)(c)$$

These elements define elements of order two in the component groups of Definition 12.4, which we write as

$$\overline{m_\alpha^{alg}} \in A({}^dT^\Gamma)^{alg,\vee G} \quad (\alpha \in {}^dR_\mathbb{R}). \quad (12.6)(d)$$

Next, suppose α is an imaginary root, and $y \in {}^dT^\Gamma - {}^dT$. Choose root vectors $X_{\pm\alpha} \in {}^\vee\mathfrak{g}$. Since θ fixes α, $\mathrm{Ad}(y)$ must send each root vector to a multiple of itself:

$$\mathrm{Ad}(y)X_{\pm\alpha} = z_{\pm\alpha}X_\alpha. \quad (12.7)(a)$$

The bracket of these two root vectors is a multiple of the derivative of the coroot $\alpha^\vee$, and is therefore fixed by $\theta = \mathrm{Ad}(y)$. It follows that

$$z_\alpha = (z_{-\alpha})^{-1}. \quad (12.7)(b)$$

Similarly, if α, β and $\alpha + \beta$ are all imaginary roots,

$$z_{\alpha+\beta} = z_\alpha z_\beta. \quad (12.7)(c)$$

Now y^2 belongs to ${}^dT$, so $\alpha(y^2)$ is defined. Clearly

$$(z_\alpha)^2 = \alpha(y^2). \quad (12.7)(d)$$

We say that α is *y-compact* if $z_\alpha = 1$, and *y-noncompact* if $z_\alpha = -1$. As a consequence of (12.7)(d), every imaginary root in the centralizer of y^2 is either compact or noncompact.

Definition 12.8. Suppose ϕ is a Langlands parameter for ${}^dT^\Gamma$. As in (5.8), we associate to ϕ elements $y = y(\phi) \in {}^dT^\Gamma - {}^dT$ and $\lambda = \lambda(\phi) \in {}^d\mathfrak{t}$. An imaginary root α is called ϕ-*compact* (respectively ϕ-*noncompact*) if it is $y(\phi)$-compact (respectively $y(\phi)$-noncompact) in the sense of (12.7). By Proposition 5.6(b) and (12.7)(d), every ϕ-singular imaginary root is either ϕ-compact or ϕ-noncompact.

Suppose next that (ϕ, τ_1) is a complete Langlands parameter for ${}^dT^\Gamma$ with respect to ${}^\vee G^\Gamma$ (Definition 12.4). We say that the pair (ϕ, τ_1) is *final* if it satisfies the following two conditions.

(a) Every ϕ-singular imaginary root of ${}^dT$ in ${}^\vee G$ is ϕ-compact.
(b) If α is a ϕ-singular real root of ${}^dT^\Gamma$ in ${}^\vee G^\Gamma$, then

$$\tau_1(\overline{m_\alpha^{alg}}) = 1$$

(notation 12.6)(c).)

Theorem 12.9. *Suppose ${}^\vee G^\Gamma$ is a weak E-group (Definition 4.3). Then there is a one-to-one correspondence between the set $\Xi({}^\vee G^\Gamma)$ of equivalence classes of complete Langlands parameters for ${}^\vee G^\Gamma$ (Definition 5.11) and the set of equivalence classes of final complete Langlands parameters for Cartan subgroups of ${}^\vee G^\Gamma$ (Definition 12.8).*

Proof. Suppose (ϕ, τ) is a complete Langlands parameter for ${}^\vee G^\Gamma$. We want to construct a Cartan subgroup ${}^dT^\Gamma$ containing the image of ϕ. To do this, use the notation of Corollary 5.9. There we defined a Levi subgroup $L(\lambda, y \cdot \lambda)$ of ${}^\vee G$, on which conjugation by y acts as an involutive automorphism θ_y. We apply to this subgroup the following well-known facts. (The slightly awkward hypotheses on K will allow us to apply the result to coverings.)

Lemma 12.10 *Suppose G is a connected complex reductive algebraic group, and θ is an involutive automorphism of G. Suppose K is an algebraic subgroup of G having the same identity component as the group of fixed points of θ, with the property that the automorphisms of G in $\mathrm{Ad}(K)$ all commute with θ. Then there is a θ-stable maximal torus $T \subset G$, determined up to conjugation by K_0 by either of the following properties. Write $\mathfrak{a}$ for the -1 eigenspace of θ on $\mathfrak{t}$, and M for the centralizer of $\mathfrak{a}$ in G.*

a) The Lie algebra $\mathfrak{a}$ is a maximal semisimple abelian subalgebra in the -1 eigenspace of θ on $\mathfrak{g}$.

b) Every root of T in M is compact (cf. (12.7); the roots in M are precisely the imaginary roots of T in G).

In addition, T has the following properties.

c) $T \cap K$ meets every connected component of K. That is, the natural map $(T \cap K)/(T \cap K)_0 \to K/K_0$ is surjective.

d) The kernel $T \cap K_0$ of the map in (c) is generated by $(T \cap K)_0$ and the elements $m_\alpha \in T$ attached to real roots α (cf. (12.6)).

In particular, the characters of the component group K/K_0 are in one-to-one correspondence with the characters of $(T\cap K)/(T\cap K)_0$ trivial on all the elements m_α (for α real).

A torus with the properties in the lemma will be called a *maximally θ-split torus for G*. We postpone a discussion of the proof for a moment, and continue with the proof of Theorem 12.9. So choose a maximally θ_y-split torus

$$^dT \subset L(\lambda, y\cdot\lambda). \qquad (12.11)(a)$$

Since $L(\lambda, y\cdot\lambda)$ is a Levi subgroup of $^\vee G$, it follows that ^dT is a maximal torus in $^\vee G$. Furthermore the Lie algebra $^d\mathfrak{t}$ must contain the central elements λ and $y\cdot\lambda$ of $\mathfrak{l}(\lambda, y\cdot\lambda)$, so

$$\phi|_{\mathbb{C}^\times} : \mathbb{C}^\times \to {}^dT.$$

Define

$$^dT^\Gamma = \text{group generated by } y \text{ and } {}^dT, \qquad (12.11)(b)$$

so that

$$\phi : W_\mathbb{R} \to {}^dT^\Gamma. \qquad (12.11)(c)$$

Since θ_y preserves ^dT, the element y normalizes ^dT. Since y^2 belongs to ^dT (cf. Proposition 5.6(b)), it follows that ^dT has index two in $^dT^\Gamma$. It follows easily that $^dT^\Gamma$ is a weak E-group, and so a Cartan subgroup of $^\vee G^\Gamma$ (Definition 12.4).

A real or imaginary root is λ-singular if and only if it is $y\cdot\lambda$-singular. It follows that the real or imaginary ϕ-singular roots are exactly the real or imaginary roots of ^dT in $L(\lambda, y\cdot\lambda)$. By Lemma 12.10(b), this means that

every ϕ-singular imaginary root of ^dT in $^\vee G$ is ϕ-compact.
$$(12.11)(d)$$

As in Corollary 5.9, write $K(y)$ for the centralizer of y in $^\vee G$. The component group of which the datum τ is a character is

$$A_\phi^{loc,alg} = [K(y) \cap L(\lambda, y\cdot\lambda)]^{alg} / \big([K(y) \cap L(\lambda, y\cdot\lambda)]^{alg}\big)_0$$

12. Langlands parameters and Cartan subgroups   145

(Corollary 5.9(c)). Lemma 12.10(c) (applied to the group $L(\lambda, y \cdot \lambda)^{alg, {}^\vee G}$) provides a surjective map

$$A({}^d T^\Gamma)^{alg, {}^\vee G} \to A_\phi^{loc, alg}. \qquad (12.11)(e)$$

We use this map to pull τ back to a character τ_1 of $A({}^d T^\Gamma)^{alg, {}^\vee G}$. By Lemma 12.10(d),

$$\tau_1(m_\alpha^{alg}) = 1 \qquad (\alpha \text{ real and } \phi\text{-singular}). \qquad (12.11)(f)$$

By (12.11)(d) and (12.11)(f), the pair (ϕ, τ_1) is a final complete Langlands parameter for ${}^d T^\Gamma$ with respect to ${}^\vee G^\Gamma$. The only choice involved is that of the maximal torus T, and Lemma 12.10 guarantees that T is unique up to conjugation by the centralizer of ϕ in ${}^\vee G$. It follows easily that the map from parameters for ${}^\vee G^\Gamma$ to parameters for Cartan subgroups is well-defined on equivalence classes.

This argument can be reversed without difficulty, to recover a unique complete parameter for ${}^\vee G^\Gamma$ from a complete final parameter for a Cartan subgroup. It follows that the correspondence is bijective. Q.E.D.

Sketch of proof of Lemma 12.10. That conditions (a) and (b) are equivalent is fairly easy; the main point is the uniqueness of T up to K_0-conjugacy. This is proved in [31], Theorem 1 (or [60], p.323). Part (c) is [31], Proposition 1 (or [60], Proposition 7).

For (d), consider the Levi subgroup

$$L = \text{centralizer of } (T \cap K)_0 \text{ in } G.$$

As the centralizer of a torus, L is connected. The roots of T in L are precisely the real roots of T in G. Since $(T \cap K)_0$ is also a torus in K_0, it follows that $L \cap K_0$ is also connected. It follows that

$$T \cap K_0 = T \cap (L \cap K_0) = T \cap (L \cap K)_0.$$

It therefore suffices to prove (d) for the subgroup L instead of for all of G; that is, under the assumption that all roots are real. It is enough to prove (d) for a finite cover of G, and then for each factor of a θ-stable direct product decomposition. After such reductions, we may assume that G is simple and simply connected, and that θ acts by -1 on $\mathfrak{t}$. Then T^θ consists precisely of the elements of order 2 in T. Since G is simply connected, $X_*(T)$ is generated by the coroot lattice. Now it follows from

Lemma 9.9 and the definitions that the elements m_α generate the full group of elements of order 2 in T. Consequently

$$T \cap K_0 \subset T^\theta = \text{ group generated by the } m_\alpha.$$

The other containment (that is, that $m_\alpha \in K_0$) follows from a standard calculation in $SL(2)$. Q.E.D.

13. Pairings between Cartan subgroups and the proof of Theorem 10.4

Theorem 12.3 describes representations in terms of Cartan subgroups of G^Γ, and Theorem 12.9 describes Langlands parameters in terms of Cartan subgroups of $^\vee G^\Gamma$. To complete the proof of Theorem 10.4, we need to relate Cartan subgroups of G^Γ and $^\vee G^\Gamma$. We begin by doing this for maximal tori. The following lemma more or less restates the definition of dual group; we leave its proof to the reader.

Lemma 13.1. *Suppose G is a complex connected reductive algebraic group, and $^\vee G$ is a dual group for G (Definition 4.2). Fix maximal tori $T \subset G$ and $^d T \subset {^\vee G}$. Write $^\vee T$ for a dual torus to T (cf. (9.3)). Suppose $\Delta \subset X^*(T)$ is a set of simple roots for T in G, and $^d \Delta \subset X^*(^d T)$ a set of simple roots for $^d T$ in $^\vee G$. Write $\Delta^\vee \subset X_*(T)$ and $^d\Delta^\vee \subset X_*(^d T)$ for the corresponding sets of simple coroots.*

a) There is a natural isomorphism

$$\zeta(\Delta, {^d\Delta}) : {^\vee T} \to {^d T}.$$

b) The induced map on one-parameter subgroup lattices

$$\zeta_*(\Delta, {^d\Delta}) : (X_*(^\vee T) = X^*(T)) \to X_*(^d T)$$

carries Δ onto $^d\Delta^\vee$.

c) The induced map (in the other direction) on character lattices

$$\zeta^*(\Delta, {^d\Delta}) : X^*(^d T) \to (X^*(^\vee T) = X_*(T))$$

carries $^d\Delta$ onto $\Delta^\vee$.

d) If sets of simple roots are not fixed, we obtain a finite family of natural isomorphisms

$$\zeta : {^\vee T} \to {^d T}.$$

Any two of these isomorphisms differ by composition with the action of a unique element of the Weyl group of $^d T$ in $^\vee G$.

The characteristic property of the isomorphism $\zeta(\Delta, {}^d\Delta)$ is that the induced maps described in (b) and (c) implement the natural isomorphism from the dual

$$ {}^\vee\Psi_0(G) = (X_*(T), \Delta^\vee, X^*(T), \Delta) $$

of the based root datum for G, onto the based root datum

$$ \Psi_0({}^\vee G) = (X^*({}^d T), {}^d\Delta, X_*({}^d T), {}^d\Delta^\vee). $$

With the rôles of G and ${}^\vee G$ reversed, (a) provides also an isomorphism

$$ \zeta({}^d\Delta, \Delta) : {}^{\vee d}T \to T. $$

It coincides with the one obtained from $\zeta(\Delta, {}^d\Delta)$ by applying the dual torus functor (cf. (9.3)). Using that functor, we can also think of the various natural isomorphisms of (d) as differing by elements of the Weyl group of T in G.

Definition 13.2 (cf. [1], Definition 9.11). Suppose G^Γ is a weak extended group, and ${}^\vee G^\Gamma$ is a weak E-group for G^Γ (Definition 4.3). Fix Cartan subgroups $T^\Gamma \subset G^\Gamma$ and ${}^d T^\Gamma \subset {}^\vee G^\Gamma$ (Definitions 12.1 and 12.4). Write σ for the real form of T defined by any element of $T^\Gamma - T$, and θ for the involutive automorphism of ${}^d T$ defined by any element of ${}^d T^\Gamma - {}^d T$. By Proposition 2.12, σ defines an involutive automorphism

$$ a_T \in \mathrm{Aut}(\Psi_0(T)) = \mathrm{Aut}(X^*(T), X_*(T)). $$

That is, a_T gives an automorphism of each of the lattices $X^*(T)$ and $X_*(T)$, respecting the pairing into $\mathbb{Z}$. Similarly (Proposition 2.11 or Proposition 4.4) the involution θ defines an involutive automorphism

$$ a_{{}^d T} \in \mathrm{Aut}({}^\vee\Psi_0({}^d T)) = \mathrm{Aut}(X_*({}^d T), X^*({}^d T)). $$

A *weak pairing* between T^Γ and ${}^d T^\Gamma$ is an identification of ${}^d T^\Gamma$ with a weak E-group for T^Γ, subject to one additional condition that we now describe. According to Definition 4.3, the identification in question amounts to an isomorphism

$$ \zeta : {}^\vee T \to {}^d T \qquad (13.2)(a) $$

13. Pairings between Cartan subgroups

of the dual torus ${}^\vee T$ for T (cf. (9.3)) with ${}^d T$. Such an identification is just an isomorphism

$$\zeta : (X^*(T), X_*(T)) \to (X_*({}^d T), X^*({}^d T)) \qquad (13.2)(b)$$

In order to make ${}^d T^\Gamma$ an E-group for T^Γ, this isomorphism must satisfy

$$\zeta \text{ carries } a_T \text{ to } a_{{}^d T}. \qquad (13.2)(c)$$

The additional requirement that we impose for a weak pairing is

$$\zeta \text{ is one of the natural isomorphisms of Lemma 13.1(d).} \qquad (13.2)(d)$$

This condition brings the groups G^Γ and ${}^\vee G^\Gamma$ into the definition.

Lemma 13.3. *Suppose we are in the setting of Definition 13.2; use the notation there.*

a) *The isomorphism ζ induces an identification of Weyl groups*

$$\zeta : W(G, T) \to W({}^\vee G, {}^d T).$$

b) *The isomorphism of (a) carries the action of the involution σ on $W(G, T)$ to the action of θ on $W({}^\vee G, {}^d T)$. In particular, we have an isomorphism*

$$\zeta : W(G, T)^\sigma \to W({}^\vee G, {}^d T)^\theta.$$

c) *Suppose ζ' is another isomorphism from ${}^\vee T$ to ${}^d T$ (Lemma 13.1(d)). Then ζ' is a weak pairing between the Cartan subgroups if and only if $\zeta' = w \circ \zeta$ for some $w \in W({}^\vee G, {}^d T)^\theta$.*

This is obvious.

Proposition 13.4 ([1], Lemma 9.16). *Suppose G^Γ is a weak extended group, and ${}^\vee G^\Gamma$ is a weak E-group for G^Γ (Definition 4.3).*

a) *Suppose T^Γ is a Cartan subgroup of G^Γ (Definition 12.1). Then there is a Cartan subgroup ${}^d T^\Gamma$ of ${}^\vee G^\Gamma$ and a weak pairing between T^Γ and ${}^d T^\Gamma$ (Definition 13.2). The Cartan subgroup ${}^d T^\Gamma$ is unique up to conjugation by ${}^\vee G$.*

b) *Suppose ${}^d T^\Gamma$ is a Cartan subgroup of ${}^\vee G^\Gamma$ (Definition 12.4). Then there is a Cartan subgroup T^Γ of G^Γ and a weak pairing between T^Γ and ${}^d T^\Gamma$ (Definition 13.2). The Cartan subgroup T^Γ is unique up to conjugation by G.*

Proof. For (a), let ^dT be any maximal torus in $^\vee G$, and y an element of $^\vee G^\Gamma - {}^\vee G$ normalizing ^dT. (To find such an element y, start with any y_0 in $^\vee G^\Gamma - {}^\vee G$, and modify it by an element of $^\vee G$ conjugating $\mathrm{Ad}(y_0)(^dT)$ back to ^dT.) Then conjugation by y defines an automorphism θ_y of ^dT, and so an automorphism a_y of $^\vee\Psi_0(^dT)$.

Fix one of the natural isomorphisms ζ from $^\vee T$ to ^dT (Lemma 13.1). This isomorphism carries the automorphism a_T to an involutive automorphism a_{dT} of $^\vee\Psi_0(^dT)$. Since $^\vee G^\Gamma$ is an E-group for G^Γ, it follows from Definition 4.3 that a_y and a_{dT} must differ by the action of an element $w \in W(^\vee G, {}^dT)$:

$$a_{dT} = w \circ a_y.$$

Now let $n \in {}^\vee G$ be any representative of w, and set $y' = ny$. Then $y' \in {}^\vee G^\Gamma - {}^\vee G$ still normalizes ^dT, and the conjugation action of y' defines the involutive automorphism θ of ^dT corresponding to a_{dT}. The group $^dT^\Gamma$ generated by y' and ^dT is therefore a Cartan subgroup of $^\vee G^\Gamma$ paired with T^Γ.

For the uniqueness, suppose $(^dT_1^\Gamma)'$ is another Cartan subgroup paired by ζ_1 with T^Γ. After conjugating by $^\vee G$, we may assume that $^dT_1 = {}^dT$ and that $\zeta_1 = \zeta$. Then the automorphism of ^dT defined by any element y of $^dT^\Gamma - {}^dT$ coincides with the automorphism defined by any element y' of $(^dT^\Gamma)' - {}^dT$. It follows that $y^dT = y'^dT$, so that $^dT^\Gamma = (^dT^\Gamma)'$ as we wished to show.

The proof of (b) is formally identical, and we omit it. Q.E.D.

Our next goal is to reduce the ambiguity in the notion of weak pairing described in Lemma 13.3(c).

Definition 13.5. Suppose $(G^\Gamma, \mathcal{W})$ is an extended group for G (Definition 1.12). A *based Cartan subgroup* of $(G^\Gamma, \mathcal{W})$ is a quadruple $\mathcal{T}^\Gamma = (T^\Gamma, \mathcal{W}(T^\Gamma), R_{i\mathbb{R}}^+, R_\mathbb{R}^+)$, subject to the following conditions.
(a) T^Γ is a Cartan subgroup of G^Γ (Definition 12.1).
(b) $\mathcal{W}(T^\Gamma)$ is an extended group structure on T^Γ (Definition 1.12). That is, $\mathcal{W}(T^\Gamma)$ is a T-conjugacy class of elements $\delta \in T^\Gamma - T$, with the property that δ^2 has finite order.
(c) $\mathcal{W}(T^\Gamma)$ is a subset of $\mathcal{W}$, in the sense that for every $\delta \in \mathcal{W}(T^\Gamma)$ there are N and χ so that $(\delta, N, \chi) \in \mathcal{W}$.
(d) $R_{i\mathbb{R}}^+$ is a set of positive imaginary roots of T in G, and $R_\mathbb{R}^+$ is a set of positive real roots.
(e) The extended group structure and the positive imaginary roots are compatible in the following sense. Fix $\delta \in \mathcal{W}(T^\Gamma)$, and N and χ so that $(\delta, N, \chi) \in \mathcal{W}$. Write $G(\mathbb{R}) = G(\mathbb{R}, \delta)$ for the corresponding

13. Pairings between Cartan subgroups

(quasisplit) real form of G. Suppose Λ^{can} is an irreducible canonical projective representation of $T(\mathbb{R})^{can,T}$ of type $zz(\rho)$, and that

$$\Lambda = (\Lambda^{can}, R_{i\mathbb{R}}^+, R_{\mathbb{R}}^+))$$

is a $G(\mathbb{R})$-limit character of type z (Definition 11.2). Then what we require is that the corresponding standard limit representation $M(\Lambda)$ admit a Whittaker model of type χ (Definition 3.1).

Two based Cartan subgroups are called *equivalent* if they are conjugate by G.

Proposition 13.6. *Suppose T^Γ is a Cartan subgroup of the extended group $(G^\Gamma, \mathcal{W})$, and $(R_{i\mathbb{R}}^+, R_{\mathbb{R}}^+)$ are systems of imaginary and real positive roots. Then there is a unique extended group structure $\mathcal{W}(T^\Gamma)$ on T^Γ making the quadruple $(T^\Gamma, \mathcal{W}(T^\Gamma), R_{i\mathbb{R}}^+, R_{\mathbb{R}}^+)$ a based Cartan subgroup.*

The proof is fairly long, so we postpone it to the next chapter.

There is an analogue of Definition 13.5 for E-groups. The key to it is the notion of "special" systems of positive imaginary roots formulated in section 6 of [1].

Definition 13.7. Suppose $({}^\vee G^\Gamma, \mathcal{S})$ is an E-group for G (Definition 4.14). A *based Cartan subgroup* of $({}^\vee G^\Gamma, \mathcal{S})$ is a quadruple ${}^d T^\Gamma = ({}^d T^\Gamma, \mathcal{S}({}^d T^\Gamma), {}^d R_{i\mathbb{R}}^+, {}^d R_{\mathbb{R}}^+)$, subject to the following conditions.

(a) ${}^d T^\Gamma$ is a Cartan subgroup of ${}^\vee G^\Gamma$ (Definition 12.4).
(b) $\mathcal{S}({}^d T^\Gamma)$ is an E-group structure on ${}^d T^\Gamma$ (Definition 4.14). That is, $\mathcal{S}({}^d T^\Gamma)$ is a ${}^d T$-conjugacy class of elements ${}^\vee \delta \in T^\Gamma - T$, with the property that ${}^\vee \delta^2$ has finite order.
(c) $\mathcal{S}({}^d T^\Gamma)$ is a subset of $\mathcal{S}$, in the sense that for every ${}^\vee \delta \in \mathcal{S}({}^d T^\Gamma)$ there is a Borel subgroup ${}^d B$ so that $({}^\vee \delta, {}^d B) \in \mathcal{S}$.
(d) ${}^d R_{i\mathbb{R}}^+$ is a set of positive imaginary roots of ${}^d T$ in ${}^\vee G$, and ${}^d R_{\mathbb{R}}^+$ is a set of positive real roots (cf. (12.5)).
(e) The E-group structure and the positive imaginary roots are compatible in the following sense. Fix ${}^\vee \delta \in \mathcal{S}({}^d T^\Gamma)$, and define ${}^d \mathcal{B}$ to be the set of Borel subgroups ${}^d B$ so that $({}^\vee \delta, {}^d B) \in \mathcal{S}$). (The set ${}^d \mathcal{B}$ is a single orbit under the action of the centralizer K of ${}^\vee \delta$ in ${}^\vee G$.) Then what we require is that ${}^d R_{i\mathbb{R}}^+$ be special with respect to ${}^d \mathcal{B}$ ([1], Definition 6.29).

Two based Cartan subgroups are called *equivalent* if they are conjugate by ${}^\vee G$.

It is possible to give a more geometric definition of "special" than the one in [1], and in fact this is crucial for the proof given there of some technical results we will use.

Here is the analogue of Proposition 13.6.

Proposition 13.8. *Suppose $^dT^\Gamma$ is a Cartan subgroup of the E-group $(^\vee G^\Gamma, \mathcal{S})$, and $(^dR_{i\mathbb{R}}^+, {}^dR_\mathbb{R}^+)$ are systems of imaginary and real positive roots. Then there is a unique E-group structure $\mathcal{S}(^dT^\Gamma)$ on $^dT^\Gamma$ making the quadruple $(^dT^\Gamma, \mathcal{S}(^dT^\Gamma), R_{i\mathbb{R}}^+, R_\mathbb{R}^+)$ a based Cartan subgroup.*

Again we postpone the proof until the next chapter.

Definition 13.9. Suppose $(G^\Gamma, \mathcal{W})$ is an extended group for G (Definition 1.12), and $(^\vee G^\Gamma, \mathcal{S})$ is a corresponding E-group (Definition 4.14). Fix based Cartan subgroups T^Γ for G^Γ (Definition 13.5) and $^dT^\Gamma$ for $^\vee G^\Gamma$ (Definition 13.7). A *pairing between* T^Γ *and* $^dT^\Gamma$ is a weak pairing

$$\zeta: {}^\vee T \to {}^dT \qquad (13.9)(a)$$

(Definition 13.2), subject to the additional conditions

ζ carries the positive imaginary roots $R_{i\mathbb{R}}^+$ onto the positive real coroots $^dR_\mathbb{R}^{\vee,+}$ $\qquad (13.9)(b)$

(cf. Lemma 13.1(b)), and

ζ carries the positive real roots $R_\mathbb{R}^+$ onto the positive imaginary coroots $^dR_{i\mathbb{R}}^{\vee,+}$. $\qquad (13.9)(c)$

In particular, this identifies $(^dT^\Gamma, \mathcal{S}(^dT^\Gamma))$ as an E-group of $(T^\Gamma, \mathcal{W}(T^\Gamma))$.

Proposition 13.10. *Suppose $(G^\Gamma, \mathcal{W})$ is an extended group for G (Definition 1.12), and $(^\vee G^\Gamma, \mathcal{S})$ is a corresponding E-group (Definition 4.14).*

a) *Suppose T^Γ is a based Cartan subgroup for G^Γ (Definition 13.5). Then there is a based Cartan subgroup $^dT^\Gamma$ for $^\vee G^\Gamma$ and a pairing between them (Definition 13.9). $^dT^\Gamma$ is uniquely determined up to conjugation by $^\vee G^\Gamma$.*

b) *Suppose $^dT^\Gamma$ is a based Cartan subgroup of $^\vee G^\Gamma$ (Definition 13.7). Then there is a based Cartan subgroup T^Γ for G^Γ and a pairing between them. T^Γ is unique up to conjugation by G.*

c) *Suppose ζ is a pairing between based Cartan subgroups as above. If ζ' is another isomorphism from $^\vee T$ to ^dT, then ζ' is a pairing if and* only if $\zeta' = w \circ \zeta$ for some $w \in W({}^\vee G, {}^d T)^\theta$ preserving ${}^d R_{i\mathbb{R}}^+$ and ${}^d R_{\mathbb{R}}^+$.

d) Suppose w is as in (c), and use the isomorphism of Lemma 13.3(b) to identify w with an element of $W(G,T)^\sigma$. If σ' is any real form of G extending σ on T, then w has a representative in $G(\mathbb{R}, \sigma')$.

Proof. For (a), use Proposition 13.4(a) to find a Cartan subgroup ${}^d T^\Gamma$ of ${}^\vee G^\Gamma$ and a weak pairing ζ between it and T^Γ. By (13.2)(c), ζ carries the positive imaginary roots $R_{i\mathbb{R}}^+$ onto some set ${}^d R_{\mathbb{R}}^{\vee,+}$ of positive real coroots for ${}^d T$; and similarly for real roots. Proposition 13.8 then guarantees that we can use these sets of positive roots to construct a based Cartan subgroup ${}^d \mathcal{T}^\Gamma$. The uniqueness of its equivalence class follows from the uniqueness in Proposition 13.4(a), Lemma 13.3(c), and the uniqueness in Proposition 13.8. Part (b) is proved in exactly the same way. Part (c) is clear from Lemma 13.3(c) and Definition 13.9. Part (d) follows from the description of real Weyl groups in [56], Proposition 4.16. Q.E.D.

We are near the end of the maze leading to Theorem 10.4 now. Before we pull all the pieces together, we need one more definition.

Definition 13.11. Suppose $\mathcal{T}^\Gamma$ is a based Cartan subgroup of an extended group $(G^\Gamma, \mathcal{W})$ (Definition 3.5). A G-limit character of T^Γ (Definition 12.1) is said to be *compatible with* $\mathcal{T}^\Gamma$ if the corresponding systems of positive real and imaginary roots agree.

Suppose ${}^d \mathcal{T}^\Gamma$ is a based Cartan subgroup of the E-group $({}^\vee G^\Gamma, \mathcal{S})$ (Definition 13.7). A complete Langlands parameter for ${}^d T^\Gamma$ (Definition 12.4) is said to be *compatible with* ${}^d \mathcal{T}^\Gamma$ if

$$\langle \alpha, \lambda \rangle \geq 0 \qquad (\alpha \in {}^d R_{\mathbb{R}}^+). \tag{13.11}$$

Here λ is constructed from the Langlands parameter as in (5.8). We do *not* assume any relationship between λ and the positive imaginary roots.

Proposition 13.12. *Suppose $\mathcal{T}^\Gamma$ is a based Cartan subgroup for an extended group $(G^\Gamma, \mathcal{W})$, ${}^d \mathcal{T}^\Gamma$ is a based Cartan subgroup for an E-group $({}^\vee G^\Gamma, \mathcal{S})$, and ζ is a pairing between them (Definition 13.9). Write z for the second invariant of the E-group (Definition 4.14). Use ζ to identify $({}^d T^\Gamma, \mathcal{S}({}^d T^\Gamma))$ as an E-group for T^Γ (of second invariant $zz(\rho)$). Then the correspondence of Corollary 10.7 induces a bijection between the set of limit characters of T^Γ compatible with $\mathcal{T}^\Gamma$ and the set of complete Langlands parameters for ${}^d T^\Gamma$ compatible with ${}^d \mathcal{T}^\Gamma$ (Definition 13.11). This bijection identifies final limit characters with final limit parameters.*

A little more precisely, each Langlands parameter ϕ for ${}^dT^\Gamma$ gives rise to a canonical projective character $\Lambda^{can}(\phi)$ of $T(\mathbb{R})$ of type $zz(\rho)$ (Proposition 10.6). Each irreducible representation τ_1 of $A({}^dT^\Gamma)^{alg}$ gives rise to a T-conjugacy class in $T^\Gamma - T$, say with a representative $\delta(\tau_1)$ (Corollary 9.12). We have

a) *The element $\delta(\tau_1)$ defines a strong real form of G — that is, $\delta(\tau_1)^2$ belongs to $Z(G)$ — if and only if τ_1 factors through the quotient group $A({}^dT^\Gamma)^{alg,\vee G}$.*

b) *The positivity requirement (12.2)(f) on $\Lambda^{can}(\phi)$ is equivalent to the compatibility requirement (13.11).*

c) *A simple imaginary root α for T is noncompact with respect to $\delta(\tau_1)$ if and only the corresponding simple real root ${}^d\alpha$ for ${}^dT$ satisfies*

$$\tau_1(\overline{m_{{}^d\alpha}^{alg}}) = 1.$$

d) *A real root β for T satisfies the parity condition for $\Lambda^{can}(\phi)$ if and only if the corresponding imaginary root ${}^d\beta$ for ${}^dT$ is ϕ-noncompact (Definition 12.8).*

Proof. Because of the definitions, it suffices to prove the assertions (a)–(d). For (a), fix an element $\delta \in W(T^\Gamma)$. The element $\delta(\tau_1)$ is obtained by multiplying δ by an appropriate element $t = e(\tau/2)$, with $\tau \in \mathfrak{t}_\mathbb{Q}^{-a}$ (Proposition 9.10); here τ represents τ_1 in the isomorphism of Proposition 9.8(c). One can check easily that τ_1 factors through $A({}^dT^\Gamma)^{alg,\vee G}$ if and only if the roots of T in G take integer values on the element τ. (To see what the root lattice has to do with the problem, recall that $\pi_1({}^dT)$ may be identified with $X_*({}^dT) \simeq X^*(T)$. The inclusion of ${}^dT$ in ${}^\vee G$ induces a surjection $\pi_1({}^dT) \to \pi_1({}^\vee G)$, and the kernel of this map is the lattice of roots of T in G.) On the other hand, we compute

$$\delta(\tau_1)^2 = t\sigma(t)\delta^2 = e(\tau)\delta^2 = e(\tau)z(\rho)$$

(cf. proof of Proposition 9.10). So $\delta(\tau_1)$ is a strong real form if and only if $e(\tau) \in Z(G)$. Of course this condition is also equivalent to the roots of T in G taking integral values on τ, proving (a).

Part (b) is clear (see Proposition 10.6). For (c), we can use the notation of (a). Inspecting the definitions in Chapter 9 and at (12.6), we find that

$$\tau_1(\overline{m_{{}^d\alpha}^{alg}}) = \exp(2\pi i \alpha(\tau)) = (-1)^{\alpha(\tau/2)}.$$

13. Pairings between Cartan subgroups   155

So the condition on τ_1 is equivalent to

$$\alpha(\tau/2) \text{ is an even integer.}$$

On the other hand, every simple imaginary root α of T in G is noncompact with respect to δ (see the proof of Proposition 13.6 in Chapter 14). Consequently α is nocompact with respect to $\delta(\tau_1) = t\delta$ if and only if $\alpha(t) = 1$. Since $t = e(\tau/2)$, this is equivalent to $\alpha(\tau/2)$ being an even integer.

The proof of (d) is similar; since we have referred to [1] for the construction of $\Lambda(\phi)$, we omit the details. Q.E.D.

Proposition 13.13. *Suppose $(G^\Gamma, \mathcal{W})$ is an extended group for G, and $({}^\vee G^\Gamma, \mathcal{S})$ is an E-group for the corresponding inner class of real forms. Write z for the second invariant of the E-group (Definition 4.14). Then the various correspondences of Proposition 13.12 induce a bijection from the set of equivalence classes of final limit characters of type z of Cartan subgroups of G^Γ (Definition 12.1) and equivalence classes of final complete limit parameters of Cartan subgroups of ${}^\vee G^\Gamma$ (Definition 12.8).*

Proof. Suppose (ϕ, τ_1) is a complete limit parameter for ${}^d T^\Gamma$. To construct the corresponding limit character, we need to make ${}^d T^\Gamma$ a based Cartan subgroup; so we must choose certain sets of positive roots. Let ${}^d R_\mathbb{R}^+$ be a set of positive real roots for ${}^d T$ making $\lambda(\phi)$ dominant (cf. (12.5) and Definition 13.11); and let ${}^d R_{i\mathbb{R}}^+$ be an arbitrary set of positive imaginary roots. Let ${}^d T^\Gamma$ be the corresponding based Cartan subgroup (Proposition 13.8). Fix a based Cartan subgroup T^Γ for G^Γ and a pairing

$$\zeta: {}^\vee T \to {}^d T \qquad (13.14)$$

with ${}^d T^\Gamma$. Proposition 13.12 gives a final limit character $(\delta(\tau_1), \Lambda(\phi))$ of T^Γ. We must show that the equivalence class of this character is independent of the two positive root systems we chose.

Consider first a second system $({}^d R_\mathbb{R}^+)'$ of positive real roots making $\lambda(\phi)$ dominant. These two positive systems differ by a Weyl group element w that is a product of simple reflections in real roots vanishing on λ. Let n be a representative of w. Changing the sytem of positive real roots changes the based Cartan subgroup ${}^d T^\Gamma$ by conjugation by n, and replaces ζ by $w \circ \zeta$. (Evidently this implies that the final limit character $(\delta(\tau_1), \Lambda(\phi))$ is unchanged.) We claim that w has a representative n fixing (ϕ, τ_1). It is enough to check this for a simple reflection s_α, with α a real root vanishing on λ. Choose a root subgroup $\phi_\alpha: SL(2) \to {}^\vee G$ as usual (cf. Definition 11.10)) with the additional property that ϕ_α carries the inverse transpose involution on $SL(2)$ to the action of θ_y. We may take

$$n_\alpha = \phi_\alpha \begin{pmatrix} 0 & 1 \\ -1 & 0 \end{pmatrix}$$

as a representative of s_α. Evidently n_α belongs to $K(y) \cap L(\lambda, y \cdot \lambda)$ (cf. Corollary 5.9); that is, to the centralizer of ϕ. To compute its effect on τ_1, recall that τ_1 is a character of (a certain quotient of) $(^d T^\theta)^{alg,\vee G}$. If t is in this group, then $\alpha(t) = \pm 1$ since α is real. Consequently

$$s_\alpha(t) = t\alpha^\vee(\alpha(t^{-1})) = t\alpha^\vee(\pm 1) = t \text{ or } tm_\alpha^{alg}.$$

Since (ϕ, τ_1) is final, $\tau_1(m_\alpha^{alg}) = 1$; so we find that $s_\alpha(\tau_1) = \tau_1$, as we wished to show.

Next, consider a second choice $(^d R_{i\mathbb{R}}^+)'$ of positive imaginary roots. This differs from the original by an element w in the Weyl group of the imaginary roots. To get the corresponding based Cartan subgroup $(^d T^\Gamma)'$, we must also conjugate $\mathcal{S}(^d T^\Gamma)$ by a representative n of w (because of Definition 13.7(e)). This conjugation has the effect of multiplying $\mathcal{S}(^d T^\Gamma)$ by an element $t = n\theta(n^{-1})$; this element satisfies $\theta(t) = t^{-1}$. Back in G, we can get the new based Cartan subgroup just by replacing the set of positive real roots by the new set $(R_{\mathbb{R}}^+)'$ corresponding to $(^d R_{i\mathbb{R}}^+)'$. The pairing ζ is then unchanged. The canonical character $(\Lambda^{can})'$ attached to these new choices differs from Λ^{can} only because of the change in the E-group structure on $^d T^\Gamma$. The proof of Lemma 9.28 of [1] shows that the effect of this change is to twist Λ^{can} by the character $\tau(R_{\mathbb{R}}^+, (R_{\mathbb{R}}^+)')$ of (11.4). By Definition 11.6 and Definition 12.1, the equivalence class of the limit character is unchanged.

That this construction actually depends only on the equivalence class of (λ, τ_1) is obvious. That it is surjective is clear from Proposition 13.12. That the inverse correspondence is well-defined may be proved in exactly the same way. Q.E.D.

Theorem 10.4 is a consequence of Proposition 13.13, Theorem 12.3, and Theorem 12.9.

14. Proof of Propositions 13.6 and 13.8

Evidently part of what a based Cartan subgroup provides is some distinguished extensions of its real form to all of G. We begin by studying such extensions.

Lemma 14.1. *Suppose T is a maximal torus in the connected reductive algebraic group G, and σ is a real form of T permuting the roots of T in G. Suppose that σ_1 and σ_2 are two real forms of G extending σ, and that they determine the same sets of compact (and noncompact) imaginary roots. Then σ_1 and σ_2 are conjugate by an element of T.*

Proof. Choose a set R^+ of positive roots of T in G having the property that σ preserves the non-imaginary positive roots. (One way to do this is to order the roots using first their restrictions to the $+1$-eigenspace of σ on the rational span of the coroots.) In the argument below, we will usually write β for an imaginary root, γ for a non-imaginary root, and α for a general root. For any root in R^+, we have either

$$\beta \text{ is imaginary, and } \sigma\beta = -\beta \qquad (14.2)(a)$$

or

$$\gamma \text{ is not imaginary, and } \sigma\gamma \text{ is positive.} \qquad (14.2)(b)$$

For simple roots in R^+, one can say even more.

Lemma 14.3. *In the setting of Lemma 14.1, suppose R^+ is a set of positive roots for T in G such that σ preserves the non-imaginary positive roots. Then every simple root of R^+ falls into exactly one of the following three categories.*

a) β is imaginary, and $\sigma\beta = -\beta$.
b) γ is not imaginary, and

$$\sigma\gamma = \gamma + \sum_{\beta \text{ simple imaginary}} n_\beta(\gamma)\beta$$

with $n_\beta(\gamma)$ a non-negative integer.

c) γ is not imaginary, and there is a distinct non-imaginary root γ' so that

$$\sigma\gamma = \gamma' + \sum_{\beta \text{ simple imaginary}} n_\beta(\gamma)\beta$$

with $n_\beta(\gamma)$ a non-negative integer. In this case

$$\sigma\gamma' = \gamma + \sum_{\beta \text{ simple imaginary}} n_\beta(\gamma)\beta.$$

Proof. Suppose γ is a non-imaginary simple root. Then $\sigma\gamma$ is a positive root by (14.2)(b), so it can be expressed as a sum of simple roots with non-negative coefficients. In this expression there must appear at least one non-imaginary simple root $\gamma' = \gamma'(\gamma)$, for otherwise $\sigma\gamma$ (and therefore γ) would be imaginary. Thus

$$\sigma\gamma = \gamma' + \text{(other non-imaginary simple roots)} + \sum n_\beta\beta.$$

Now apply σ to this expression. On the right we get at least one non-imaginary simple root from the first term, more from the second if it is non-zero, and various imaginary simple roots. On the left we get γ, since σ is an involution. It follows that the second term is zero, and that $\sigma\gamma'$ involves γ. The remaining assertions of the lemma are now clear. Q.E.D.

We continue now with the proof of Lemma 14.1. Fix root vectors $\{X_\alpha\}$ for all the roots. As usual we may choose these so that

$$[X_\alpha, X_{-\alpha}] = H_\alpha \qquad (14.4)(a)$$

(the usual coroot), and

$$[X_\alpha, X_{\alpha'}] = q(\alpha, \alpha') X_{\alpha+\alpha'} \qquad (14.4)(b)$$

(with $q(\alpha, \alpha')$ rational) whenever $\alpha + \alpha'$ is a root. Define complex constants $c_i(\alpha)$ by

$$\sigma_i(X_\alpha) = c_i(\alpha) X_{\sigma\alpha}. \qquad (14.4)(c)$$

Since σ_1 and σ_2 agree on T, they differ by the adjoint action of an element s of T:

$$c_2(\alpha)/c_1(\alpha) = \alpha(s). \qquad (14.4)(d)$$

14. Proof of Propositions 13.6 and 13.8

In particular,

c_2/c_1 extends to a multiplicative character of the root lattice. \hfill (14.4)(e)

Replacing σ_1 by $t\sigma_1 t^{-1}$ replaces $c_1(\alpha)$ by

$$[(\sigma\alpha)(t)/\overline{\alpha(t)}]c_1(\alpha). \qquad (14.5)(a)$$

It follows that t conjugates σ_1 to σ_2 if and only if

$$(\sigma\alpha)(t)/\overline{\alpha(t)} = c_2(\alpha)/c_1(\alpha) \qquad (14.5)(b)$$

for every root α. Because of (14.4)(e), it suffices to verify (14.5)(b) for any set of roots α generating the root lattice. What we must show is that our hypotheses on σ_i provide enough control on c_i to guarantee the existence of an element t satisfying (14.5)(b). We choose a generating set S of roots as follows: S is the union of

$S_1 =$ simple imaginary roots;
$S_2 =$ simple roots as in Lemma 14.3(b);
$S_3 =$ a simple root γ from each pair (γ, γ') in Lemma 14.3(c); and
$S_4 =$ the roots $\sigma\gamma$, with $\gamma \in S_3$.
\hfill (14.6)

Lemma 14.3 shows that S is actually a basis of the root lattice. We may therefore choose t so that $\alpha(t)$ has any value we specify for $\alpha \in S$; we must show that this can be done so that (14.5)(b) holds for $\alpha \in S$.

Applying the antiholomorphic involution σ_i to (14.4)(c), we find that

$$\overline{c_i(\alpha)}c_i(\sigma\alpha) = 1. \qquad (14.7)(a)$$

Applying σ_i to (14.4)(b) gives

$$c_i(\alpha)c_i(\alpha') = q(\alpha, \alpha')c_i(\alpha + \alpha') \qquad (14.7)(b)$$

whenever $\alpha + \alpha'$ is a root.

Suppose now that β is an imaginary root. Then $\sigma H_\beta = -H_\beta$. Applying σ_i to (14.4)(a) gives

$$c_i(\beta)c_i(-\beta) = 1. \qquad (14.8)(a)$$

On the other hand, (14.7)(a) gives

$$\overline{c_i(\beta)}c_i(-\beta) = 1. \qquad (14.8)(b)$$

Consequently

$$c_i(\beta) \text{ is real, and } c_i(-\beta) = c_i(\beta)^{-1}. \qquad (14.8)(c)$$

Calculation in $SL(2)$ shows that

$$c_i(\beta) \text{ is positive if and only if } \beta \text{ is noncompact for } \sigma_i. \qquad (14.8)(d)$$

The hypothesis of Lemma 14.1 therefore guarantees that $c_1(\beta)/c_2(\beta)$ is positive for every imaginary root β. We therefore require of our element t that

$$\beta(t) \text{ is a square root of } c_1(\beta)/c_2(\beta) \qquad (14.8)(e)$$

for every β in S_1. Condition (14.5)(b) follows for such β, and then for all imaginary roots by (14.4)(e).

Suppose next that γ belongs to S_2. The requirement (14.5)(b) on t may be written as

$$(c_2/c_1)(\gamma) = (\gamma(t)/\overline{\gamma(t)})(\sigma\gamma - \gamma)(t). \qquad (14.9)(a)$$

This is the same as

$$\gamma(t)/\overline{\gamma(t)} = (c_2/c_1)(\gamma)(\gamma - \sigma\gamma)(t). \qquad (14.9)(b)$$

We can choose $\gamma(t)$ so that this equation is satisfied if and only if the right side has absolute value 1. To see that this is the case, divide (14.7)(a) for σ_2 by the same equation for σ_1, to obtain

$$\begin{aligned} 1 &= \overline{(c_2/c_1)(\gamma)}(c_2/c_1)(\sigma\gamma) \\ &= \overline{(c_2/c_1)(\gamma)}(c_2/c_1)(\gamma)(c_1/c_2)(\gamma - \sigma\gamma) \\ &= |(c_2/c_1)(\gamma)|^2(c_1/c_2)(\gamma - \sigma\gamma). \end{aligned} \qquad (14.9)(c)$$

Now Lemma 14.3 shows that $\gamma - \sigma\gamma$ is a sum of imaginary roots. It therefore follows from (14.8) that the last factor on the right here is the positive real number $(\gamma - \sigma\gamma)(t)^2$. Consequently

$$1 = |(c_2/c_1)(\gamma)|^2(\gamma - \sigma\gamma)(t)^2. \qquad (14.9)(d)$$

This guarantees the existence of a solution $\gamma(t)$ to (14.9)(b).

Finally, suppose $\gamma \in S_3$. We require of t that

$$\gamma(t) = 1, \qquad \sigma\gamma(t) = (c_2/c_1)(\gamma). \qquad (14.10)$$

14. Proof of Propositions 13.6 and 13.8

Then (14.5)(b) is automatically satisfied for $\alpha = \gamma$. To check it for $\alpha = \sigma\gamma$, just apply (14.7)(a). This completes the construction of t satisfying (14.5)(b), and so the proof of Lemma 14.1. Q.E.D.

Proof of Proposition 13.6. The real form σ_T of T (defined by conjugation by any element y of $T^\Gamma - T$) is an antiholomorphic involution preserving the roots. It may therefore be extended to a quasisplit real form σ_G of G, with the property that every simple root of T in $R_{i\mathbb{R}}^+$ is noncompact. (A proof of this fact may be found in [56], Lemma 10.9.) Clearly σ_G belongs to the inner class defined by G^Γ, so by Proposition 2.12 there is a triple $(\delta_1, N_1, \chi_1) \in \mathcal{W}$ with $\mathrm{Ad}(\delta_1) = \sigma_G$. In particular, δ_1 and y both act by σ_T on T, so $\delta_1 \in yT = T^\Gamma - T$. If we write $\mathcal{W}_1(T^\Gamma)$ for the T-conjugacy class of δ_1, then conditions (a) – (d) of Definition 13.5 are satisfied. The interesting and subtle point is arranging (e). This is essentially in [30]; we sketch the argument.

Fix (δ_1, N, χ_1) as above, and write $G(\mathbb{R}, \delta_1)$ for the corresponding real form. Define $A_\mathbb{R} \subset T(\mathbb{R})$ as in (11.3) (the identity component of the split part of $T(\mathbb{R})$), and put

$$M = \text{centralizer of } A_\mathbb{R} \text{ in } G.$$

Fix a σ_G-stable parabolic subgroup $P = MU$ of G with Levi subgroup M. After replacing (N, χ_1) by a conjugate under $G(\mathbb{R}, \delta_1)$, we may assume that

$$PN \text{ is open in } G.$$

It follows that $N_M = N \cap M$ is a maximal unipotent subgroup of M normalized by δ_1, and that $\chi_{M,1} = \chi_1|_{N_M(\mathbb{R})}$ is a non-degenerate unitary character (Definition 1.12). We apply to this situation the following lemma.

Lemma 14.11. *Suppose σ is a quasisplit real form of a complex connected reductive algebraic group G, and $P = MU$ is a σ-stable Levi decomposition of a parabolic subgroup of G. Suppose (N, χ) is the data for a Whittaker model for G (Definition 3.1). Suppose also that PN is open in G, so that if we write $N_M = N \cap M$, $\chi_M = \chi|_{N_M(\mathbb{R})}$, then (N_M, χ_M) is the data for a Whittaker model for M.*

Suppose π_M is a finite length admissible Hilbert space representation of $M(\mathbb{R})$, and $\pi = \mathrm{Ind}_{P(\mathbb{R})}^{G(\mathbb{R})} \pi_M$. Then π admits a Whittaker model of type χ if and only if π_M admits a Whittaker model of type χ_M.

Proof. The "only if" assertion is a special case of Theorem 1 of [21]. The "if" part is a formal consequence of the "only if," Harish-Chandra's

subquotient theorem, and the fact that a principal series representation has exactly one Whittaker model of a given type ([30], Theorem 6.6.2).
Q.E.D.

We continue with the argument for Proposition 13.6. Now the standard limit representations in Definition 13.5(e) may be constructed by induction from limits of discrete series on M (see the discussion at (11.2)). By Lemma 14.11, they will admit Whittaker models of type χ_1 if and only if these limits of discrete series admit Whittaker models of type $\chi_{M,1}$. Because all of the limits of discrete series in question correspond to a single system of positive imaginary roots, it is possible to pass from any one to any other by tensoring with finite-dimensional representations of $M(\mathbb{R})$ (more precisely, of its canonical covering). This does not affect the existence of Whittaker models (cf. [30], proof of Theorem 6.6.2).

We may therefore confine our attention to a single limit of discrete series representation π_M of $M(\mathbb{R})$. By the construction of σ_G, every simple imaginary root of T in M is noncompact. It follows that the annihilator of π_M in the enveloping algebra of $\mathfrak{m}$ is a minimal primitive ideal ([52], Theorem 6.2). By Theorem 6.7.2 of [30], there is a non-degenerate unitary character χ_M of $N_M(\mathbb{R})$ so that π_M admits a Whittaker model of type χ_M. It is clear from Definition 3.1 that there is a non-degenerate unitary character χ of $N(\mathbb{R})$ restricting to χ_M on M. By Lemma 14.11, our standard limit representations all admit Whittaker models of type χ. By Lemma 3.2, there is a $t \in G$ normalizing N, so that $\mathrm{Ad}(t)$ is defined over $\mathbb{R}$, and $t \cdot \chi_1 = \chi$. Consider now the triple

$$(t\delta_1 t^{-1}, N, t \cdot \chi_1) = (\delta, N, \chi).$$

Since $\mathcal{W}$ is a G-orbit, this triple belongs to $\mathcal{W}$. Since $\mathrm{Ad}(t)$ is defined over $\mathbb{R}$, conjugation by δ defines the same real form σ_G as δ_1. Define

$$\mathcal{W}(T^\Gamma) = T\text{-conjugacy class of } \delta.$$

This satisfies the requirements of Definition 13.5.

For the uniqueness, suppose $\mathcal{W}'(T^\Gamma)$ is another extended group structure on T^Γ satisfying the requirements of Definition 13.5. Fix a triple $(\delta', N', \chi') \in \mathcal{W}$ so that δ' belongs to $\mathcal{W}'(T^\Gamma)$. The argument in the first half of the proof may be reversed to deduce from the assumed existence of Whittaker models (Definition 13.5)(e)) that every simple root of $R_{i\mathbb{R}}^+$ must be noncompact for the real form σ_G' defined by conjugation by δ'. By the construction of σ_G, it follows that σ_G and σ_G' define exactly the same sets of compact and noncompact imaginary roots. By Lemma 14.1, they are conjugate by an element of T. Since $\mathcal{W}'(T^\Gamma)$ is a

14. Proof of Propositions 13.6 and 13.8

T-conjugacy class, we may therefore assume that δ' is chosen so that

$$\text{the conjugation action } \sigma'_G \text{ of } \delta' \text{ is equal to } \sigma_G. \qquad (14.12)(a)$$

This means that the real form $G(\mathbb{R})$ defined by δ' coincides with the one defined by δ. The groups N and N' are therefore conjugate by $G(\mathbb{R})$; after applying such a conjugation to N' (which does not change δ') we may assume that $N = N'$. Fix a maximal torus $T_s \subset M$ normalizing N and defined over $\mathbb{R}$; we write

$$B_s = T_s N, \qquad B_s(\mathbb{R}) = T_s(\mathbb{R}) N_s(\mathbb{R}). \qquad (14.12)(b)$$

Of course T_s contains $Z(G)$, so the set of elements of B_s for which the conjugation action on G is defined over $\mathbb{R}$ may be decomposed as

$$(B_s/Z(G))(\mathbb{R}) = (T_s/Z(G))(\mathbb{R}) N_s(\mathbb{R}). \qquad (14.12)(c)$$

Recall now that (δ, N, χ) and (δ', N, χ') both belong to $\mathcal{W}$, and so are conjugate by an element t of G. This element normalizes N, and so belongs to B_s. By (14.12), the element may be chosen to belong to T_s (where it must represent a class in $(T_s/Z(G))(\mathbb{R})$). In particular,

$$t \cdot \chi = \chi', \qquad t\delta t^{-1} = \delta'. \qquad (14.13)$$

On the other hand, Definition 13.5(e) and Lemma 14.11 provide a single limit of discrete series representation for $M(\mathbb{R})$ admitting Whittaker models of types χ_M and χ'_M. To this situation we can apply

Lemma 14.14. *Suppose $G(\mathbb{R})$ is a quasisplit real form of a complex connected reductive algebraic group, and π is an irreducible representation in the limits of (relative) discrete series. If π admits Whittaker models of types χ and χ', then χ and χ' must be conjugate under $G(\mathbb{R})$.*

Proof. Fix a Borel subgroup $B_s = T_s N$ defined over $\mathbb{R}$, and a maximally compact maximal torus $T(\mathbb{R})$. We may as well assume that χ and χ' are non-degenerate unitary characters of $N(\mathbb{R})$. Consider the group $(\operatorname{Ad} G)(\mathbb{R})$ of inner automorphisms of G defined over $\mathbb{R}$. It contains the image $(\operatorname{Ad}(G(\mathbb{R}))$ of $G(\mathbb{R})$ as a subgroup of finite index. Define

$$Q(G(\mathbb{R})) = (\operatorname{Ad} G)(\mathbb{R})/(\operatorname{Ad}(G(\mathbb{R}))). \qquad (14.15)(a)$$

This finite group acts on the set of equivalence classes of representations of $G(\mathbb{R})$, and on the set of conjugacy classes of Whittaker models.

It follows easily from the essential uniqueness of B_s that every coset in $Q(G(\mathbb{R}))$ meets T_s. It therefore follows from Lemma 3.2 that

$Q(G(\mathbb{R}))$ acts simply transitively on conjugacy classes of Whittaker models for $G(\mathbb{R})$. \hfill (14.15)(b)

At the same time, the existence of representatives in T_s shows that

$Q(G(\mathbb{R}))$ acts trivially on the set of principal series representations for $G(\mathbb{R})$. \hfill (14.15)(c)

On the other hand, the essential uniqueness of the maximally compact torus T shows that every coset in $Q(G(\mathbb{R}))$ meets the normalizer of T in G. Define

$$W_2(G, T) = \text{ Weyl group of } T \text{ in } (\operatorname{Ad} G)(\mathbb{R}).$$

Since $T(\mathbb{R})$ is connected, it follows that

$$Q(G(\mathbb{R})) \simeq W_2(G,T)/W(G(\mathbb{R}), T(\mathbb{R})).$$

From this it follows that

$Q(G(\mathbb{R}))$ acts without fixed points on limits of discrete series representations of $G(\mathbb{R})$. \hfill (14.15)(d)

To prove the lemma, realize π as a quotient of a principal series representation ρ. By (14.15)(b), there is an element $q \in Q(G(\mathbb{R}))$ with $q \cdot \chi$ is conjugate by $G(\mathbb{R})$ to χ'. It follows formally that $\pi' = q \cdot \pi$ has a Whittaker model of type χ'. By (14.15)(c), $q \cdot \rho$ is equivalent to ρ, so π' is also a quotient of ρ. The quotient representations π and π' of ρ now both have Whittaker models of type χ'. Because ρ has a unique Whittaker model ([30], Theorem 6.6.2) it follows that π must be equivalent to π'. By (14.15)(d), $q = 1$, so χ is conjugate to χ'. Q.E.D.

Returning to the proof of Proposition 13.6, we deduce that χ_M is conjugate to χ'_M by an element of $M(\mathbb{R})$; by Lemma 3.2, we may as well choose this element t_M in $T_s(\mathbb{R})$. We can replace χ' by $t_M \cdot \chi'$, and t by tt_M; then (14.13) remains true, and we have in addition that $\chi_M = \chi'_M$. From this we deduce that $\operatorname{Ad}(t)$ acts trivially on each of the simple restricted root spaces in $\mathfrak{m}$. Consequently t is central in M, so it belongs also to the maximal torus T of M. The second equation of (14.13) now shows that δ and δ' are conjugate by T, as we wished to show. Q.E.D.

We turn now to the proof of Proposition 13.8. Formally the argument is quite similar, and most of the subtleties have been dealt with in [1]. We will therefore omit some details. Again the first point is

14. Proof of Propositions 13.6 and 13.8

to understand extensions of involutions from maximal tori to reductive groups.

Lemma 14.16. *Suppose T is a maximal torus in the connected reductive algebraic group G, and θ is a (holomorphic) involutive automorphism of T permuting the roots of T in G. Suppose θ_1 and θ_2 are two involutive automorphisms of G extending θ, and that they determine the same sets of compact (and noncompact) imaginary roots. Then θ_1 and θ_2 are conjugate by an element of T.*

(Recall from (12.5) – (12.7) that a root is called imaginary if it is fixed by θ, and compact if the corresponding root vector is also fixed.)

The proof is exactly parallel to (and perhaps slightly simpler than) that of Lemma 14.1, so we omit it.

Proof of Proposition 13.8. By Lemma 9.17 of [1], there is an element $^{\vee}\delta_1 \in \mathcal{S} \cap {^dT}^{\Gamma}$. Write K_1 for the centralizer of $^{\vee}\delta_1$ in $^{\vee}G$, and

$$^d\mathcal{B}_1 = \{{^d}B | ({^{\vee}}\delta_1, {^d}B) \in \mathcal{S}\}$$

This is evidently a single K_1-orbit of Borel subgroups. By Proposition 6.30 of [1], there is a set $^dR_{i\mathbb{R},1}^+$ of positive imaginary roots special with respect to $^d\mathcal{B}_1$. Now there is a unique element w of the Weyl group of imaginary roots carrying $^dR_{i\mathbb{R},1}^+$ to $^dR_{i\mathbb{R}}^+$. The action of w on ^dT (as a product of reflections in imaginary roots) commutes with the action of $^{\vee}\delta_1$ on ^dT. Choose an element n of the normalizer of ^dT in $^{\vee}G$ representing w, and define $^{\vee}\delta = n({^{\vee}}\delta_1)n^{-1}$. Clearly $^{\vee}\delta$ belongs to $\mathcal{S}$. The action of $^{\vee}\delta$ on ^dT agrees with that of $^{\vee}\delta_1$ (since the latter commutes with w); so $^{\vee}\delta = {^{\vee}}\delta_1 t$ for some $t \in {^dT}$; so $^{\vee}\delta$ belongs to $\mathcal{S} \cap {^dT}^{\Gamma}$. The corresponding set $^d\mathcal{B}$ of Borel subgroups is obtained from $^d\mathcal{B}_1$ by conjugating by n; so it follows immediately from the definition that $^dR_{i\mathbb{R}}^+ = w({^dR}_{i\mathbb{R},1}^+)$ is special with respect to $^d\mathcal{B}$. The ^dT-conjugacy class $\mathcal{S}({^dT}^{\Gamma})$ of $^{\vee}\delta$ therefore satisfies the requirements of the proposition.

For the uniqueness, suppose $\mathcal{S}'({^dT}^{\Gamma})$ is another E-group structure satisfying the requirements of the proposition. Fix $^{\vee}\delta' \in \mathcal{S}'({^dT}^{\Gamma})$; what we are trying to show is that $^{\vee}\delta'$ is conjugate to $^{\vee}\delta$ by ^dT. Write θ and θ' for the involutive automorphisms of $^{\vee}G$ defined by conjugation by $^{\vee}\delta$ and $^{\vee}\delta'$. By Proposition 6.30(a) of [1], every simple root in $^dR_{i\mathbb{R}}^+$ is noncompact with respect both to θ and to θ'. By Lemma 14.16, it follows that θ and θ' are conjugate by ^dT. After replacing $^{\vee}\delta'$ by a conjugate, we may therefore assume that $\theta = \theta'$.

Now $^{\vee}\delta$ and $^{\vee}\delta'$ belong to $\mathcal{S}$, which is a single orbit of $^{\vee}G$. It follows that there is an element x of $^{\vee}G$ conjugating $^{\vee}\delta$ to $^{\vee}\delta'$. We can multiply

x on the right or left by elements of K without affecting this property; what we need to show is that this may be done so as to put x in ^dT. The automorphism $\mathrm{Ad}(x)$ commutes with θ, so the coset $xZ(^\vee G)$ belongs to the group $\overline{K}$ of fixed points of θ in $\mathrm{Ad}(^\vee G)$. Now $\mathrm{Ad}(x)$ carries ^dT to another θ-stable maximal torus $^dT'$, so these two tori are conjugate by $\overline{K}$. We apply to $\mathrm{Ad}(^\vee G)$ the following easy lemma.

Lemma 14.17. *Suppose G is a connected reductive algebraic group, and θ is an involutive automorphism of G. Write K for the group of fixed points of θ. Suppose T and T' are θ-stable maximal tori in G, conjugate by K. Then they are conjugate by the identity component K_0.*

We omit the elementary proof.

In our situation, we conclude that there is an element $x_0 \in K$ conjugating $^dT'$ to ^dT. Multiplying x by x_0 on the left, we may therefore assume that $^dT' = {}^dT$; that is, that x normalizes ^dT. In particular, the action of x defines an element $w \in W(^\vee G, {}^dT)$ commuting with θ. Proposition 3.12 of [56] describes the Weyl group elements commuting with θ. By Proposition 4.16 of [56], we may modify x by a representative in K of an appropriate Weyl group element, and arrange for w to be in the imaginary Weyl group. Since x carries $^\vee\delta$ to $^\vee\delta'$, it follows that $w(^dR_{i\mathbb{R}}^+)$ must (like $^dR_{i\mathbb{R}}^+$) be special with respect to $^d\mathcal{B}'$. By Proposition 6.30(c) of [1], it follows that w has a representative in K. After multiplying x by the inverse of such a representative, we get $x \in {}^dT$, as we wished to show. Q.E.D.

15. Multiplicity formulas for representations

Our next goal is Theorem 1.24 of the introduction, relating the geometric invariants discussed in Chapters 7 and 8 to representation theory. We begin by discussing a little more carefully the definition of the representation-theoretic multiplicity and character matrices (cf. (1.21)). For the same reasons as in Chapter 11, we work at first in the setting of (11.1). Recall from Definition 11.13 the set $L^z(G_\mathbb{R})$ of equivalence classes of final limit characters of type z. For $\Lambda \in L^z(G_\mathbb{R})$, write

$$M(\Lambda) = \text{standard limit representation attached to } \Lambda \qquad (15.1)(a)$$

(cf. (11.2)), and

$$\pi(\Lambda) = \text{Langlands quotient of } M(\Lambda). \qquad (15.1)(b)$$

(We leave open the question of which form of the representation to use — Harish-Chandra module or some topological version. Several reasonable possibilities are discussed below.) By Theorem 11.14, $\pi(\Lambda)$ is irreducible, and this correspondence establishes a bijection

$$L^z(G_\mathbb{R}) \leftrightarrow \Pi^z(G_\mathbb{R}), \qquad \Lambda \leftrightarrow \pi(\Lambda). \qquad (15.1)(c)$$

There are several ways to fit these representations into a nice abelian category. By far the simplest approach, due to Harish-Chandra, is to choose a maximal compact subgroup $K_\mathbb{R}$ of $G_\mathbb{R}$. (This is unique up to conjugation by $G_\mathbb{R}$.) One can then consider

$$\mathcal{M}^z(\mathfrak{g}, K_\mathbb{R}), \qquad (15.2)(a)$$

the category of finite-length canonical projective $(\mathfrak{g}, K_\mathbb{R})$-modules of type z, or *Harish-Chandra modules*. (Such a module is a representation simultaneously of $\mathfrak{g}$ and of $K_\mathbb{R}^{can,G}$ (cf. (11.1)(e)), with $\pi_1(G)^{can}$ acting by the character parametrized by z.) Sometimes it is convenient to obscure the choice of $K_\mathbb{R}$, writing instead

$$\mathcal{M}_{HC}^z(G_\mathbb{R}) \qquad (15.2)(b)$$

for this category. Alternatively, one can consider the category

$$\mathcal{M}^z_\infty(G_\mathbb{R}) \qquad (15.2)(c)$$

for which a typical object is the space of smooth vectors in a finite-length canonical projective representation of $G_\mathbb{R}$ of type z on a Hilbert space. That this is a nice category is a deep theorem of Casselman and Wallach; in fact they show that it is equivalent (by taking $K_\mathbb{R}$-finite vectors) to $\mathcal{M}^z(\mathfrak{g}, K_\mathbb{R})$. Instead of smooth vectors, one can consider distribution vectors, analytic vectors, or hyperfunction vectors; all of these choices lead to equivalent categories (designated with a subscript $-\infty$, ω, or $-\omega$). For our purposes in this paper, the category of Harish-Chandra modules is sufficient; but the aesthetic advantages of the other possibilities (such as the elimination of the choice of $K_\mathbb{R}$) are significant. Recall also that the critical Definition 13.5(e) really makes sense only on $\mathcal{M}^z_\infty$ (cf. Definition 3.1).

In any case, the categories are all canonically equivalent, so we can safely define

$$K\Pi^z(G_\mathbb{R}) = \text{Grothendieck group of } \mathcal{M}^z_{HC}(G_\mathbb{R}). \qquad (15.2)(d)$$

Suppose Θ and Λ belong to $L^z(G_\mathbb{R})$. As in (1.21), define

$$m_r(\Theta, \Lambda) = \text{multiplicity of } \pi(\Theta) \text{ in } M(\Lambda). \qquad (15.3)(a)$$

We call m_r the *representation-theoretic multiplicity matrix*. In the Grothendieck group this definition amounts to

$$M(\Lambda) = \sum_{\Theta \in L^z(G_\mathbb{R})} m_r(\Theta, \Lambda)\pi(\Theta). \qquad (15.3)(b)$$

Entries of m_r corresponding to limit characters of distinct infinitesimal character are zero, so m_r is "block-diagonal" with finite blocks. In an appropriate ordering of the basis, each block is upper triangular with one's on the diagonal ([54], Lemma 6.6.6). Consequently the multiplicity matrix is invertible; its inverse c_r is called the *representation-theoretic character matrix*. Explicitly,

$$\pi(\Theta) = \sum_{\Lambda \in L^z(G_\mathbb{R})} c_r(\Lambda, \Theta)M(\Lambda) \qquad (15.3)(c)$$

in $K\Pi^z(G_\mathbb{R})$. In particular, the final standard limit representations $M(\Theta)$ form a basis of the Grothendieck group. (The distribution characters of the standard representations are relatively simple. The equation

15. Multiplicity formulas for representations

(15.3)(c) therefore provides a fairly good formula for the distribution character of the irreducible representation $\pi(\Theta)$; hence the term "character matrix." In the case of regular infinitesimal character, the standard representation $M(\Lambda)$ is characterized by the appearance of one particular term in the formula for its distribution character (on the Cartan subgroup $T_{\mathbb{R}}$ corresponding to Λ). The entries $c_r(\Lambda, \Theta)$ may therefore be interpreted as certain coefficients in the character formula for $\pi(\Theta)$.) In analogy with (7.11)(f), we define the *Bruhat order on* $L^z(G_{\mathbb{R}})$ to be the smallest partial order with the property that

$$m_r(\Theta, \Lambda) \neq 0 \text{ only if } \Theta \leq \Lambda. \qquad (15.3)(d)$$

We could replace m_r by c_r without changing the order. The Bruhat order makes tempered final limit characters minimal, since the corresponding standard representations are irreducible. (Of course there are non-tempered minimal elements as well.)

We will need to recall a little about the Kazhdan-Lusztig algorithm for computing the matrices m_r and c_r (for linear groups). Just as in the geometric case (Proposition 8.8) the first step is a reduction to the regular case using a translation principle. We begin with an easy and well-known result.

Lemma 15.4. *Suppose G is a complex connected reductive algebraic group, and $^\vee G$ is a dual group for G. Write $\mathfrak{g}$ for the Lie algebra of G, and*

$$\mathfrak{z}(\mathfrak{g}) = \text{ center of } U(\mathfrak{g}).$$

Then there is a natural one-to-one correspondence between the set of algebra homomorphisms

$$\chi : \mathfrak{z}(\mathfrak{g}) \to \mathbb{C}$$

and the set of semisimple orbits $\mathcal{O}$ of $^\vee G$ on $^\vee \mathfrak{g}$.

Proof. Fix maximal tori $T \subset G$ and $^d T \subset {}^\vee G$. Harish-Chandra's theorem (see [22]) parametrizes the characters of $\mathfrak{z}(\mathfrak{g})$ by orbits of the Weyl group on the dual $\mathfrak{t}^*$ of the Lie algebra of T. By (9.3)(b), $\mathfrak{t}^*$ is naturally isomorphic to the Lie algebra $^\vee \mathfrak{t}$ of the dual torus $^\vee T$. By Lemma 13.1, there is an isomorphism of $^\vee T$ with $^d T$ determined uniquely up to the action of W; so Weyl group orbits on the respective Lie algebras are canonically identified. But every semisimple orbit of $^\vee G$ on $^\vee \mathfrak{g}$ meets $^d \mathfrak{t}$ in a unique Weyl group orbit. Q.E.D.

We sometimes write

$$\chi_{\mathcal{O}} : \mathfrak{z}(\mathfrak{g}) \to \mathbb{C} \qquad (15.5)(a)$$

for the character corresponding to the semisimple orbit $\mathcal{O}$ in the bijection of Lemma 15.4. If T is a maximal torus in G and $\lambda \in \mathfrak{t}^*$, then we may also write

$$\chi_\lambda : \mathfrak{z}(\mathfrak{g}) \to \mathbb{C} \qquad (15.5)(b)$$

using Harish-Chandra's theorem more directly. The various definitions in (15.2) may be restricted to a single infinitesimal character, as in

$$L^z(\mathcal{O}, G_\mathbb{R}) = \{ \Lambda \in L^z(G_\mathbb{R}) \mid \pi(\Lambda) \text{ has infinitesimal character } \chi_{\mathcal{O}} \} \qquad (15.5)(c)$$

$$\mathcal{M}^z_{HC}(\mathcal{O}, G_\mathbb{R}) = \text{Harish-Chandra modules of generalized infinitesimal character } \chi_{\mathcal{O}} \qquad (15.5)(d)$$

$$K\Pi^z(\mathcal{O}, G_\mathbb{R}) = \text{Grothendieck group of } \mathcal{M}^z_{HC}(\mathcal{O}, G_\mathbb{R}). \qquad (15.5)(e)$$

The set $L^z(\mathcal{O}, G_\mathbb{R})$ is finite, so $K\Pi^z(\mathcal{O}, G_\mathbb{R})$ is a lattice of finite rank.

It is convenient to include here the notation we will use in the extended group setting, even though we have a little more work to do with $G_\mathbb{R}$. So suppose for a moment that $(G^\Gamma, \mathcal{W})$ is an extended group for G. As in Lemma 1.15, choose a set $\{ \delta_s \mid s \in \Sigma \}$ of representatives for the equivalence classes of strong real forms of G. Recall from Definition 12.1 that $L^z(G/\mathbb{R})$ is the set of G-conjugacy classes of pairs (Λ, δ), with δ a strong real form of G (Definition 1.13) and $\Lambda \in L^z(G(\mathbb{R}, \delta))$ a final limit character. As in Lemma 1.15, there is a natural identification

$$L^z(G/\mathbb{R}) \simeq \coprod_{s \in \Sigma} L^z(G(\mathbb{R}, \delta_s)). \qquad (15.6)(a)$$

Theorem 12.3 provides a bijection between these parameters and the set $\Pi^z(G/\mathbb{R})$ of Definition 10.3; and the proof of Lemma 1.15 shows that

$$\Pi^z(G/\mathbb{R}) \simeq \coprod_{s \in \Sigma} \Pi^z(G(\mathbb{R}, \delta_s)). \qquad (15.6)(b)$$

Just as in (15.5), we can restrict attention to a single infinitesimal character corresponding to a ${}^\vee G$-orbit $\mathcal{O}$, writing for example $L^z(\mathcal{O}, G/\mathbb{R})$.

15. Multiplicity formulas for representations

Finally, let $({}^\vee G^\Gamma, \mathcal{S})$ be an E-group for G^Γ with second invariant z. Theorem 10.4 identifies $\Pi^z(G/\mathbb{R})$ with

$$\Xi^z(G/\mathbb{R}) = \Xi({}^\vee G^\Gamma). \qquad (15.6)(c)$$

This set also decomposes according to the semisimple orbits of ${}^\vee G$ on ${}^\vee \mathfrak{g}$, and we have

$$\Xi^z(\mathcal{O}, G/\mathbb{R}) = \{\text{ complete geometric parameters for}$$
$${}^\vee G^{alg} \text{ acting on } X(\mathcal{O}, {}^\vee G^\Gamma)\} \qquad (15.6)(d)$$

It is clear from the definitions that Theorem 10.4 provides an identification

$$\Pi^z(\mathcal{O}, G/\mathbb{R}) \simeq \Xi^z(\mathcal{O}, G/\mathbb{R}). \qquad (15.6)(e)$$

In analogy with Definition 7.13, we can form the direct sum of abelian categories

$$\mathcal{M}_{HC}^z(G/\mathbb{R}) = \bigoplus_{s \in \Sigma} \mathcal{M}_{HC}^z(G(\mathbb{R}, \delta_s)). \qquad (15.7)(a)$$

(Recall that the notation conceals a choice of maximal compact subgroup for each strong real form.) Every object in this category has finite length, and we have irreducible and standard representations parametrized by $L^z(G/\mathbb{R})$ or $\Xi^z(G/\mathbb{R})$. For $\xi \in \Xi^z(\mathcal{O}, G/\mathbb{R})$, we may write as in (15.3)

$$M(\xi) = \sum_{\gamma \in \Xi^z(\mathcal{O}, G/\mathbb{R})} m_r(\gamma, \xi) \pi(\gamma) \qquad (15.7)(b)$$

and so forth; this identity is in the Grothendieck group

$$K\Pi^z(\mathcal{O}, G/\mathbb{R}) = \bigoplus_{s \in \Sigma} K\Pi^z(\mathcal{O}, G(\mathbb{R}, \delta_s)). \qquad (15.7)(c)$$

In order to define the pairing of Theorem 1.24, we need one more definition.

Definition 15.8. Suppose ${}^\vee G^\Gamma$ is a weak E-group. Recall from Definition 4.9 the element $z(\rho) \in Z({}^\vee G^{alg})$; it has order 2. Because it is canonically defined, it is fixed by the conjugation action of any element of ${}^\vee G^\Gamma$. If $\xi = (\phi, \tau) \in \Xi({}^\vee G^\Gamma)$, it follows that $z(\rho)$ belongs to ${}^\vee G_\phi^{alg}$. Write $\overline{z(\rho)}$ for its image in the universal component group $A_\phi^{loc,alg}$; this is a central element of order 2. It therefore acts by ± 1 in the irreducible representation τ, and we define

$$e(\xi) = \tau(\overline{z(\rho)}) = \pm 1.$$

We also recall from (1.22) the notation

$$d(\xi) = \dim S_\xi,$$

with S_ξ the ${}^\vee G$-orbit on $X({}^\vee G^\Gamma)$ corresponding to ξ (Definition 7.6).

Lemma 15.9. *In the setting of Theorem 10.4, suppose $\xi \in \Xi^z(G/\mathbb{R})$ corresponds to an irreducible representation π of a real form $G(\mathbb{R}, \delta)$. Then the sign $e(\xi)$ of Definition 15.8 is equal to the sign $e(G(\mathbb{R}, \delta))$ defined in [32]. In particular, it is equal to 1 if $G(\mathbb{R}, \delta)$ is quasisplit.*

We will give a proof at the end of Chapter 17.

Definition 15.11. Suppose we are in the setting (15.6). The Grothendieck groups $K\Pi^z(\mathcal{O}, G/\mathbb{R})$ (cf. (15.7)(c)) and $KX(\mathcal{O}, {}^\vee G^\Gamma)$ (Definition 7.13) are both free $\mathbb{Z}$-modules on bases parametrized by $\Xi^z(G/\mathbb{R})$. It therefore makes sense to define the *canonical perfect pairing*

$$\langle , \rangle : K\Pi^z(\mathcal{O}, G/\mathbb{R}) \times KX(\mathcal{O}, {}^\vee G^\Gamma) \to \mathbb{Z}$$

by the requirement

$$\langle M(\xi), \mu(\gamma) \rangle = e(\xi)\delta_{\xi,\gamma}.$$

The last term on the right is a Kronecker delta.

We can now restate Theorem 1.24 using E-groups instead of L-groups.

Theorem 15.12. *Suppose $(G^\Gamma, \mathcal{W})$ is an extended group for G (Definition 1.12), and $({}^\vee G^\Gamma, \mathcal{S})$ is an E-group for the corresponding inner class of real forms, having second invariant z (Definition 4.14). Fix a semisimple orbit $\mathcal{O}$ of ${}^\vee G$ on ${}^\vee \mathfrak{g}$. Write $K\Pi^z(\mathcal{O}, G/\mathbb{R})$ for the Grothendieck group of canonical projective representations of type z and infinitesimal character $\chi_\mathcal{O}$ of strong real forms of G (cf. (15.7)(c)), and $KX(\mathcal{O}, {}^\vee G^\Gamma)$ for the Grothendieck group of ${}^\vee G^{alg}$-equivariant constructible sheaves on the geometric parameter space $X(\mathcal{O}, {}^\vee G^\Gamma)$ (Definitions 6.9 and 7.13). Fix complete geometric parameters $\xi, \gamma \in \Xi^z(\mathcal{O}, G/\mathbb{R})$ (cf. (15.6)(d)), and write $\pi(\xi), P(\gamma)$ for the corresponding*

15. Multiplicity formulas for representations

irreducible representation and perverse sheaf (Theorem 10.4 and Definition 7.13). Then the canonical pairing of Definition 15.11 satisfies

$$\langle \pi(\xi), P(\gamma)\rangle = e(\xi)(-1)^{d(\xi)}\delta_{\xi,\gamma}.$$

Here $e(\xi)$ is defined in (15.8).

We will prove this result in the next two chapters. The following relationship between the geometric and representation-theoretic multiplicity matrices is just a reformulation.

Corollary 15.13. *In the setting of Theorem 15.12, fix η and γ in $\Xi^z(G/\mathbb{R})$. Then the geometric and representation-theoretic multiplicity and character matrices (cf. (7.11) and (15.3)) satisfy*
a) $c_g(\eta,\gamma)(-1)^{d(\eta)-d(\gamma)} = m_r(\gamma,\eta)$.
b) $c_r(\eta,\gamma)(-1)^{d(\eta)-d(\gamma)} = m_g(\gamma,\eta)$.

Proof. Using (15.3)(c) and (7.11)(c), we can rewrite the pairing in Theorem 15.12 in terms of standard representations and elementary constructible sheaves. The result is

$$\langle \pi(\xi), P(\gamma)\rangle = \langle \sum_{\xi'} c_r(\xi',\xi)M(\xi'), \sum_{\gamma'} c_g(\gamma',\gamma)\mu(\gamma')(-1)^{d(\gamma')}\rangle$$
$$= \sum_{\xi',\gamma'} c_r(\xi',\xi)c_g(\gamma',\gamma)(-1)^{d(\gamma')}\langle M(\xi'),\mu(\gamma')\rangle.$$

By Definition 15.11, only the terms with $\gamma' = \xi'$ contribute; we get

$$\sum_{\eta} c_r(\eta,\xi)c_g(\eta,\gamma)(-1)^{d(\eta)}e(\eta).$$

Now the central element $z(\rho)$ clearly acts by $e(\gamma)$ on all the stalks of the perverse sheaf $P(\gamma)$. So whenever the second factor is non-zero, we must have $e(\eta) = e(\gamma)$. This leads to

$$e(\gamma)(-1)^{d(\gamma)} \sum_{\eta} c_r(\eta,\xi)c_g(\eta,\gamma)(-1)^{d(\eta)-d(\gamma)}.$$

Comparing with the formula in Theorem 15.12, we find

$$\sum_{\eta} c_r(\eta,\xi)c_g(\eta,\gamma)(-1)^{d(\eta)-d(\gamma)} = \delta_{\xi,\gamma}.$$

Since the inverse of the matrix $c_r(\gamma,\xi)$ is by definition $m_r(\gamma,\xi)$, it follows that
$$m_r(\gamma,\eta) = c_g(\eta,\gamma)(-1)^{d(\eta)-d(\gamma)},$$
which is (a).

For (b), write
$$c_g(\eta,\gamma) = m_r(\gamma,\eta)(-1)^{d(\eta)-d(\gamma)}.$$
Multiply by $m_g(\xi,\eta)$ and sum over η. Since c_g is by definition the inverse of m_g, we get $\delta_{\xi,\gamma}$ on the left. The right side can be written as
$$(-1)^{d(\gamma)-d(\xi)} \sum_\eta m_r(\gamma,\eta) m_g(\xi,\eta)(-1)^{d(\xi)-d(\eta)}.$$
Since the left side is zero unless $\gamma = \xi$, we can drop the sign in front of the summation. Since c_r is by definition the inverse of m_r, it follows that
$$c_r(\eta,\xi) = m_g(\xi,\eta)(-1)^{d(\xi)-d(\eta)}.$$
This is (b). Q.E.D.

We leave to the reader the straightforward verification that the argument given for Corollary 15.13(a) can be reversed, so that

Theorem 15.12 is equivalent to Corollary 15.13(a). (15.14)

16. The translation principle and the Kazhdan-Lusztig algorithm

In this chapter we will begin the proof of Theorem 15.12 (which is essentially Theorem 1.24). The first step is a reduction to the case of regular infinitesimal character. For this, we reformulate the translation principle for representations so as to emphasize the connection with the geometric translation principle of Proposition 8.8.

Definition 16.1. In the setting of Lemma 15.4, suppose $\mathcal{O}$ and $\mathcal{O}'$ are semisimple orbits of ${}^\vee G$ on ${}^\vee \mathfrak{g}$, and $\mathcal{T}$ is a translation datum from $\mathcal{O}$ to $\mathcal{O}'$ (Definition 8.6). Fix maximal tori $T \subset G$ and ${}^d T \subset {}^\vee G$, and construct $\mu = \lambda' - \lambda \in X_*({}^d T)$ as in Definition 8.6(e); this element is defined up to the Weyl group. By Lemma 13.1(b), μ corresponds to a weight (also called μ) in $X^*(T)$, defined up to the action of the Weyl group. By the Cartan-Weyl theory, there is a unique finite-dimensional irreducible algebraic representation $F_\mathcal{T}$ of G having extremal weight $-\mu$. Write $P_\mathcal{O}$ for the functor of projection on the generalized infinitesimal character $\chi_\mathcal{O}$; this is defined on all representations of $\mathfrak{g}$ on which $\mathfrak{z}(\mathfrak{g})$ acts in a locally finite way.

The *Jantzen-Zuckerman translation functor*

$$\psi_\mathcal{T} : \mathcal{M}^z(\mathcal{O}', G_\mathbb{R}) \to \mathcal{M}^z(\mathcal{O}, G_\mathbb{R})$$

is defined by

$$\psi_\mathcal{T}(M) = P_\mathcal{O}(M \otimes F_\mathcal{T}).$$

(Of course it is also defined on $\mathcal{M}^z(\mathcal{O}', G/\mathbb{R})$.) In the notation of (11.15), this is $\psi_{\lambda_a+\mu_a}^{\lambda_a}$. Our present hypotheses are somewhat weaker. Because of the definition of translation datum, this functor is a translation "to the wall." A complete development of the theory also requires the adjoint translation functor ("away from the wall")

$$\phi_\mathcal{T} : \mathcal{M}^z(\mathcal{O}, G_\mathbb{R}) \to \mathcal{M}^z(\mathcal{O}', G_\mathbb{R})$$

defined by

$$\phi_\mathcal{T}(M) = P_{\mathcal{O}'}(M \otimes F_\mathcal{T}^*).$$

176   The Langlands Classification and Irreducible Characters

The most basic properties of these functors (that they are well-defined, covariant, exact, and adjoint to each other) are established in [54], Proposition 4.5.8. Before we recall anything deeper, it may be helpful to reformulate the definition of translation datum so as to put Definition 16.1 in a more familiar setting.

Lemma 16.2. *Suppose G is a complex connected reductive algebraic group, $^\vee G$ is a dual group, and $\mathcal{O}$ and $\mathcal{O}'$ are semisimple orbits of $^\vee G$ on $^\vee \mathfrak{g}$. Then a translation datum T from $\mathcal{O}$ to $\mathcal{O}'$ (Definition 8.6) may be identified with a G-conjugacy class of triples (T, λ, λ') subject to the conditions below.*

i) T *is a maximal torus in G, and λ and λ' belong to $\mathfrak{t}^*$.*

ii) *The infinitesimal characters χ_λ and χ'_λ are equal to $\chi_\mathcal{O}$ and $\chi_{\mathcal{O}'}$ respectively (notation (15.5)).*

iii) *Suppose α is a root of T in $\mathfrak{g}$, and $\langle \lambda, \alpha^\vee \rangle$ is a positive integer. Then $\langle \lambda', \alpha^\vee \rangle$ is a positive integer.*

iv) *The weight $\mu = \lambda' - \lambda$ belongs to $X^*(T)$.*

Any triple (T, λ, λ') satisfying (i)–(iv) determines a unique translation datum.

This is clear from Definition 8.6 and Lemma 13.1. In the language of Definition 4.5.7 of [54], our $(\mathfrak{t}, \lambda, \lambda')$ corresponds to $(\mathfrak{h}^s, \xi + \overline{\lambda_2}, \xi + \overline{\lambda_1})$. (Various extra complications appear in [54] because the finite-dimensional representations are not assumed to come from an ambient algebraic group. On the other hand, the definition in [54] does not require any condition like (iii) of Lemma 16.2; such conditions appear only when there are theorems to be proved.) Our functor ψ_T is called $\psi^{\xi+\lambda_1}_{\xi+\lambda_2}$ in [54], and our ϕ_T is $\psi^{\xi+\lambda_2}_{\xi+\lambda_1}$.

In the setting of Definition 16.1, suppose $\Lambda' = ((\Lambda^{can})', R^+_{i\mathbb{R}}, R^+_\mathbb{R})$ is a limit character of infinitesimal character $\mathcal{O}'$ (cf. (11.2)). When G is linear we can define a new limit character

$$\Lambda = \psi_T(\Lambda') \qquad (16.3)(a)$$

as follows (cf. (11.17)). First, the Cartan subgroup $T_\mathbb{R}$ and the positive root systems are unchanged. Identify the translation datum T as in Lemma 16.2, and choose a triple (T, λ, λ') with T the complexification of $T_\mathbb{R}$ and λ' the differential of $(\Lambda')^{can}$; this is possible by our hypothesis on the infinitesimal character of Λ'. By the definition of a translation datum (see Lemma 16.2(iii)), the stabilizer of λ' in the Weyl group $W(G, T)$ also stabilizes λ. It follows that λ is uniquely determined by T and λ'; so the weight $\mu = \lambda' - \lambda \in X^*(T)$ is well-defined. (This step is not symmetric in Λ and Λ'; we cannot reverse it to define "$\phi_T(\Lambda)$.")

16. The translation principle and the Kazhdan-Lusztig algorithm   177

Via the map (11.1)(b), we can regard μ as a character of $T_{\mathbb{R}}$; so we can define

$$\Lambda^{can} = ((\Lambda)^{can})' \otimes \mu^{-1}. \qquad (16.3)(b)$$

The positivity assumption in Lemma 16.2(iii) guarantees that Λ inherits from Λ' the requirement (11.2)(d) in the definition of a limit character. (This is where the linearity of G enters: it forces the imaginary roots to be integral, so that Lemma 16.2(iii) imposes some restrictions on their positivity. This point introduces substantial difficulties in the detailed character theory of non-linear groups. In Chapter 11, we avoided these problems by assuming μ dominant.)

Theorem 16.4. *Suppose G is a complex connected reductive algebraic group, and $G_{\mathbb{R}}$ is a linear real Lie group as in (11.1). Fix infinitesimal characters $\mathcal{O}$ and $\mathcal{O}'$ (Lemma 15.4) and a translation datum $\mathcal{T}$ from $\mathcal{O}$ to $\mathcal{O}'$.*

a) Suppose Λ' is a limit character with infinitesimal character $\mathcal{O}'$, and write $\Lambda = \psi_{\mathcal{T}}(\Lambda')$ (cf. (16.3)). Then

$$\psi_{\mathcal{T}}(M(\Lambda')) = M(\Lambda)$$

(notation (11.2) and (16.1).)

b) Suppose $\pi' \in \Pi^z(G_{\mathbb{R}})$ is an irreducible canonical projective representation of infinitesimal character $\mathcal{O}'$. Then $\psi_{\mathcal{T}}(\pi')$ is irreducible or zero.

c) In the setting of (a), suppose in addition that Λ' is final and that $\psi_{\mathcal{T}}(\pi(\Lambda')) \neq 0$. Then Λ is also final, and

$$\psi_{\mathcal{T}}(\pi(\Lambda')) = \pi(\Lambda).$$

In this way $\psi_{\mathcal{T}}$ defines a bijection from a subset of $\Pi^z(\mathcal{O}', G_{\mathbb{R}})$ onto $\Pi^z(\mathcal{O}, G_{\mathbb{R}})$.

This is another version of the results in Propositions 11.16, 11.18, and 11.20; the arguments and references given there apply here as well. Write

$$\psi_{\mathcal{T}}^{-1} : \Pi^z(\mathcal{O}, G_{\mathbb{R}}) \to \Pi^z(\mathcal{O}', G_{\mathbb{R}}) \qquad (16.5)(a)$$

for the injective map inverting the correspondence of (c). This is not so easy to describe explicitly; one characterization is

$$\psi_{\mathcal{T}}^{-1}(\pi) = \text{unique irreducible subrepresentation of } \phi_{\mathcal{T}}(\pi). \qquad (16.5)(b)$$

Because of Theorem 11.14, there is a corresponding injective map on equivalence classes of final limit characters

$$\phi_T : L^z(\mathcal{O}, G_\mathbb{R}) \to L^z(\mathcal{O}', G_\mathbb{R}). \qquad (16.5)(c)$$

As a consequence of Theorem 16.4, this map respects the representation-theoretic multiplicity matrix:

$$m_r(\phi_T(\Theta), \phi_T(\Lambda)) = m_r(\Theta, \Lambda) \qquad (\Theta, \Lambda \in L^z(\mathcal{O}, G_\mathbb{R})). \qquad (16.5)(d)$$

Because of Lemma 8.7, this reduces the calculation of the multiplicity matrix m_r to the case of regular infinitesimal character. In connection with Theorem 1.24, we need to know that this reduction is compatible with the corresponding one on the geometric side.

Proposition 16.6. *In the setting of (15.6), suppose T is a translation datum from $\mathcal{O}$ to $\mathcal{O}'$. Then the diagram*

$$\begin{array}{ccc} L^z(\mathcal{O}, G/\mathbb{R}) & \xrightarrow{\phi_T} & L^z(\mathcal{O}', G/\mathbb{R}) \\ \downarrow & & \downarrow \\ \Xi^z(\mathcal{O}, {}^\vee G^\Gamma) & \xrightarrow{f_T^*} & \Xi^z(\mathcal{O}', {}^\vee G^\Gamma) \end{array}$$

commutes. Here the vertical arrows are the bijections of Theorem 10.4.

Proof. The map in the top row has been described rather explicitly in terms of Cartan subgroups of G^Γ in the course of the proof of Proposition 11.20. (The hypotheses there were slightly different, but the same ideas apply.) The sets in the bottom row have been parametrized in terms of Cartan subgroups of ${}^\vee G^\Gamma$ (Theorem 12.9). The vertical maps are defined in terms of Cartan subgroups (Propositions 13.12 and 13.13). All that remains is to check that the geometrically defined map in the bottom row (cf. (7.16)) can be computed in terms of Cartan subgroups. This is an elementary exercise, and we leave it to the reader. (Notice that there are no perverse sheaves in this calculation — just homogeneous spaces and representations of component groups.) Q.E.D.

To get the reduction of Theorem 15.12 to the case of regular infinitesimal character, we use the equivalent form Corollary 15.13(a) (cf. (15.14)). For that, we need only compare Proposition 7.15(c) with (16.5)(d). (The orbit correspondence of Proposition 7.15 does not preserve dimensions of orbits, but it changes all of them by the same constant d. Since only differences of dimensions appear in Corollary 15.13(a), this suffices.)

16. The translation principle and the Kazhdan-Lusztig algorithm

For the balance of this chapter, we may therefore assume that

$$\mathcal{O} \subset {}^\vee\mathfrak{g} \text{ is a regular semisimple orbit of } {}^\vee G. \qquad (16.7)(a)$$

In order to describe the proof of Theorem 15.12, we need to recall in some detail the structure of the Kazhdan-Lusztig algorithm that computes the various character matrices. Suppose $({}^dT_1, \lambda_1)$ and $({}^dT_2, \lambda_2)$ are pairs with ${}^dT_i$ a maximal torus in ${}^\vee G$ and $\lambda_i \in {}^dT_i \cap \mathcal{O}$. Just as in (11.17), there is a unique isomorphism

$$j(\lambda_1, \lambda_2) : {}^dT_1 \to {}^dT_2 \qquad (16.7)(b)$$

induced by an element of ${}^\vee G$ and carrying λ_1 to λ_2. (It is the assumption that $\mathcal{O}$ is regular that makes the isomorphism unique.) Define

$$({}^\vee T_\mathcal{O}, \lambda_\mathcal{O}) = \lim\, ({}^dT, \lambda), \qquad (16.7)(c)$$

the projective limit taken over pairs as above using the isomorphisms of (16.7)(b). Then any such pair is canonically isomorphic to $({}^dT_\mathcal{O}, \lambda_\mathcal{O})$, say by

$$j(\lambda_\mathcal{O}, \lambda) : {}^\vee T_\mathcal{O} \to {}^dT \qquad (16.7)(d)$$

Write

$$T_\mathcal{O} = \text{dual torus to } {}^\vee T_\mathcal{O}. \qquad (16.8)(a)$$

The dual of the Lie algebra of $T_\mathcal{O}$ may be identified with ${}^\vee\mathfrak{t}_\mathcal{O}$, so we may write

$$\lambda_\mathcal{O} \in \mathfrak{t}_\mathcal{O}^*. \qquad (16.8)(b)$$

By inspection of the proof of Lemma 15.4, it is clear that we may identify

$$(T_\mathcal{O}, \lambda_\mathcal{O}) = \lim\,(T, \lambda). \qquad (16.8)(c)$$

The projective limit is taken over pairs (T, λ) with T a maximal torus in G, and $\lambda \in \mathfrak{t}^*$ a weight defining the infinitesimal character $\chi_\mathcal{O}$, using isomorphisms analogous to those of (16.7)(b). In particular, we get isomorphisms

$$i(\lambda_\mathcal{O}, \lambda) : T_\mathcal{O} \to T \qquad (16.8)(d)$$

as in (16.7)(d).

The torus ${}^\vee T_\mathcal{O}$ inherits a root system $R({}^\vee G, {}^\vee T_\mathcal{O}) \subset X^*({}^\vee T_\mathcal{O})$, a Weyl group $W({}^\vee G, {}^\vee T_\mathcal{O})$, and so on. Similarly (using (16.8)(c)) we can define $R(G, T_\mathcal{O})$. The identification (16.8)(a) makes $R(G, T_\mathcal{O})$ and $R({}^\vee G, {}^\vee T_\mathcal{O})$ into dual root systems, so that the roots for one may be regarded as coroots for the other. We define the set of $\mathcal{O}$-*integral coroots*

$$R^\vee(\mathcal{O}) = \{\, \alpha^\vee \in R({}^\vee G, {}^\vee T_\mathcal{O}) | \alpha^\vee(\lambda_\mathcal{O}) \in \mathbb{Z}\,\}. \qquad (16.9)(a)$$

Equivalently,

$$R^\vee(\mathcal{O}) = \{\, \alpha^\vee \in R^\vee(G, T_\mathcal{O}) | \lambda_\mathcal{O}(\alpha^\vee) \in \mathbb{Z}\,\}. \qquad (16.9)(b)$$

Clearly an isomorphism $j(\lambda_\mathcal{O}, \lambda)$ as in (16.7)(d) identifies $R^\vee(\mathcal{O})$ with the set of roots of ${}^d T$ in ${}^\vee G(\lambda)$ (cf. (6.2)(b)). In G, $i(\lambda_\mathcal{O}, \lambda)$ identifies $R(\mathcal{O})$ with the set of λ-integral roots of T in G. There is a natural positive root system

$$R^+(\mathcal{O}) = \{\, \alpha \in R(\mathcal{O}) | \alpha^\vee(\lambda_\mathcal{O}) > 0\,\}; \qquad (16.9)(c)$$

$j(\lambda_\mathcal{O}, \lambda)$ identifies the positive coroots with the roots of ${}^d T$ in the Borel subgroup $P(\lambda)$ of ${}^\vee G(\lambda)$ (cf. (6.2)(e)). We write

$$\begin{aligned} R(\lambda) &= i(\lambda_\mathcal{O}, \lambda)(R(\mathcal{O})) = \{\, \alpha \in R(G, T) | \alpha^\vee(\lambda) \in \mathbb{Z}\,\}, \\ R^+(\lambda) &= i(\lambda_\mathcal{O}, \lambda)(R^+(\mathcal{O})), \end{aligned} \qquad (16.9)(d)$$

and similarly for $R^\vee$. Define

$$\Delta(\mathcal{O}) = \text{set of simple roots of } R^+(\mathcal{O}), \quad W(\mathcal{O}) = \text{Weyl group of } R(\mathcal{O}). \qquad (16.9)(e)$$

We can regard $W(\mathcal{O})$ as a group of automorphisms of $T_\mathcal{O}$ or of ${}^\vee T_\mathcal{O}$. The corresponding set of simple reflections is written

$$S(\mathcal{O}) = \{\, s_\alpha | \alpha \in \Delta(\mathcal{O})\,\}. \qquad (16.9)(f)$$

Of course $(W(\mathcal{O}), S(\mathcal{O}))$ is a Coxeter group. We may therefore attach to it a *Hecke algebra* $\mathcal{H}(\mathcal{O})$ (cf. [27]). The Hecke algebra is a free $\mathbb{Z}[u^{1/2}, u^{-1/2}]$-algebra with basis

$$\{\, T_w | w \in W(\mathcal{O})\,\}. \qquad (16.10)(a)$$

16. The translation principle and the Kazhdan-Lusztig algorithm

It is characterized by the relations

$$T_x T_y = T_{xy} \qquad (x, y \in W(\mathcal{O}), l(x) + l(y) = l(xy)) \qquad (16.10)(b)$$

and

$$(T_s + 1)(T_s - u) = 0 \qquad (s \in S(\mathcal{O})). \qquad (16.10)(c)$$

(Here $l(x)$ is the length function on the Coxeter group $(W(\mathcal{O}), S(\mathcal{O}))$; it need not be the restriction of any length function on the larger Weyl group $W(G, T_\mathcal{O})$.) It follows that the operators T_s generate the Hecke algebra. The specialization to $u^{1/2} = 1$ of $\mathcal{H}(\mathcal{O})$ — that is, its quotient by the ideal generated by $u^{1/2} - 1$ — is naturally isomorphic to the group algebra of $W(\mathcal{O})$.

This is the basic structure required for the Kazhdan-Lusztig algorithms, and we will make no explicit use of any more natural descriptions of it. As a hint about the geometry underlying the algorithms, however, we mention a (well-known) geometric description of $W(\mathcal{O})$. Recall the map e from ${}^\vee\mathfrak{g}$ to ${}^\vee G$ (cf. (6.2)(a)). We use the notation of (6.10). Thus $\mathcal{F}(\mathcal{O})$ is the set of canonical flats in $\mathcal{O}$, $\mathcal{C}(\mathcal{O})$ is the corresponding conjugacy class in ${}^\vee G$, and e is a smooth projective algebraic morphism from $\mathcal{F}(\mathcal{O})$ to $\mathcal{C}(\mathcal{O})$. Define

$$Z(\mathcal{O}, {}^\vee G) = \mathcal{F}(\mathcal{O}) \times_{\mathcal{C}(\mathcal{O})} \mathcal{F}(\mathcal{O}) \qquad (16.11)(a)$$

This definition is analogous to that of the geometric parameter space in Proposition 6.16, but it is substantially simpler. An analysis along the lines of that proposition shows that there is a natural bijection

$$\{ {}^\vee G\text{-orbits on } Z(\mathcal{O}) \} \leftrightarrow W(\mathcal{O}). \qquad (16.11)(b)$$

Here the diagonal orbit (which is just $\mathcal{F}(\mathcal{O})$) corresponds to the identity element of $W(\mathcal{O})$. The length function on $W(\mathcal{O})$ corresponds to the dimension of orbits (shifted by the dimension of $\mathcal{F}(\mathcal{O})$). Now one can realize the group algebra of $W(\mathcal{O})$ (and even the Hecke algebra $\mathcal{H}(\mathcal{O})$) as an algebra of correspondences, endowed with a natural geometric action on $KX(\mathcal{O}, {}^\vee G^\Gamma)$. For a sketch of this, we refer to the end of [40]. (Our construction of geometric translation functors in Chapter 8 was of essentially the same nature; recall that a translation datum was a very special kind of ${}^\vee G$-orbit on $\mathcal{F}(\mathcal{O}) \times_{\mathcal{C}(\mathcal{O})} \mathcal{F}(\mathcal{O}')$.)

We return now to our description of the Kazhdan-Lusztig algorithms. Suppose $G_\mathbb{R}$ is a linear Lie group as in (15.1)–(15.2). Define

$$\mathcal{K}\Pi^z(\mathcal{O}, G_\mathbb{R}) = K\Pi^z(\mathcal{O}, G_\mathbb{R}) \otimes_\mathbb{Z} \mathbb{Z}[u^{1/2}, u^{-1/2}], \qquad (16.12)(a)$$

(cf. (15.7)), the free $\mathbb{Z}[u^{1/2}, u^{-1/2}]$-module with basis the irreducible (or standard) representations of $G_\mathbb{R}$. We may refer to this as the *mixed Grothendieck group of representations*. (In a sense the definition is misleading: an ordinary Grothendieck group should be naturally the specialization to $u = 1$ of a mixed one, but a mixed one does not arise naturally by extension of scalars from an ordinary one. In our case there is little to be done, because of the lack of a category of "mixed representations.") In the setting of (15.6), we will write

$$\mathcal{K}\Pi^z(\mathcal{O}, G/\mathbb{R}) = K\Pi^z(\mathcal{O}, G/\mathbb{R}) \otimes_\mathbb{Z} \mathbb{Z}[u^{1/2}, u^{-1/2}], \quad (16.12)(b)$$

Similarly, we define

$$\mathcal{K}X(\mathcal{O}, {}^\vee G^\Gamma) = KX(\mathcal{O}, {}^\vee G^\Gamma) \otimes_\mathbb{Z} \mathbb{Z}[u^{1/2}, u^{-1/2}], \quad (16.12)(c)$$

the *mixed Grothendieck group of perverse sheaves*. In this case a more geometric interpretation is available ([40], Definition 2.2). Finally, we will need an analogue of the definition in Corollary 1.26:

$$\overline{\mathcal{K}}\Pi^z(\mathcal{O}, G/\mathbb{R}) = \mathbb{Z}[u^{1/2}, u^{-1/2}]\text{-linear combinations of}$$
$$\text{elements of } \Pi^z(\mathcal{O}, G/\mathbb{R}); \quad (16.12)(d)$$

here infinite linear combinations are allowed. (Infinite combinations can arise only for non-semisimple groups, because of the possibility of infinitely many strong real forms.) We call $\overline{\mathcal{K}}$ the *formal mixed Grothendieck group of representations*.

Proposition 16.13. *In the setting (16.7) – (16.12), there are natural actions of the Hecke algebra $\mathcal{H}(\mathcal{O})$ on the mixed Grothendieck groups $\mathcal{K}\Pi^z(\mathcal{O}, G_\mathbb{R})$ (or $\mathcal{K}\Pi^z(\mathcal{O}, G/\mathbb{R})$, or $\overline{\mathcal{K}}\Pi^z(\mathcal{O}, G/\mathbb{R})$) and $\mathcal{K}X(\mathcal{O}, {}^\vee G^\Gamma)$.*

Proof. By (8.9), the $\mathbb{Z}[u^{1/2}, u^{-1/2}]$-module $\mathcal{K}X(\mathcal{O}, {}^\vee G^\Gamma)$ is a direct sum of copies of Hecke algebra modules constructed in [40]. Explicit formulas for the action of the generators T_s appear in Lemma 3.5 of [40]. For $\mathcal{K}\Pi^z(\mathcal{O}, G_\mathbb{R})$, the action is constructed in [56], section 12. Explicit formulas may be found there or in [55], Definition 6.4. (There is a small but dangerous subtlety concealed in the details omitted here. The basis for $\mathcal{K}\Pi^z(\mathcal{O}, G_\mathbb{R})$ used in the references corresponds after Beilinson-Bernstein localization to a local system on an orbit, placed in degree 0. This differs by something like $(-1)^d$, with d the orbit dimension, from a standard representation. One effect appears in the formula for C_r in Proposition 16.20 below, which contains a sign absent from the references. Of course the sign could be absorbed in the definition of the representation-theoretic Kazhdan-Lusztig polynomials, and this is probably where it belongs.) Q.E.D.

16. The translation principle and the Kazhdan-Lusztig algorithm

In addition to the Hecke algebra module structure, there are two more structures that are needed for the Kazhdan-Lusztig algorithms.

Definition 16.14 (see [27]). *Verdier duality* is the unique algebra automorphism

$$D : \mathcal{H}(\mathcal{O}) \to \mathcal{H}(\mathcal{O})$$

satisfying

$$Du^{1/2} = u^{-1/2}, DT_s = u^{-1}(T_s + (1-u)) \quad (s \in S(\mathcal{O})).$$

The terminology arises from the geometric interpretation of $\mathcal{H}(\mathcal{O})$. Using the defining relations (16.10), one can easily verify that D is also characterized by

$$Du^{1/2} = u^{-1/2}, DT_w = (T_{w^{-1}})^{-1} \quad (w \in W(\mathcal{O})).$$

Another useful characterization is

$$Du^{1/2} = u^{-1/2}, D(T_s + 1) = u^{-1}(T_s + 1) \quad (s \in S(\mathcal{O})).$$

It follows from either characterization that $D^2 = 1$. If $\mathcal{M}$ is a module for $\mathcal{H}(\mathcal{O})$, then a *Verdier duality for* $\mathcal{M}$ is a **Z**-linear involution

$$D_\mathcal{M} : \mathcal{M} \to \mathcal{M}$$

satisfying

$$D_\mathcal{M}(a \cdot m) = (Da) \cdot (D_\mathcal{M} m) \quad (a \in \mathcal{H}(\mathcal{O}), m \in \mathcal{M}).$$

Again several equivalent characterizations are possible, notably

$$D_\mathcal{M}(u^{1/2} \cdot m) = u^{-1/2}(D_\mathcal{M} m), \quad D_\mathcal{M}((T_s+1) \cdot m) = u^{-1}(T_s+1) \cdot (D_\mathcal{M} m)$$
$$(m \in \mathcal{M}, s \in S(\mathcal{O})).$$

Proposition 16.15 ([40], Theorem 1.10, and [55], Lemma 6.8). *In the setting of Proposition 16.13, there is a natural Verdier duality D_X on the Hecke algebra module $KX(\mathcal{O}, {}^\vee G^\Gamma)$. It is characterized uniquely by the requirement that the matrix of D_X with respect to the basis $\{\mu(\xi)\}$ of (7.10)(c) be upper triangular in the Bruhat order (cf. (7.11)(f)) with $u^{-d(\xi)}$ on the diagonal.*

The duality D_X is constructed in [40] directly from Verdier duality for complexes of sheaves. (The powers of u on the diagonal here differ from those in [55] by a constant. This modifies D_X by a power of u, but has no other effect.) Since we are interested in using this duality to compute the geometric character matrix, there is a small problem of circularity: the characterization of D_X involves the Bruhat order, which in turn is defined using the geometric character matrix. This problem is circumvented in [55] by defining in elementary terms a weaker preorder relation, then proving a stronger uniqueness result for D_X, involving only the weaker preorder. In any case, the proof of the uniqueness of D_X provides an algorithm for computing it.

To formulate an analogous result for the mixed Grothendieck group of representations, we need a function on $L^z(G_\mathbb{R})$ analogous to the function $d(\xi)$ (dimension of the orbit) on $\Xi({}^\vee G^\Gamma)$. In the cases of interest Theorem 10.4 provides a bijection between these two sets, so we could simply use $d(\xi)$; but this is a little unsatisfactory aesthetically. What we want is essentially the "integral length" of [54], Definition 8.1.4; we will normalize it in a slightly different way, however.

Definition 16.16. Let $G_\mathbb{R}$ be a linear Lie group as in (11.1). Fix a Langlands decomposition $M_\mathbb{R} A_\mathbb{R} N_\mathbb{R}$ of a minimal parabolic subgroup of $G_\mathbb{R}$, and let B_M be a Borel subgroup of M. Define

$$c_0(G_\mathbb{R}) = 1/2(\dim B_M). \qquad (16.16)(a)$$

Suppose that Λ is a $G_\mathbb{R}$-limit character of a Cartan subgroup $T_\mathbb{R}$, with differential $\lambda \in \mathfrak{t}_\mathbb{R}$ (Definition 11.2). Assume that Λ has infinitesimal character corresponding to $\mathcal{O}$ as in (16.7), so that we have a well-defined set $R^+(\lambda)$ of positive integral roots (cf. (16.9)(d)). Choose a Cartan involution θ of $G_\mathbb{R}$ preserving $T_\mathbb{R}$. Then θ acts on the roots of T in G, and this action preserves the integral roots. (For this we need the linearity of $G_\mathbb{R}$.) We can therefore define the *integral length of* Λ by

$$l^I(\Lambda) = -1/2(|\{\,\alpha \in R^+(\lambda) | \theta\alpha \in R^+(\lambda)\,\}| + \dim(T^\theta_\mathbb{R})) + c_0(G_\mathbb{R}).$$
$$(16.16)(b)$$

Clearly this differs by a constant (depending on $\mathcal{O}$ and $G_\mathbb{R}$) from the definition in [56], Definition 12.1 or [54], Definition 8.1.4. It is also evident from the discussion in [56] that l^I takes non-positive integral values. In the setting of (15.6), we will write

$$l^I(\xi) = l^I(\Lambda) \qquad (16.16)(c)$$

whenever $\xi \in \Xi^z(G/\mathbb{R})$ corresponds to $\Lambda \in L^z(G(\mathbb{R}, \delta_s))$.

16. The translation principle and the Kazhdan-Lusztig algorithm

Proposition 16.17 ([56], Lemma 12.14). *In the setting of Proposition 16.13, there is natural Verdier duality D_Π on the Hecke algebra modules $\mathcal{K}\Pi^z(\mathcal{O}, G/\mathbb{R})$. It is characterized uniquely by the requirement that the matrix of D_Π with respect to the basis $\{M(\Lambda)\}$ of (15.1)(a) be upper triangular in the Bruhat order (cf. (15.3)(d)) with $u^{-l^I(\Lambda)}$ on the diagonal.*

The construction of D_Π in [56] is a very complicated reduction to the case of integral infinitesimal character. In that case the Beilinson-Bernstein localization theory provides a further reduction to the case of Proposition 16.15. Again it is important to establish a stronger uniqueness theorem (not involving the Bruhat order) so that D_Π is computable.

We now have all the ingredients needed to define the Kazhdan-Lusztig polynomials.

Proposition 16.18 ([55], Theorem 7.1). *In the setting (16.7)–(16.12), there is for every geometric parameter $\gamma \in \Xi(\mathcal{O}, {}^\vee G^\Gamma)$ an element*

$$C_g(\gamma) = \sum_\xi P_g(\xi, \gamma)(u)\mu(\xi) \qquad (P_g(\xi, \gamma) \in \mathbb{Z}[u^{1/2}, u^{-1/2}])$$

of $\mathcal{K}X(\mathcal{O}, {}^\vee G^\Gamma)$ characterized by the following properties.

i) $D_X C_g(\gamma) = u^{-d(\gamma)} C_g(\gamma)$ (cf. Proposition 16.15).
ii) $P_g(\gamma, \gamma) = 1$.
iii) $P_g(\xi, \gamma) \neq 0$ only if $\xi \leq \gamma$ in the Bruhat order (cf. (7.11)(f)).
iv) If $\xi \neq \gamma$, then $P_g(\xi, \gamma)$ is a polynomial in u of degree at most $1/2(d(\gamma) - d(\xi) - 1)$.

As in Proposition 16.15, there is actually a stronger uniqueness theorem, and an algorithm for computing the polynomials. We call the P_g *geometric Kazhdan-Lusztig polynomials.* Here is the main result of [40].

Theorem 16.19 ([40], Theorem 1.12). *In the setting of Proposition 16.18, fix $\gamma, \xi \in \Xi(\mathcal{O}, {}^\vee G^\Gamma)$. Recall from (7.10)–(7.11) the perverse sheaf $P(\gamma)$ and the local system V_ξ on S_ξ.*

a) $H^{-d(\gamma)+i} P(\gamma) = 0$ if i is odd.
b) The multiplicity of V_ξ in $H^{-d(\gamma)+i} P(\gamma)|_{S_\xi}$ is the coefficient of $u^{i/2}$ in $P_g(\xi, \gamma)$.
c) $c_g(\xi, \gamma) = (-1)^{d(\xi)-d(\gamma)} P_g(\xi, \gamma)(1)$.
d) The specialization $C_g(\gamma)(1) \in \mathcal{K}X(\mathcal{O}, {}^\vee G^\Gamma)$ is equal to $(-1)^{d(\gamma)} P(\gamma)$.

Here the first two assertions are in [40] (although our present definition of $P(\gamma)$, which follows [9], differs by a degree shift from the one in

[40]). The third assertion is immediate from the second and (7.11)(d), and the last follows from (7.11)(c).

On the representation-theoretic side, things are formally quite similar.

Proposition 16.20 ([56], Lemma 12.15). *Suppose we are in the setting (16.7)–(16.12) (so that in particular $G_\mathbb{R}$ is a linear group). Then there is for every element $\Theta \in L^z(\mathcal{O}, G_\mathbb{R})$ an element*

$$C_r(\Theta) = \sum_\Lambda (-1)^{l^I(\Theta) - l^I(\Lambda)} P_r(\Lambda, \Theta)(u) M(\Lambda)$$

$$(P_r(\Lambda, \Theta) \in \mathbb{Z}[u^{1/2}, u^{-1/2}])$$

of $\mathcal{K}\Pi^z(\mathcal{O}, G_\mathbb{R})$ characterized by the following properties.

i) $D_\Pi C_r(\Theta) = u^{-l^I(\Theta)} C_r(\Theta)$ *(cf. Proposition 16.17).*
ii) $P_r(\Theta, \Theta) = 1$.
iii) $P_r(\Lambda, \Theta) \neq 0$ *only if $\Lambda \leq \Theta$ in the Bruhat order (cf. (15.3)(d)).*
iv) *If $\Lambda \neq \Theta$, then $P_r(\Lambda, \Theta)$ is a polynomial in u of degree at most $1/2(l^I(\Theta) - l^I(\Lambda) - 1)$.*

Again there is a better uniqueness theorem and an algorithm for computing the polynomials. We call the P_r *representation-theoretic Kazhdan-Lusztig polynomials*. To state a result completely analogous to Theorem 16.19, we need to compute entries of the representation-theoretic character matrix as Euler characteristics.

Proposition 16.21. *Suppose $\Lambda = (\Lambda^{can}, R^+_{i\mathbb{R}}, R^+_\mathbb{R})$ is a $G_\mathbb{R}$-limit character of $T_\mathbb{R}$ of infinitesimal character $\mathcal{O}$ (cf. Definition 11.2 and (16.7)). Choose a Cartan involution θ for $G_\mathbb{R}$ preserving $T_\mathbb{R}$. Write λ for the differential of Λ^{can}, so that $R^+(\lambda)$ is the corresponding set of positive integral roots (cf. (16.9)(d)). After replacing Λ by an equivalent limit character, assume that $R^+_\mathbb{R}(\lambda) \subset R^+_\mathbb{R}$ (Definition 11.6). Fix a system of positive roots*

$$R^+ \supset R^+(\lambda) \cup R^+_\mathbb{R}$$

for T in G. Define

$$b = 1/2 |\{\alpha \in R^+ - R^+(\lambda) \mid \theta\alpha \in R^+\}|,$$

a non-negative integer. Let $\mathfrak{n}$ be the nilpotent subalgebra spanned by the negative root vectors for R^+, so that $\Lambda^{can} \otimes \rho(\mathfrak{n})$ is a character of $T_\mathbb{R}^{can, G}$ of type z. Recall from Definition 11.2 the dual standard representation $\check{M}(\Lambda)$.

16. The translation principle and the Kazhdan-Lusztig algorithm   187

a) *The weight $\Lambda^{can} \otimes \rho(\mathfrak{n})$ occurs in $H_i(\mathfrak{n}, \tilde{M}(\Lambda))$ exactly once, in degree $b - l^I(\Lambda)$.*
b) *If Θ is another standard limit character not equivalent to Λ, then $\Lambda^{can} \otimes \rho(\mathfrak{n})$ does not occur in $H_i(\mathfrak{n}, \tilde{M}(\Theta))$.*
c) *For any standard limit character Θ, the representation-theoretic character matrix is given by*

$$c_r(\Lambda, \Theta) = (-1)^{l^I(\Lambda)+b} \sum_i (-1)^i \quad \text{multiplicity of}$$

$$\Lambda^{can} \otimes \rho(\mathfrak{n}) \; in H_i(\mathfrak{n}, \pi(\Theta)).$$

We use here the Harish-Chandra module version of the standard representations (cf. (15.2), taking for $K_\mathbb{R}$ the fixed points of the Cartan involution θ chosen at the beginning of the proposition. The Lie algebra $\mathfrak{t}$ and the group $T_\mathbb{R} \cap K_\mathbb{R}$ both act on the Lie algebra homology groups, and the multiplicities are to be interpreted in the category of $(\mathfrak{t}, T_\mathbb{R} \cap K_\mathbb{R})$-modules. It is not difficult to reformulate the result without a choice of θ, using instead the complex conjugation on T coming from $T_\mathbb{R}$. In this form it is probably true for the various smooth forms of the standard modules discussed in (15.2), but we do not know how to prove it.

Proof. Since $G_\mathbb{R}$ is assumed to be linear, all the imaginary roots are integral. The set of roots appearing in the definition of b therefore consists of certain pairs $\alpha, \theta\alpha$ of complex roots. Consequently b is an integer.

In the case of integral infinitesimal character, parts (a) and (b) are Corollary 4.7 of [55]. The proof given there carries through essentially without change in general. (It consists mostly of references to [54], where there is no assumption of integrality.) Part (c) is a formal consequence of (a) and (b), together with the fact that the Euler characteristic of $\mathfrak{n}$ homology is a well-defined map from the Grothendieck group $K\Pi^z(G_\mathbb{R})$ to $K\Pi^z(T_\mathbb{R})$.

Theorem 16.22. *In the setting of Proposition 16.20, fix Λ and Θ in $L^z(\mathcal{O}, G_\mathbb{R})$, and choose $\mathfrak{n}$ as in Proposition 16.21.*
a) *$\Lambda^{can} \otimes \rho(\mathfrak{n})$ does not occur in $H_{b-l^I(\Theta)+i}(\mathfrak{n}, \pi(\Theta))$ if i is odd.*
b) *The multiplicity of $\Lambda^{can} \otimes \rho(\mathfrak{n})$ in $H_{b-l^I(\Theta)+i}(\mathfrak{n}, \pi(\Theta))$ is the coefficient of $u^{i/2}$ in $P_r(\Lambda, \Theta)$.*
c) *$c_r(\Lambda, \Theta) = (-1)^{l^I(\Lambda)-l^I(\Theta)} P_r(\Lambda, \Theta)(1)$.*
d) *The specialization $C_r(\Theta)(1) \in K\Pi^z(\mathcal{O}, G_\mathbb{R})$ is equal to $\pi(\Theta)$.*

We postpone a discussion of the proof to the next chapter.

Definition 16.23. Suppose we are in the setting (15.6). The mixed Grothendieck groups $\mathcal{K}\Pi^z(\mathcal{O}, G/\mathbb{R})$ and $\mathcal{K}X(\mathcal{O}, {}^\vee G^\Gamma)$ (cf. (16.12)) are both free $\mathbb{Z}[u^{1/2}, u^{-1/2}]$-modules on bases parametrized by $\Xi^z(\mathcal{O}, G/\mathbb{R})$. It therefore makes sense to define the *canonical perfect pairing*

$$\langle , \rangle : \mathcal{K}\Pi^z(\mathcal{O}, G/\mathbb{R}) \times \mathcal{K}X(\mathcal{O}, {}^\vee G^\Gamma) \to \mathbb{Z}[u^{1/2}, u^{-1/2}]$$

to be the $\mathbb{Z}[u^{1/2}, u^{-1/2}]$-linear map satisfying

$$\langle M(\xi), \mu(\gamma) \rangle = e(\xi)\delta_{\xi,\gamma} u^{1/2(d(\gamma)+l^I(\xi))}$$

(cf. Definition 15.8, Definition 16.16) Notice that the specialization to $u = 1$ of this pairing is the canonical pairing of Definition 15.11.

Theorem 16.24. *Suppose we are in the setting (15.6). With respect to the pairing of Definition 16.23, the elements $C_r(\xi)$ and $C_g(\gamma)$ of Propositions 16.18 and 16.20 satisfy*

$$\langle C_r(\xi), C_g(\gamma) \rangle = e(\xi)\delta_{\xi,\gamma} u^{1/2(d(\gamma)+l^I(\xi))}.$$

The geometric and representation-theoretic Kazhdan-Lusztig matrices are essentially inverse transposes of each other:

$$\sum_\eta (-1)^{l^I(\xi)-l^I(\eta)} P_r(\eta, \xi) P_g(\eta, \gamma) = \delta_{\xi,\gamma}.$$

We will discuss the proof of the first assertion in Chapter 17. The second is a formal consequence, as in the proof of Corollary 15.13 from Theorem 15.12.

Theorem 15.12 follows from Theorem 16.24 by specializing to $u = 1$ (see Theorem 16.19(d) and Theorem 16.22(d)). Q.E.D.

17. Proof of Theorems 16.22 and 16.24

When the infinitesimal character is integral, Theorem 16.22 is the main result (Theorem 7.3) of [55]; it is a more or less straightforward consequence of [8] and [40]. The proofs in [55] can be modified easily to cover the case when the simple root system $\Delta(\mathcal{O})$ of (16.9) is contained in a set of simple roots for $R(G, T_\mathcal{O})$. Unfortunately this is not always the case. That the general case can be treated has been known to various experts for many years, but there does not seem to be an account of it in print. The geometric part of the argument for the case of Verma modules may be found in the first chapter of [37]. The outline below is gleaned from conversations with Bernstein, Brylinski, Kashiwara, and Lusztig; it is due to them and to Beilinson. To simplify the notation, we take the central element z in $Z(^\vee G)^{\theta_z}$ to be trivial; this changes nothing.

To begin, we must choose a system of positive roots

$$R^+(G, T_\mathcal{O}) \supset R^+(\mathcal{O}) \qquad (17.1)(a)$$

for the root system $R(G, T_\mathcal{O})$. There is no distinguished choice for this positive system, and in fact the argument will use several different ones. We write

$$\rho_\mathcal{O} = 1/2 \sum_{\alpha \in R^+(G, T_\mathcal{O})} \alpha. \qquad (17.1)(b)$$

Now classical intertwining operator methods (as for example in [49]) show that $c_r(\Lambda, \Theta)$ is unchanged by a small modification of Θ that does not affect the integral roots. (Implicit here is the assertion that it is possible to make a corresponding modification of Λ.) After making such a modification, we may assume that the infinitesimal character is *rational*; that is, (in the notation of (16.8)) that there is a positive integer n with

$$n(\lambda_\mathcal{O} - \rho_\mathcal{O}) = \gamma_\mathcal{O} \in X^*(T_\mathcal{O}). \qquad (17.1)(c)$$

Define

$$\mathcal{B} = \text{variety of Borel subalgebras of } \mathfrak{g}. \qquad (17.2)(a)$$

If $\mathfrak{b} \in \mathcal{B}$ is any Borel subalgebra with Cartan subalgebra $\mathfrak{t}$, then there is a natural isomorphism $j : \mathfrak{t}_{\mathcal{O}} \to \mathfrak{t}$ carrying $R^+(G, T_{\mathcal{O}})$ to the roots of $\mathfrak{t}$ in $\mathfrak{g}/\mathfrak{b}$. Using these isomorphisms, and the character $\gamma_{\mathcal{O}}$ of $T_{\mathcal{O}}$, we can define an algebraic line bundle

$$\mathcal{L} \to \mathcal{B}. \qquad (17.2)(b)$$

(If the infinitesimal character is integral, then $\mathcal{L}$ has non-trivial sections.) The complexification $K_{\mathbb{C}}$ of a maximal compact subgroup $K_{\mathbb{R}}$ of $G_{\mathbb{R}}$ acts algebraically on $\mathcal{B}$ and $\mathcal{L}$, with a finite number of orbits on $\mathcal{B}$. We make the multiplicative group $\mathbb{C}^\times$ act on $\mathcal{L}$ one fiber at a time, by

$$z \cdot \xi = z^n \xi \qquad (z \in \mathbb{C}^\times, \xi \in \mathcal{L}). \qquad (17.2)(c)$$

Then the product group $H = K_{\mathbb{C}} \times \mathbb{C}^\times$ acts on

$$\mathcal{L}^\times = \mathcal{L} - \text{ zero section }, \qquad (17.2)(d)$$

with finitely many orbits corresponding precisely to the orbits of $K_{\mathbb{C}}$ on $\mathcal{B}$.

It is convenient to develop this situation a little more generally at first. So suppose Y is a smooth complex algebraic variety, and $\mathcal{L}$ is an algebraic line bundle. Write $\mathcal{L}^\times$ for $\mathcal{L}$ with the zero section removed:

$$\pi : \mathcal{L}^\times \to Y \qquad (17.3)(a)$$

a principal $\mathbb{C}^\times$ bundle. We make $\mathbb{C}^\times$ act on $\mathcal{L}^\times$ as in (17.2)(c):

$$z \cdot \xi = z^n \xi \qquad (z \in \mathbb{C}^\times, \xi \in \mathcal{L}^\times). \qquad (17.3)(b)$$

We may therefore speak of $\mathbb{C}^\times$-*equivariant* $\mathcal{D}_{\mathcal{L}^\times}$-*modules* on $\mathcal{L}^\times$. If $\mathcal{M}$ is such a module, then its direct image on Y is graded by the action of $\mathbb{C}^\times$:

$$\pi_*(\mathcal{M}) = \sum_{k \in \mathbb{Z}} \pi_*(\mathcal{M})(k) \qquad (17.3)(c)$$

We say that $\mathcal{M}$ is *genuine* if the group of nth roots of unity in $\mathbb{C}^\times$ acts by the inverse of the tautological character:

$$\omega \cdot m = \omega^{-1} m \qquad (\omega^n = 1, m \in \mathcal{M}) \qquad (17.3)(d)$$

17. Proof of Theorems 16.22 and 16.24

Evidently this is equivalent to

$$\pi_*(\mathcal{M}) = \sum_{k \equiv -1 \pmod{n}} \pi_*(\mathcal{M})(k) \qquad (17.3)(e)$$

One can define a sheaf of algebras $\mathcal{D}_Y(\mathcal{L}^{1/n})$ on Y, as follows. Informally, $\mathcal{D}_Y(\mathcal{L}^{1/n})$ is the algebra of differential operators on sections of the $(1/n)$th power of $\mathcal{L}$. The difficulty is that there is usually no such line bundle. Now if k is an integer, the space of sections of $\mathcal{L}^{\otimes k}$ may be identified with the space of functions on $\mathcal{L}^\times$ homogeneous of degree $-k$ in the fiber variable; that is, with functions satisfying

$$f(z \cdot \xi) = z^{-nk} f(\xi). \qquad (17.4)(a)$$

Write E for the (Euler) vector field on $\mathcal{L}^\times$, induced by the vector field $z\frac{d}{dz}$ on $\mathbb{C}^\times$ and the action (17.2)(c). This is a globally defined vector field on $\mathcal{L}^\times$. Clearly

$$\{\text{ functions homogeneous of degree } -k\}$$
$$= \{\text{ functions killed by } E + nk\}. \qquad (17.4)(b)$$

It follows easily that the differential operators on sections of $\mathcal{L}^{\otimes k}$ may be identified with the differential operators on $\mathcal{L}$ commuting with E, modulo the ideal generated by $E + nk$. This suggests defining (over every open set $U \subset Y$)

$$\mathcal{D}_Y(\mathcal{L}^{1/n})(U) = (\mathcal{D}_{\mathcal{L}^\times}(\pi^{-1}U))^E / (E + 1) \qquad (17.4)(c)$$

Here on the right we are dividing by the ideal generated by $E + 1$.

Proposition 17.5. *In the setting (17.3)-(17.4), there is a natural equivalence of categories*

$$(\text{genuine } \mathbb{C}^\times\text{-equivariant } \mathcal{D}_{\mathcal{L}^\times}\text{-modules on } \mathcal{L}^\times) \leftrightarrow$$
$$(\mathcal{D}_Y(\mathcal{L}^{1/n})\text{-modules on } Y).$$

In the notation of (17.3)(e), the equivalence is

$$\mathcal{M} \leftrightarrow \pi_*(\mathcal{M})(-1).$$

This is easy general nonsense.

We now return to the setting of (17.1) and (17.2). The action of G on $\mathcal{B}$ lifts to $\mathcal{L}^\times$, and so defines an operator representation

$$\psi_{\mathcal{L}^\times} : U(\mathfrak{g}) \to \mathcal{D}_{\mathcal{L}^\times}. \qquad (17.6)(a)$$

The action of G commutes with that of $\mathbb{C}^\times$, so the image of $U(\mathfrak{g})$ commutes with E. By (17.4)(c), we get

$$\psi_B(\mathcal{L}^{1/n}) : U(\mathfrak{g}) \to \mathcal{D}_B(\mathcal{L}^{1/n}). \qquad (17.6)(b)$$

We write

$$I_B(\mathcal{L}^{1/n}) = \ker \psi_B(\mathcal{L}^{1/n}) \subset U(\mathfrak{g}). \qquad (17.6)(c)$$

Theorem 17.7 (Beilinson-Bernstein localization theorem — see [8]). *Suppose we are in the setting (17.1)–(17.2) (so that in particular the weight $\lambda_\mathcal{O}$ is regular). Use the notation of (17.3)–(17.4) and (17.6).*

a) The operator representation (17.6)(b) is surjective, with kernel equal to the ideal generated by the kernel of the infinitesimal character $\chi_\mathcal{O}$.

b) The global sections and localization functors provide an equivalence of categories between quasicoherent sheaves of $\mathcal{D}_B(\mathcal{L}^{1/n})$-modules on $\mathcal{B}$ and $\mathfrak{g}$-modules of infinitesimal character $\chi_\mathcal{O}$.

Corollary 17.8. *In the setting of Theorem 17.7, there is a (contravariant) equivalence of categories between finite-length $(\mathfrak{g}, K_\mathbb{C})$-modules of infinitesimal character $\chi_\mathcal{O}$, and genuine H-equivariant perverse sheaves on $\mathcal{L}^\times$.*

Here $H = K_\mathbb{C} \times \mathbb{C}^\times$ acts on $\mathcal{L}^\times$ as in (17.2)(c), and "genuine" is explained in (17.3)(d). To get this, we first apply Proposition 17.5 to pass to genuine H-equivariant coherent $\mathcal{D}_{\mathcal{L}^\times}$-modules, then use the Riemann-Hilbert correspondence of Theorem 7.9. To get the equivalence to be contravariant, we must use the solution sheaf version of the Riemann-Hilbert correspondence rather than the DeRham functor; this is the approach used in [55].

According to (7.10), the genuine irreducible H-equivariant perverse sheaves on $\mathcal{L}^\times$ are parametrized by genuine H-equivariant local systems on orbits; so we must study such local systems.

Lemma 17.9. *In the setting of Theorem 17.7, there is a natural bijection between genuine H-equivariant local systems on H-orbits on $\mathcal{L}^\times$, and equivalence classes of limit characters in $L(\mathcal{O}, G_\mathbb{R})$ (cf. (15.5)). Suppose the limit character Λ corresponds to a local system on an orbit S. In the notation of Definition 16.16 and Propostion 16.21, the codimension of the orbit S is equal to $b - l^I(\Lambda)$.*

Proof. Fix a $K_\mathbb{C}$-orbit S_0 on $\mathcal{B}$, and write S for its preimage in $\mathcal{L}^\times$

17. Proof of Theorems 16.22 and 16.24

(an orbit of H). We can find a Borel subalgebra

$$\mathfrak{b} = \mathfrak{t} + \mathfrak{n} \in S_0 \qquad (17.10)(a)$$

with the property that $\mathfrak{t}$ is both θ-stable and defined over $\mathbb{R}$. The stabilizer B_K of $\mathfrak{b}$ in K has a Levi decomposition

$$B_K = T_K N_K; \qquad (17.10)(b)$$

the first factor is reductive, with compact real form

$$T_{K,\mathbb{R}} = T_K \cap K_\mathbb{R} = T_\mathbb{R} \cap K_\mathbb{R}. \qquad (17.10)(c)$$

There is a natural map from B_K into the Borel subgroup B corresponding to $\mathfrak{b}$ (induced by restricting (11.1)(a) to $K_\mathbb{R}$ and complexifying). On the other hand, the line bundle $\mathcal{L}$ gives a character γ of B, and so of B_K. The stabilizer in H of any point ξ of $\mathcal{L}^\times$ over $\mathfrak{b}$ is then

$$\widetilde{B_K} = \{ (b, z) \mid b \in B_K, z \in \mathbb{C}^\times, \gamma(b) = z^{-n} \}. \qquad (17.10)(d)$$

By projection on the first factor, we see that $\widetilde{B_K}$ is a central extension of B_K by the nth roots of one. Projection on the second factor defines a genuine character of $\widetilde{B_K}$ of differential equal to $\rho - \lambda$. Tensoring with this character therefore defines a bijection

$$\text{(characters of } B_K \text{ with differential } \lambda - \rho) \leftrightarrow$$
$$\text{(genuine characters of } \widetilde{B_K}/(\widetilde{B_K})_0). \qquad (17.10)(e)$$

The objects on the right are essentially the genuine geometric parameters for the orbit S. Using (17.10)(b) and (17.10)(c), we may identify the characters on the left with characters of $T_\mathbb{R}$ with differential $\lambda - \rho$. (The point is that $T_\mathbb{R}$ is a direct product of $T_{K,\mathbb{R}}$ with a vector group, so that characters with specified differential are determined by their restrictions to $T_{K,\mathbb{R}}$.) Such characters in turn give rise (by twisting by ρ as in (11.3)) to limit characters of differential λ, and so to classes in $L(\mathcal{O}, G_\mathbb{R})$. We leave to the reader the easy verification that this construction establishes the bijection we need. The last assertion is an elementary calculation, which we omit. Q.E.D

Proposition 17.11. *In the setting of Theorem 17.7, suppose V is a finite-length Harish-Chandra module for $G_\mathbb{R}$ of infinitesimal character $\chi_\mathcal{O}$. Let P be the corresponding perverse sheaf on $\mathcal{L}^\times$ (Theorem 17.7). Fix an orbit S and a genuine H-equivariant irreducible local system $\mathcal{V}$ on S. Choose corresponding $T_\mathbb{R}$, $\mathfrak{n}$, and limit character Λ as in the proof* of Lemma 17.9. Then the multiplicity of $\mathcal{V}$ in $H^i P|_S$ is equal to the multiplicity of the character $\Lambda^{can} \otimes \rho(\mathfrak{n})$ in $H_{i+\dim \mathcal{L}^\times}(\mathfrak{n}, V)$.

The proof is parallel to that of Proposition 4.1 of [55], and we omit it.

The main assertion in Theorem 16.22 is (b); for (a) is a special case, and (c) and (d) follow from (b) and Proposition 16.21. So we need to calculate certain Lie algebra homology groups with coefficients in irreducible representations. Proposition 17.11 reduces this to calculating the stalks of some irreducible perverse sheaves. Assembling all of this (and the dimension calculation in Lemma 17.9) we see that Theorem 16.22 is equivalent to

Theorem 17.12. *In the setting of Theorem 17.7, suppose $P(\Theta)$ is the irreducible perverse sheaf on $\mathcal{L}^\times$ corresponding to the limit character $\Theta \in L(\mathcal{O}, G_\mathbb{R})$. Write d for the dimension of the underlying orbit. Suppose Λ is another limit character, corresponding to the local system $\mathcal{V}$ on the orbit S. Then the multiplicity of $\mathcal{V}$ in $H^{-d+i} P(\Theta)|_S$ is the coefficient of $u^{i/2}$ in $P_r(\Lambda, \Theta)$.*

The analogy with Theorem 16.19 is now immediately apparent.

To prove Theorem 17.12, we can proceed as follows. The corresponding calculation of Lie algebra homology groups is carried out in [54], under the hypothesis that certain representations $U_\alpha(X)$ are completely reducible (cf. [55], section 7). Here X is an irreducible representation of infinitesimal character $\chi_\mathcal{O}$, and $\alpha \in \Delta(\mathcal{O})$ is a simple integral root (cf. (16.9)). The idea is to prove that complete reducibility more or less geometrically. Write P and $U_\alpha(P)$ for the perverse sheaves corresponding to X and $U_\alpha(X)$ (Corollary 17.8). We may assume by induction that Theorem 17.12 is true for P. We would like to find a geometric description of $U_\alpha(P)$. Unfortunately, no such description is available in general. To get one, we must first replace our choice of positive root system in (17.1) by a new one

$$(R^+)'(G, T_\mathcal{O}) \supset R^+(\mathcal{O}) \qquad (17.13)(a)$$

with the additional property that

$$\alpha \text{ is a simple root in } (R^+)'(G, T_\mathcal{O}). \qquad (17.13)(b)$$

Such a choice certainly exists; the difficulty is that we cannot make one choice for all α. This change replaces all the nilpotent algebras $\mathfrak{n}$ whose homology we considered by slightly different algebras $\mathfrak{n}'$; but Proposition 4.3 of [55] guarantees that the homology changes only by a

17. Proof of Theorems 16.22 and 16.24

shift in degree. Next, we must replace the infinitesimal character $\lambda_{\mathcal{O}}$ by some

$$\lambda_{\mathcal{O}'} = \lambda_{\mathcal{O}} + \mu, \qquad (17.13)(c)$$

with $\mu \in X^*(T_{\mathcal{O}})$. What we require of the new infinitesimal character is

$$\lambda_{\mathcal{O}'} \text{ is regular and dominant for } R^+(\mathcal{O}), \text{ and } \alpha^\vee(\lambda_{\mathcal{O}'}) = 1. \quad (17.13)(d)$$

(To get the second condition, we may have to replace G by some finite cover.) The translation principle identifies the Lie algebra homology groups we want with some corresponding ones for representations of infinitesimal character attached to $\mathcal{O}'$.

After making all of these adjustments, we are reduced to the case when our simple integral root α is simple in $R^+(G,T)$, and the coroot $\alpha^\vee$ is one on $\lambda_{\mathcal{O}}$. Because of (17.1)(c), we have

$$\alpha^\vee(\gamma_{\mathcal{O}}) = 0 \qquad (17.14)(a)$$

Write

$$\mathcal{P}_\alpha = \text{variety of parabolic subalgebras of type } \alpha \qquad (17.14)(b)$$

(as for example in [40], (3.2)). The condition (17.14)(a) is what is needed for the weight $\gamma_{\mathcal{O}}$ to define an equivariant algebraic line bundle

$$\mathcal{L}_\alpha \to \mathcal{P}_\alpha. \qquad (17.14)(c)$$

The natural projection π_α from $\mathcal{B}$ to $\mathcal{P}_\alpha$ pulls $\mathcal{L}_\alpha$ back to $\mathcal{L}$; so it defines a (proper equivariant) morphism

$$\pi_\alpha : \mathcal{L}^\times \to \mathcal{L}_\alpha^\times. \qquad (17.14)(d)$$

The fibers of π_α are one-dimensional projective lines. The *geometric U_α functor* on the derived category of H-equivariant constructible sheaves on $\mathcal{L}^\times$ is given by

$$U_{\alpha,g}(C) = \pi_\alpha^*(\pi_\alpha)_*(C) \qquad (17.14)(e)$$

(cf. [40], Definition 3.1, and [37], Lemma 1.19).

Suppose now that P is an irreducible genuine H-equivariant perverse sheaf on $\mathcal{L}^\times$ as above. By the (very deep!) decomposition theorem of Beilinson, Bernstein, Deligne, and Gabber ([9], Theorem 6.2.5), the direct image $(\pi_\alpha)_* P$ (in an appropriate derived category) is a direct sum of irreducible (genuine H-equivariant) perverse sheaves on $\mathcal{L}_\alpha^\times$, with various degree shifts. By Proposition 7.15, $\pi_\alpha^*[1]$ carries irreducible perverse sheaves on $\mathcal{L}_\alpha^\times$ to irreducible perverse sheaves on $\mathcal{L}^\times$. It follows that $U_{\alpha,g}(P)$ is a direct sum in the derived category of irreducible perverse sheaves, with degree shifts. Since we know the cohomology sheaves of P by induction, and since they vanish in every other degree, it is possible to compute the cohomology sheaves of $U_{\alpha,g}(P)$ completely. The result agrees with the calculation of the Lie algebra homology of $U_\alpha(X)$ made in [54] (compare the argument at (7.9) – (7.12) in [55]). It follows first of all that $U_\alpha(P)$ and $U_{\alpha,g}(P)$ have the same image in the Grothendieck group. Now an equivariant perverse sheaf in our setting that is *not* completely reducible must have some stalks of its homology sheaves strictly smaller than those for the direct sum of its composition factors. (This is elementary; it is also not difficult to prove a corresponding statement on the level of Harish-Chandra modules.) Consequently $U_\alpha(P)$ is completely reducible, as we wished to show. Q.E.D.

We turn now to the proof of Theorem 16.24. Of course this is essentially taken from [56]. The idea is to define a certain natural dual of the Hecke algebra module $\mathcal{K}X(\mathcal{O},{}^\vee G^\Gamma)$, then to identify this dual with $\mathcal{K}\Pi^z(\mathcal{O},G/\mathbb{R})$ (or rather an appropriate completion) using the pairing of Definition 16.23. The dual module has a basis dual to the basis $\{C_g(\gamma)\}$; an element of this basis must for formal reasons satisfy the requirements of Definition 16.20 characterizing the $C_r(\xi)$. The first point is therefore to define a dual Hecke module.

Definition 17.15. In the setting of (16.10), suppose $\mathcal{M}$ is a module for $\mathcal{H}(\mathcal{O})$. Set

$$\mathcal{M}^* = \operatorname{Hom}_{\mathbb{Z}[u^{1/2},u^{-1/2}]}(\mathcal{M},\mathbb{Z}[u^{1/2},u^{-1/2}]). \qquad (17.15)(a)$$

Any $\mathbb{Z}[u^{1/2},u^{-1/2}]$-linear map A on $\mathcal{M}$ defines a $\mathbb{Z}[u^{1/2},u^{-1/2}]$-linear map A^t on $\mathcal{M}^*$, by the requirement

$$(A^t \mu)(m) = \mu(Am) \qquad (m \in \mathcal{M}, \mu \in \mathcal{M}^*). \qquad (17.15)(b)$$

As usual, we have $(AB)^t = B^t A^t$. Since the Hecke algebra is non-commutative, we cannot make $\mathcal{M}^*$ into a module for it just by transposing the action on $\mathcal{M}$; we must twist by an anti-automorphism of $\mathcal{H}(\mathcal{O})$. Such an anti-automorphism may be defined by sending T_w to $(-u)^{l(w)} T_w^{-1}$. (The inverse of T_s may be computed from (16.10)(c); it is

$$T_s^{-1} = u^{-1}(T_s + (1-u)). \qquad (17.15)(c)$$

17. Proof of Theorems 16.22 and 16.24

The invertibility of the other T_w then follows from (16.10)(b).) We can therefore make $\mathcal{M}^*$ a module for the Hecke algebra by defining

$$T_w \cdot \mu = (-u)^{l(w)} (T_w^{-1})^t \cdot \mu. \qquad (17.15)(d)$$

Using (17.15)(b)–(d), we can get three equivalent formulations of this definition:

$$\begin{aligned}(T_s \cdot \mu)(m) &= \mu((-T_s - 1 + u) \cdot m), \text{ or} \\ ((T_s + 1) \cdot \mu)(m) &= -\mu((T_s - u) \cdot m), \text{ or} \\ ((T_s - u) \cdot \mu)(m) &= -\mu((T_s + 1) \cdot m).\end{aligned} \qquad (17.15)(e)$$

Suppose that $\mathcal{M}$ admits a Verdier duality $D_\mathcal{M}$. We can define a $\mathbb{Z}$-linear involution $D_{\mathcal{M}^*}$ on $\mathcal{M}^*$ by

$$(D_{\mathcal{M}^*}\mu)(m) = \overline{\mu(D_\mathcal{M} m)} \qquad (17.15)(f)$$

Here bar denotes the automorphism of $\mathbb{Z}[u^{1/2}, u^{-1/2}]$ defined by $\overline{u^{1/2}} = u^{-1/2}$. It is straightforward to check that $D_{\mathcal{M}^*}$ is a Verdier duality for $\mathcal{M}^*$ (Definition 16.14).

Proposition 17.16. *Suppose we are in the setting of (15.6); use the notation of (16.12). Then the pairing of Definition 16.23 extends to*

$$\langle,\rangle : \overline{\mathcal{K}\Pi}^z(\mathcal{O}, G/\mathbb{R}) \times \mathcal{K}X(\mathcal{O}, {}^\vee G^\Gamma) \to \mathbb{Z}[u^{1/2}, u^{-1/2}]$$

On this level it provides an identification

$$\overline{\mathcal{K}\Pi}^z(\mathcal{O}, G/\mathbb{R}) \simeq \mathrm{Hom}_{\mathbb{Z}[u^{1/2}, u^{-1/2}]}(\mathcal{K}X(\mathcal{O}, {}^\vee G^\Gamma),$$
$$\mathbb{Z}[u^{1/2}, u^{-1/2}]) = \mathcal{K}X(\mathcal{O}, {}^\vee G^\Gamma)^*.$$

The Hecke algebra actions provided by Proposition 16.13 and Definition 17.15 are identified by this isomorphism.

Proof. Because the representation-theoretic character matrix is a direct sum of invertible finite blocks, the formal mixed Grothendieck group $\overline{\mathcal{K}\Pi}^z(\mathcal{O}, G/\mathbb{R})$ may be identified with formal sums of standard representations. Now the first assertion is obvious from Definition 16.23: the dual of a free module is a direct product over a dual basis. For the second, we must show (according to Definition 17.15(e))

$$\langle (T_s - u)M(\xi_1), \mu(\xi_2)\rangle = -\langle M(\xi_1), (T_s + 1)\mu(\xi_2)\rangle \qquad (17.17)$$

for every $\xi_1, \xi_2 \in \Xi^z(\mathcal{O}, G/\mathbb{R})$, and every simple reflection $s = s_\alpha \in S(\mathcal{O})$ (cf. (16.9)(f)). The proof of this is essentially identical to the proof of Proposition 13.10(b) in [56]. We will examine carefully one easy case and one difficult case.

So suppose ξ_1 corresponds to a complete Langlands parameter (ϕ_1, τ_1) for a Cartan subgroup ${}^d T_1^\Gamma$ of ${}^\vee G^\Gamma$ (Definition 12.4 and Theorem 12.9). Choose a compatible based Cartan subgroup structure

$$ {}^d T_1^\Gamma = ({}^d T_1^\Gamma, \mathcal{S}({}^d T_1^\Gamma), {}^d R_{i\mathbb{R},1}^+, {}^d R_{\mathbb{R},1}^+) \qquad (17.18)(a) $$

(Definition 13.11). Choose a based Cartan subgroup

$$ T_1^\Gamma = (T_1^\Gamma, \mathcal{W}(T_1^\Gamma), R_{i\mathbb{R},1}^+, R_{\mathbb{R},1}^+) \qquad (17.18)(b) $$

for G^Γ and a pairing

$$ \zeta_1 : {}^\vee T_1 \to {}^d T_1 \qquad (17.18)(c) $$

between them (Definition 13.9 and Proposition 13.10(b)). Construct a limit character (δ_1, Λ_1) for T_1^Γ (Definition 12.1) as in Proposition 13.12. By the definition in Proposition 13.13, the standard representation $M(\xi_1)$ is attached to the real form δ_1 of G and the limit character Λ_1. The parameter $\lambda_1 \in {}^d T_1$ attached to ϕ_1 (Proposition 5.6) belongs to $\mathcal{O}$, so (16.7)(d) provides natural isomorphisms

$$ j(\lambda_\mathcal{O}, \lambda_1) : {}^\vee T_\mathcal{O} \to {}^d T_1, \qquad i(\lambda_\mathcal{O}, \lambda_1) : T_\mathcal{O} \to T_1. \qquad (17.18)(d) $$

Under these maps, the simple root $\alpha \in \Delta(\mathcal{O})$ of (17.17) corresponds to

$$ \alpha_1 \in \Delta(\lambda_1) \qquad (17.18)(e) $$

(a root of T_1 in G), and $\alpha^\vee$ to a root

$$ {}^d \alpha_1 \in {}^d \Delta(\lambda_1) \qquad (17.18)(f) $$

of ${}^d T_1$ in ${}^\vee G$.

For our easy case, assume that α_1 is a compact imaginary root of T_1 in $G(\mathbb{R}, \delta_1)$, and that $\xi_1 = \xi_2$. By [56], Definition 12.3,

$$ T_s M(\xi_1) = u M(\xi_1). \qquad (17.19)(a) $$

17. Proof of Theorems 16.22 and 16.24

By Proposition 13.12(c), $^d\alpha_1$ is a real root failing to satisfy the parity condition. By [40], Lemma 3.5(e),

$$T_s\mu(\xi_2) = -\mu(\xi_2). \qquad (17.19)(b)$$

It follows from (17.19) that both sides of (17.17) are zero.

For a hard case, assume that α_1 is noncompact imaginary in $G(\mathbb{R}, \delta_1)$, and that the reflection s_{α_1} does *not* belong to the Weyl group of $T_1(\mathbb{R})$ in $G(\mathbb{R}, \delta_1)$. (In the terminology of [54], α_1 is "type I." This case arises in $SL(2, \mathbb{R})$ but not in $PGL(2, \mathbb{R})$. Our assumption on ξ_2 will be formulated after (17.20).) Define $G(\alpha_1)$ to be the subgroup generated by T_1 and the image of the root subgroup ϕ_{α_1} (see (11.10)). The assumption that α_1 is imaginary implies that $G(\alpha_1)$ is normalized by T_1^Γ, and in fact

$$G(\alpha_1)^\Gamma = G(\alpha_1)T_1^\Gamma \qquad (17.20)(a)$$

is a weak extended group. By the proof of Proposition 13.6 (after (14.10)), α_1 is also noncompact imaginary with respect to the real forms in $\mathcal{W}(T_1^\Gamma)$. We can therefore find an element

$$\delta_1^0 \in \mathcal{W}(T_1^\Gamma) \qquad (17.20)(b)$$

that acts as δ_1 does on $G(\alpha_1)$. Up to center, the common real form $G(\alpha_1, \delta_1) = G(\alpha_1, \delta_1^0)$ is locally isomorphic to $SL(2, \mathbb{R})$, and $T_1(\mathbb{R})$ corresponds to a compact Cartan subgroup. Now let $T_2(\mathbb{R})$ be another Cartan subgroup of $G(\alpha_1, \delta_1^0)$, corresponding to a split Cartan subgroup of $SL(2, \mathbb{R})$. Define

$$T_2^\Gamma = \text{group generated by } T_2 \text{ and } \delta_1^0, \quad \mathcal{W}(T_2^\Gamma) = \text{Ad}(T_2) \cdot \delta_1^0. \qquad (17.20)(c)$$

This is a Cartan subgroup of G^Γ, the *Cayley transform of T_1^Γ through α_1.* (Compare [54], Definition 8.3.4.) We want to define on T_2^Γ a structure of based Cartan subgroup. To do that, we first fix an isomorphism

$$c : T_1 \to T_2 \qquad (17.20)(d)$$

given by conjugation by an element of $G(\alpha_1)$. This choice carries α_1 to a real root α_2 of T_2 in G. It identifies the imaginary roots in T_2 with the imaginary roots in T_1 orthogonal to α_1, so we can define

$$R_{i\mathbb{R},2}^+ = c(R_{i\mathbb{R},1}^+) \cap R_{i\mathbb{R},2}. \qquad (17.20)(e)$$

On the other hand, c^{-1} carries the real roots of T_2 orthogonal to α_2 onto the real roots of T_1. We choose a set of positive real roots subject to the two conditions

$$R^+_{\mathbb{R},2} \supset c(R^+_{\mathbb{R},1}), \qquad \alpha_2 \text{ is simple in } R^+_{\mathbb{R},2}. \qquad (17.20)(f)$$

This is certainly possible. We claim that

$$T_2^\Gamma = (T_2^\Gamma, \mathcal{W}(T_2^\Gamma), R^+_{i\mathbb{R},2}, R^+_{\mathbb{R},2}) \qquad (17.20)(g)$$

is a based Cartan subgroup of G^Γ. The only difficult condition is (e) of Definition 13.5. To check it, suppose Λ_1^0 is a limit character for $G(\mathbb{R}, \delta_1^0)$ as in Definition 13.5(e). Form the Cayley transform of this limit character through the noncompact root α_1 ([54], Definition 8.3.6). This is a limit character Λ_2^0 for the Cartan subgroup $T_2(\mathbb{R})$ of $G(\mathbb{R}, \delta_1^0)$, again of the type considered in Definition 13.5(e). By a theorem of Hecht and Schmid ([54], Proposition 8.4.5), the standard representation $M(\Lambda_2^0)$ contains $M(\Lambda_1^0)$ as a subrepresentation. Since the latter is assumed to admit a Whittaker model, the former must as well; so T_2^Γ is a based Cartan subgroup. We define

$$\Lambda_2 = (\Lambda_2^{can}, R^+_{i\mathbb{R},2}, R^+_{\mathbb{R},2}) \qquad (17.20)(h)$$

to be the Cayley transform of Λ_1 through the root α_1 ([54], Definition 8.3.6).

We assume (back in (17.17)) that ξ_2 is the geometric parameter corresponding to (Λ_2, δ_1). Definition 12.3(c) of [56] expresses $(T_s - u)M(\xi_1)$ in terms of standard representations. The coefficient of $M(\xi_2)$ is 1. By Definition 16.23, the left side of (17.17) in this case is therefore $e(\xi_2)u^{1/2(d(\xi_2)+l^I(\xi_2))}$.

To continue with our calculation, we need to describe a based Cartan subgroup for $^\vee G^\Gamma$ paired with T_2^Γ. Since α_1 was assumed to be real, the corresponding root $^d\alpha_1$ for $^dT_1$ is a real root satisfying the parity condition (Proposition 13.12(c)). We may therefore proceed in analogy with (17.20) to construct a Cayley transform $^dT_2^\Gamma$ of $^dT_1^\Gamma$ through $^d\alpha_1$. Specifically, the map ϕ_1 of (17.18) provides an element

$$y \in {}^dT_1^\Gamma - {}^dT_1, \qquad (17.21)(a)$$

and we can choose

$$^\vee\delta_1 \in \mathcal{S}(^dT_1^\Gamma) \qquad (17.21)(b)$$

17. Proof of Theorems 16.22 and 16.24

acting on $^\vee G(^d\alpha_1)$ as y does. We define

$$\mathcal{S}(^dT_2^\Gamma) = \mathrm{Ad}(^dT_2) \cdot {}^\vee\delta_1, \qquad (17.21)(c)$$

and fix an isomorphism ^dc between the maximal tori inner for $^\vee G(^d\alpha_1)$. As in (17.20)(e), this choice provides a noncompact imaginary root $^d\alpha_2$, and a natural choice $^dR_{\mathbb{R},2}^+$ of positive real roots. It also carries $\lambda_1 \in {}^dt_1$ to $\lambda_2 \in {}^dt_2$, and the pair (y, λ_2) defines a Langlands parameter

$$\phi_2 : W_\mathbb{R} \to {}^dT_2^\Gamma. \qquad (17.21)(d)$$

We want to define a pairing between T_2 and $^dT_2$. We have a pairing ζ_1 from (17.18)(c), and the map c of (17.20)(d) defines an isomorphism $^\vee c$ from $^\vee T_2$ to $^\vee T_1$. Put

$$\zeta_2 = {}^dc \circ \zeta_1 \circ {}^\vee c : {}^\vee T_2 \to {}^dT_2, \qquad (17.21)(e)$$

Finally, let $^dR_{i\mathbb{R},2}^+$ be the set of roots corresponding to $R_{\mathbb{R},2}^+$ under this pairing. By construction and (17.20)(f),

$$R_{i\mathbb{R},2}^+ \supset {}^dc(R_{i\mathbb{R},1}^+), \qquad {}^d\alpha_2 \text{ is simple in } R_{i\mathbb{R},2}^+. \qquad (17.21)(f)$$

We claim that $R_{i\mathbb{R},2}^+$ is special ([1], Definition 6.29.) This follows from the appendix to [1] in essentially the same way as we checked Definition 13.5(e) after (17.20)(g). Granted this, it is now clear that

$$^dT_2^\Gamma = ({}^dT_2^\Gamma, \mathcal{S}({}^dT_2^\Gamma), {}^dR_{i\mathbb{R},2}^+, {}^dR_{\mathbb{R},2}^+) \qquad (17.21)(g)$$

is a based Cartan subgroup (Definition 13.7) paired with T_2^Γ by ζ_2. The strong real form δ_1 of T_2 (cf. (17.20)) defines a character τ_2 of the canonical component group for $^dT_2^\Gamma$ with respect to $^\vee G^\Gamma$ (Definition 12.4 and Proposition 13.12(a)). By Proposition 13.12, the pair (ϕ_2, τ_2) corresponds to the complete geometric parameter ξ_2.

By inspection of the definitions (particularly the proofs in Chapter 9) one can find the following more direct description of τ_2. (Additional details may be found in [54], section 8.3). Write

$$A_i = ({}^dT_i^y)^{alg, {}^\vee G} / ({}^dT_i^y)_0^{alg, {}^\vee G} \qquad (17.22)(a)$$

By calculation in $SL(2)$, one sees that

$$({}^dT_1^y)_0^{alg, {}^\vee G} \subset ({}^dT_2^y)_0^{alg, {}^\vee G}. \qquad (17.22)(b)$$

(That is, the compact part of the more compact Cartan subalgebra is larger.) Set

$$A_{12} = ({}^dT_1^y \cap {}^dT_2^y)^{alg, \vee G} / ({}^dT_1^y)_0^{alg, \vee G}. \qquad (17.22)(c)$$

By definition A_{12} is a subgroup of A_1; in fact it is precisely the kernel of the character defined by the real root ${}^d\alpha_1$ (which takes values in $\{\pm 1\}$). Similarly, the natural map from A_{12} to A_2 is surjective; the kernel is generated by the element $m_{{}^d\alpha_1}^{alg}$. Now τ_1 is a representation of A_1. We may therefore restrict it to a representation τ_{12} of A_{12}. Because of Proposition 13.12(c), τ_{12} descends to a representation of A_2, and this is the one we want.

We can now consider the right side of (17.17). Lemma 3.5(c2) of [40] expresses $(T_s + 1)\mu(\xi_2)$ in terms of other extensions by zero of sheaves on orbits. The coefficient of $\mu(\xi_1)$ is -1. By Definition 16.23, the right side of (17.17) is therefore $e(\xi_1)u^{1/2(d(\xi_1)+l^I(\xi_1))}$. To compare this with the formula for the left side obtained after (17.20), notice first that τ_1 and τ_2 agree on the central element $z(\rho)$ by (17.22). By Definition 15.8, $e(\xi_1) = e(\xi_2)$. Next, the geometric description of orbits in [40] shows that $d(\xi_1) = d(\xi_2) + 1$. Finally, it is elementary to show from Definition 16.16 and (17.20) that $l^I(\xi_1) = l^I(\xi_2) - 1$. It follows that the exponents of u on the two sides of (17.17) agree as well.

This completes our hard case; the similar arguments for other cases are left to the reader. Q.E.D.

Corollary 17.23. *The identification*

$$\overline{\mathcal{K}\Pi}^z(\mathcal{O}, G/\mathbb{R}) \simeq \mathcal{K}X(\mathcal{O}, {}^\vee G^\Gamma)^*$$

of Proposition 17.16 identifies the Verdier duality D_Π of Proposition 16.17 with the "conjugate transpose" of D_X (Definition 17.15). Explicitly,

$$\langle D_\Pi M, \mu \rangle = \overline{\langle M, D_X \mu \rangle}$$

for $M \in \overline{\mathcal{K}\Pi}^z(\mathcal{O}, G/\mathbb{R})$ and $\mu \in \mathcal{K}X(\mathcal{O}, {}^\vee G^\Gamma)$.

Proof. Proposition 17.16 guarantees that the conjugate transpose of D_X is a Verdier duality. Proposition 16.17 provides a characterization of D_Π in terms of its action on the basis $\{M(\xi)\}$. Definition 16.23 allows us to check very easily that the conjugate transpose of D_X has the required characteristic properties. (We refer to [56], Lemma 13.4 for more details.) Q.E.D.

17. Proof of Theorems 16.22 and 16.24

To complete the proof of Theorem 16.24, notice that Proposition 16.18 guarantees that the elements $C_g(\xi)$ form a basis of $\mathcal{K}X(\mathcal{O}, {}^\vee G^\Gamma)$. By Proposition 17.16, we may therefore define elements $C'_r(\xi)$ of $\mathcal{K}\Pi^z(\mathcal{O}, G/\mathbb{R})$ by the requirement

$$\langle C'_r(\xi), C_g(\gamma) \rangle = e(\xi)\delta_{\xi,\gamma} u^{1/2(d(\gamma)+l^I(\xi))}.$$

By Corollary 17.23 and Proposition 16.18(i), these elements satisfy Proposition 16.20(i). Slightly more calculation (using also Definition 16.23) shows that the remaining conditions (ii)–(iv) of Proposition 16.18 guarantee those in Proposition 16.20. By Proposition 16.20, $C'_r(\xi) = C_r(\xi)$, as we wished to show. (We refer to Lemma 13.7 of [56] for more details.) Q.E.D.

We conclude this chapter with the proof of Lemma 15.9. Applying the translation principle as in Proposition 16.6, we can reduce to the case when ξ has regular infinitesimal character. We may therefore place ourselves in the setting of (12.4)–(12.6), with ξ corresponding to some complete Langlands parameter (ϕ, τ_1) for ${}^dT^\Gamma$. We can choose a set ${}^dR^+$ of positive roots for ${}^dT$ in ${}^\vee G$ with the property that

$$ {}^dR^+ - {}^dR_\mathbb{R}^+ \text{ is preserved by } \theta. \qquad (17.24)(a)$$

For every root α, we can define an element

$$m_\alpha = \alpha^\vee(-1) \qquad (17.24)(b)$$

as in (12.6). For any positive root system, we have by (4.9)(b)

$$z(\rho) = \prod_{\alpha \in {}^dR^+} m_\alpha. \qquad (17.24)(c)$$

By (17.24),

$$m_\alpha m_{\theta\alpha} = m_\alpha(\theta m_\alpha) \in {}^dT_0^\theta. \qquad (17.24)(d)$$

Since the character τ_1 is trivial on the identity component of ${}^dT^\theta$, it follows from (17.24)(c)–(d) that

$$\tau_1(z(\rho)) = \prod_{\alpha \in {}^dR_\mathbb{R}^+} \tau_1(m_\alpha). \qquad (17.24)(e)$$

Proposition 13.12(c) allows us to translate this back to $G(\mathbb{R}, \delta)$ and a Cartan subgroup $T(\mathbb{R})$ dual to ${}^d T^\Gamma$. Here is the result. Let χ be the character of the lattice of imaginary roots for $T(\mathbb{R})$ in $G(\mathbb{R}, \delta)$ which is 1 on noncompact simple roots and -1 on compact simple roots. Then

$$e(\xi) = \prod_{\beta \in R^+_{i\mathbb{R}}} \chi(\beta). \qquad (17.24)(f)$$

At this point there are several ways to proceed. For one of them, let $A(\mathbb{R})$ be the maximal $\mathbb{R}$-split torus in $T(\mathbb{R})$, and $M(\mathbb{R})$ its centralizer in $G(\mathbb{R}, \delta)$. We have shown that $e(\xi)$ may be calculated in $M(\mathbb{R})$; since the same is true of Kottwitz's sign ([32], Corollary (6) on page 295) we may assume $M = G$. Let χ_0 be the character of the imaginary root lattice that is -1 on each simple root. Evidently $\chi\chi_0$ is -1 exactly on the noncompact roots, so

$$e(\xi) = (-1)^{(\text{number of noncompact positive roots})} \cdot e_0,$$

where e_0 does not depend on the real form. The exponent of -1 here is just $q(G(\mathbb{R}, \delta))$, half the dimension of the symmetric space attached to $G(\mathbb{R})$; so

$$e(\xi) = (-1)^{q(G(\mathbb{R}, \delta))} \cdot e_0.$$

On the other hand, if $G(\mathbb{R}, \delta')$ is quasisplit, then we can choose the positive root system so that all simple imaginary roots are noncompact. It follows from (17.24)(f) that $e(\xi) = 1$ in that case. Comparing with the last formula gives $e_0 = (-1)^{q(G(\mathbb{R}, \delta'))}$, so finally

$$e(\xi) = (-1)^{q(G(\mathbb{R}, \delta)) - q(G(\mathbb{R}, \delta'))}, \qquad (17.24)(g)$$

which agrees with Kottwitz's definition. Q.E.D.

18. Strongly stable characters and Theorem 1.29

We begin by recalling Langlands' notion of stable characters. This seems to make sense only for linear groups, so we will not try to define it in the setting of (11.1). Suppose therefore that we are in the setting of Definition 10.3, and that η is a finite-length canonical projective representation of type z of a strong real form δ of G^Γ. Thus η is in particular a representation of $G(\mathbb{R}, \delta)^{can}$. The *character of η* is a generalized function $\Theta(\eta)$ on $G(\mathbb{R}, \delta)^{can}$, defined as follows. Suppose f is a compactly supported smooth density on $G(\mathbb{R}, \delta)^{can}$. Then $\eta(f)$ is a well-defined operator on the space of η. (If we write f as a compactly supported smooth function times a Haar measure, $f(g)dg$, then $\eta(f)$ is given by the familiar formula

$$\eta(f) = \int_{G(\mathbb{R},\delta)^{can}} f(g)\eta(g)dg.)$$

The operator $\eta(f)$ is trace class, and the value of the generalized function $\Theta(\eta)$ on the test density f is by definition the trace of this operator:

$$\Theta(\eta)(f) = \operatorname{tr}(\eta(f)) \qquad (18.1)(a)$$

Of course we can immediately define the trace of any virtual canonical projective representation (cf. (15.5)):

$$\Theta : K\Pi^z(G(\mathbb{R}, \delta)) \to (\text{ generalized functions on } G(\mathbb{R}, \delta)^{can}). \quad (18.1)(b)$$

This map is injective (since characters of inequivalent irreducible representations are linearly independent). Recall that an element g of $G(\mathbb{R}, \delta)^{can}$ is called *strongly regular* if its centralizer is a Cartan subgroup. (This condition is slightly stronger than regular semisimple, which asks only that the centralizer in the Lie algebra be a Cartan subalgebra.) Write

$$G(\mathbb{R}, \delta)^{can}_{SR} \subset G(\mathbb{R}, \delta)^{can} \qquad (18.1)(c)$$

for the set of strongly regular elements; it is a dense open subset of full measure. Harish-Chandra's regularity theorem for invariant eigen-distributions shows that the trace of a virtual representation is determined by its restriction to this subset, and that this restriction is an analytic function. The map of (18.1)(b) therefore becomes

$$\Theta_{SR} : K\Pi^z(G(\mathbb{R},\delta)) \hookrightarrow C^\infty(G(\mathbb{R},\delta)_{SR}^{can}). \qquad (18.1)(d)$$

Of course the resulting functions are constant on conjugacy classes of $G(\mathbb{R},\delta)^{can}$.

Definition 18.2. ([35]). In the setting of (18.1), the virtual representation η is said to be *stable* if whenever g and g' are strongly regular elements of $G(\mathbb{R},\delta)^{can}$ conjugate under the complex group G^{can}, we have

$$\Theta_{SR}(\eta)(g) = \Theta_{SR}(\eta)(g').$$

Now we have in Corollary 1.26 a formal parametrization of all virtual representations in terms of L-group data, specifically as linear functionals on a category of perverse sheaves. It is natural to look for a characterization of the subset of stable virtual characters. As we remarked after the formulation of Theorem 1.29, it is quite difficult to find such a characterization. Fortunately the ideas of Langlands and Shelstad about stable characters admit a slightly different formulation. To begin, define

$$G(\mathbb{R},*)^{can} = \{\,(g,\delta) \mid \delta \in G^\Gamma - G \text{ a strong real form}, g \in G(\mathbb{R},\delta)^{can}\,\} \qquad (18.3)(a)$$

This set is analogous to $P(^\vee G^\Gamma)$ (Proposition 5.6), or to various fiber products appearing in Chapter 6. The group G^{can} acts on it by conjugation. More generally, whenever $\mathcal{D}$ is a set of strong real forms of G, we can define

$$G(\mathbb{R},\mathcal{D})^{can} = \{\,(g,\delta) \mid \delta \in \mathcal{D}, g \in G(\mathbb{R},\delta)^{can}\,\} \qquad (18.3)(b)$$

Suppose for example that $\mathcal{D}_0$ is the equivalence class of the strong real form δ_0. Then

$$G(\mathbb{R},\mathcal{D}_0)^{can} \simeq G^{can} \times_{G(\mathbb{R},\delta_0)^{can}} G(\mathbb{R},\delta_0)^{can}, \qquad (18.3)(c)$$

with $G(\mathbb{R},\delta_0)^{can}$ acting by conjugation to define the induced bundle. (The base space of the bundle is $G^{can}/G(\mathbb{R},\delta_0)^{can}$, which is just $\mathcal{D}_0$.) The singular space $G(\mathbb{R},*)^{can}$ is undoubtedly an interesting place to do analysis, but we will confine our attention to the smooth open subset

$$G(\mathbb{R},*)_{SR}^{can} = \{\,(g,\delta) \in G(\mathbb{R},*)^{can} \mid g \text{ is strongly regular}\,\}. \qquad (18.3)(d)$$

18. Strongly stable characters and Theorem 1.29

This is a disjoint union of open subsets corresponding to equivalence classes of strong real forms; each open subset is invariant under conjugation by G^{can}, and may be described as a smooth induced bundle in analogy with (18.3)(c).

Lemma 18.4. *Suppose G^Γ is a weak extended group (Definition 2.13). In the setting (18.3), to specify a smooth G^{can}-invariant function on $G(\mathbb{R},*)_{SR}^{can}$ is equivalent to specifying (for one representative δ_0 of each equivalence class of strong real forms of G) a smooth $G(\mathbb{R},\delta_0)^{can}$-invariant function on $G(\mathbb{R},\delta_0)_{SR}^{can}$.*

This is obvious from the definition. Using this idea, we can reformulate Langlands' definition of stability.

Lemma 18.5. *Suppose $\mathcal{D}_0$ is an equivalence class of strong real forms of the weak extended group G^Γ, $\delta_0 \in \mathcal{D}_0$, and η is a virtual representation of $G(\mathbb{R},\delta_0)^{can}$. Use the character of η to define a smooth G^{can}-invariant function $\Theta_{SR}(\eta,\mathcal{D}_0)$ on $G(\mathbb{R},\mathcal{D}_0)_{SR}^{can}$ ((18.3)(b) and Lemma 18.4). Then η is stable (Definition 18.2) if and only if $\Theta_{SR}(\eta,\mathcal{D}_0)$ is constant on the fibers of the first projection*

$$p_1 : G(\mathbb{R},\mathcal{D}_0)_{SR}^{can} \to G^{can}, \qquad p_1(g,\delta) = g.$$

Now Corollary 1.26 suggests considering families of virtual representations of all strong real forms. Such families are going to give rise to smooth functions on $G(\mathbb{R},*)_{SR}^{can}$, by means of (18.1)(d) and Lemma 18.4. Here is the appropriate class of virtual representations.

Definition 18.6 (cf. Definition 1.27). Suppose G^Γ is a weak extended group (Definition 2.13). As in Corollary 1.26, write $\overline{K}^z(G/\mathbb{R})$ for the set of formal infinite integral combinations of irreducible canonical projective representations of strong real forms of G, the group of *formal virtual representations*. Such a formal virtual representation

$$\eta = \sum_{\xi \in \Xi^z(G/\mathbb{R})} n(\xi)\pi(\xi)$$

is called *locally finite* if for each fixed strong real form δ there are only finitely many ξ with $n(\xi) \neq 0$ and $\delta(\xi)$ (the strong real form parametrized by ξ) equivalent to δ. Write

$$\overline{K}_f \Pi^z(G/\mathbb{R}) \qquad (18.6)(a)$$

for the set of locally finite formal virtual representations. Evidently the map of (18.1) extends to

$$\Theta(\cdot,\delta) : \overline{K}_f \Pi^z(G/\mathbb{R}) \to (\text{ generalized functions on } G(\mathbb{R},\delta)^{can}). \qquad (18.6)(b)$$

Finally, these maps can be assembled into a single map

$$\Theta_{SR}(\cdot,*) : \overline{K}_f \Pi^z(G/\mathbb{R}) \to C^\infty(G(\mathbb{R},*)^{can}_{SR})) \qquad (18.6)(c)$$

defined by

$$\Theta_{SR}(\eta,*)(g,\delta) = \Theta(\eta,\delta)(g). \qquad (18.6)(d)$$

That is, given a strong real form δ and a (strongly regular) element g of it, we find the (finitely many) irreducible representations of $G(\mathbb{R},\delta)^{can}$ appearing in η; add their distribution characters; and evaluate the sum at g.

The locally finite formal virtual representation η is said to be *strongly stable* if the function $\Theta_{SR}(\eta,*)$ is constant on the fibers of the first projection

$$p_1 : G(\mathbb{R},*)^{can}_{SR} \to G^{can}, \qquad p_1(g,\delta) = g.$$

By Lemma 18.5, the restriction of a strongly stable virtual representation to a single strong real form must be stable. The converse is false, however, since a strongly regular element g of G^{can} may belong to several inequivalent strong real forms of G. In fact g must belong to a quasisplit real form (if it belongs to any at all); we have already encountered this elementary fact in the proof of Proposition 13.6. It follows that a strongly stable virtual representation is determined by its restriction to any quasisplit real form of G. The results of Shelstad in [47] can be formulated as a converse to this statement, as follows.

Theorem 18.7 ([47]). *Suppose G^Γ is a weak extended group for G (Definition 2.13), and δ_0 is a strong real form of G. Suppose η_0 is a stable virtual representation of $G(\mathbb{R},\delta_0)^{can}$ (Definition 18.2). Then there is a strongly stable locally finite formal virtual representation η for G (Definition 18.6) with the property that*

$$\Theta_{SR}(\eta,*)(g,\delta_0) = \Theta_{SR}(\eta_0)(g)$$

for every strongly regular element g of $G(\mathbb{R},\delta_0)^{can}$ (cf. (18.6)(c) and (18.1)(d)). If δ_0 is quasisplit, then η is unique.

18. Strongly stable characters and Theorem 1.29

We will need a more explicit statement than this; formulating it will also allow us to explain the relationship between Theorem 18.7 and the superficially rather different results of [47]. We begin with something in terms of the traditional Langlands classification. For the rest of this chapter we will fix an E-group

$$(^\vee G^\Gamma, \mathcal{D}) \tag{18.8}$$

for G^Γ with second invariant z (Definition 4.6).

Definition 18.9 ([47], section 3). In the setting of (18.1) and (18.8), suppose $\phi \in P(^\vee G^\Gamma)$ is a Langlands parameter (Definition 5.2). The *stable standard (virtual) representation attached to* ϕ is by definition the sum of all the (inequivalent) standard final limit representations of $G(\mathbb{R}, \delta)^{can}$ parametrized by ϕ (Proposition 13.12); to simplify later definitions we insert the sign $e(G(\mathbb{R}, \delta))$ attached to the real form by Kottwitz (Definition 15.8). It is written as $\eta_\phi^{loc}(\delta)$. (Equivalently, $\eta_\phi^{loc}(\delta)$ may be defined as the sum of the inequivalent standard representations induced from discrete series parametrized by ϕ.) In the notation of (1.19),

$$\eta_\phi^{loc}(\delta) = e(G(\mathbb{R}, \delta)) \sum_{\substack{\xi=(\phi,\tau)\in\Xi^z(G/\mathbb{R}) \\ \delta(\xi)=\delta}} M(\xi);$$

the sum is over complete Langlands parameters (Definition 5.11), with ϕ fixed.

The terminology of the definition is justified by

Lemma 18.10 ([47], Lemma 5.2). *In the setting of Definition 18.9, the virtual representation $\eta_\phi^{loc}(\delta)$ is stable.*

Shelstad proves this under the assumption that ϕ is tempered, but the argument is unchanged in general. Although she does not state the result explicitly, her argument also proves

Lemma 18.11. *In the setting of (18.1), the lattice of stable virtual representations has as a basis the set of non-zero $\eta_\phi^{loc}(\delta)$ (as ϕ varies over $\Phi(^\vee G^\Gamma)$).*

To prove Theorem 18.7, we need also a way to compare different real forms.

Theorem 18.12 ([47], Theorem 6.3). *In the setting of Definition 18.9, suppose that δ and δ' are strong real forms of G, and that*

$$g \in G(\mathbb{R}, \delta)^{can} \cap G(\mathbb{R}, \delta')^{can}$$

is strongly regular. Then

$$\Theta_{SR}(\eta_\phi^{loc}(\delta))(g) = \Theta_{SR}(\eta_\phi^{loc}(\delta'))(g).$$

Here Θ_{SR} is as in (18.1).

Definition 18.13. In the setting of Definitions 18.6 and 18.9, suppose $\phi \in P(^\vee G^\Gamma)$ is a Langlands parameter (Definition 5.2). The *strongly stable standard formal virtual representation attached to ϕ* is by definition the sum of all the (inequivalent) standard final limit representations parametrized by ϕ (Proposition 13.12), normalized by the signs $e(\xi)$ (Definition 15.8). It is written as η_ϕ^{loc}. In the notation of (1.19),

$$\eta_\phi^{loc} = \sum_{\xi=(\phi,\tau)\in\Xi^z(G/\mathbb{R})} e(\xi)M(\xi);$$

the sum is over complete Langlands parameters (Definition 5.11). Equivalently,

$$\eta_\phi^{loc} = \sum_\delta \eta_\phi^{loc}(\delta);$$

the sum is over strong real forms. Finally, we can also write

$$\eta_\phi^{loc} = \sum_{\tau \in \widehat{(A_\phi^{loc,alg})}} e(\phi,\tau)M(\phi,\tau),$$

a sum over the (necessarily one-dimensional) irreducible representations of the universal component group.

The terminology is justified by

Theorem 18.14. *In the setting of Definition 18.12, the formal virtual representation η_ϕ^{loc} is strongly stable. The lattice of strongly stable formal virtual representations has as a basis the set of η_ϕ^{loc} (as ϕ varies over $\Phi(^\vee G^\Gamma)$).*

Theorems 18.7 and 18.14 follow from Lemmas 18.10 and 18.11, and Theorem 18.12.

To complete the proof of Theorem 1.29, we need to identify the formal virtual representation η_ϕ^{loc} in the isomorphism of Corollary 1.26; that is, as a $\mathbb{Z}$-linear functional on the Grothendieck group $KX(^\vee G^\Gamma)$.

18. Strongly stable characters and Theorem 1.29

Lemma 18.15. *In the setting of Definition 18.13, write S for the orbit of $^\vee G$ on $X(^\vee G^\Gamma)$ corresponding to ϕ (Proposition 6.17). Then*

$$\eta_\phi^{loc} = \sum_{\xi=(S,\mathcal{V})\in\Xi^z(G/\mathbb{R})} e(\xi)M(\xi).$$

Here we identify complete geometric parameters as in Definition 7.6; the sum is therefore over irreducible equivariant local systems $\mathcal{V}$ on S. In the isomorphism of Corollary 1.26, this formal virtual representation corresponds to the map

$$\chi_S^{loc} : \mathrm{Ob}\,\mathcal{C}(X(^\vee G^\Gamma), {^\vee G}^{alg}) \to \mathbb{N}$$

that assigns to a constructible equivariant sheaf the dimension of its fiber over S (Definition 1.28).

Proof. The first formula is immediate from Definition 18.13 and the identification of complete Langlands parameters with complete geometric parameters (Definition 7.6). For the second, notice first of all that all the irreducible equivariant local systems on S have fiber dimension 1. (This is equivalent to the assertion that the group $A_S^{loc,alg}$ is abelian. The proof of Theorem 10.4 exhibits $A_S^{loc,alg}$ as a subquotient of a Cartan subgroup of the connected reductive complex group $^\vee G^{alg}$ (Definition 12.4 and Theorem 12.9), proving the assertion.) We may therefore write

$$\chi_S^{loc}(C) = \sum_{\mathcal{V}} (\text{ multiplicity of } \mathcal{V} \text{ in } C \text{ restricted to } S\ ), \qquad (18.16)$$

the sum extending over irreducible equivariant local systems on S. Comparing (18.16) with the first formula of the lemma, and using Corollary 1.26(a), we get the second formula. Q.E.D.

Theorem 1.29 now follows from Lemma 18.15 and Theorem 18.14.

19. Characteristic cycles, micro-packets, and Corollary 1.32

In this chapter we will repeat in a little more detail the argument given in the introduction for defining the new strongly stable virtual representations η_S^{mic}. It is convenient to work at first in the setting of Definition 7.7; that is, that Y is a smooth complex algebraic variety on which the pro-algebraic group H acts with finitely many orbits. At each point y of Y, the differential of this action defines a map

$$a_y : \mathfrak{h} \to T_y Y \qquad (19.1)(a)$$

from the Lie algebra $\mathfrak{h}$ of H into the tangent space of Y at y. If y belongs to the H-orbit S, then

$$a_y(\mathfrak{h}) = T_y S \qquad (19.1)(b)$$

If we regard $\mathfrak{h} \times Y$ as a trivial bundle over Y, we get a bundle map

$$a : \mathfrak{h} \times Y \to TY \qquad (19.1)(c)$$

Of course the cotangent bundle of Y is dual to the tangent bundle, so we may define the *conormal bundle to the H action* as the annihilator of the image of a. Explicitly,

$$T_H^*(Y) = \{\, (\lambda, y) \mid \lambda \in T_y^*(Y), \lambda(a_y(\mathfrak{h})) = 0 \,\} \qquad (19.1)(d)$$

It is clear from this definition that $T_H^*(Y)$ is a closed cone in the cotangent bundle. In the situation of (19.1)(b), the fiber of $T_H^*(Y)$ at y is the conormal bundle to S at y (that is, the annihilator of $T_y S$ in $T_y^*(Y)$):

$$T_{H,y}^*(Y) = T_{S,y}^*(Y). \qquad (19.1)(e)$$

Consequently

$$T_H^*(Y) = \bigcup_{H\text{-orbits } S} T_S^*(Y). \qquad (19.1)(f)$$

That is, the conormal bundle to the action is the union of the conormal bundles to the orbits.

Lemma 19.2. *Suppose the proalgebraic group H acts on the smooth variety Y with finitely many orbits; use the notation of (19.1).*

a) The conormal bundle $T_H^*(Y)$ is a closed Lagrangian cone in $T^*(Y)$.
b) The H-components (that is, the smallest H-invariant unions of irreducible components) of $T_H^*(Y)$ are the closures $\overline{T_S^*(Y)}$ of conormal bundles of H-orbits S in Y.

Proof. We have already seen that $T_H^*(Y)$ is a closed cone. Since the conormal bundle to a locally closed smooth subvariety is automatically Lagrangian, part (a) follows from (19.1)(f). The conormal bundle $T_S^*(Y)$ to an H-orbit S has dimension equal to the dimension of Y, and is H-irreducible (that is, H permutes the irreducible components transitively). Part (b) follows. Q.E.D.

We turn now to a discussion of the characteristic variety. We follow [12], VI,1.9, which the reader may consult for more details. On any smooth algebraic variety Y, the sheaf $\mathcal{D}_Y$ of algebraic differential operators has a natural filtration by order

$$\mathcal{O}_Y = \mathcal{D}_Y(0) \subset \mathcal{D}_Y(1) \subset \cdots \qquad (19.3)(a)$$

The terms here are coherent $\mathcal{O}_Y$-modules, and the multiplication of differential operators satisfies

$$\mathcal{D}_Y(p)\mathcal{D}_Y(q) \subset \mathcal{D}_Y(p+q), \qquad (19.3)(b)$$

Write $p : T^*(Y) \to Y$ for the natural bundle map. The symbol calculus for differential operators is an isomorphism of graded sheaves of algebras

$$\operatorname{gr} \mathcal{D}_Y \simeq p_*(\mathcal{O}_{T^*(Y)}). \qquad (19.3)(c)$$

Because p is an affine map, p_* defines an equivalence of categories

$$\text{(quasicoherent sheaves of } \mathcal{O}_{T^*(Y)} - \text{modules)}$$
$$\to (\text{ quasicoherent sheaves of } p_*\mathcal{O}_{T^*(Y)} - \text{modules }) \qquad (19.3)(d)$$

([20], Exercise II.5.17(e)).

Suppose now that $\mathcal{M}$ is a quasicoherent $\mathcal{D}_Y$-module. A *filtration* on $\mathcal{M}$ is a sequence

$$\cdots \subset \mathcal{M}(p) \subset \mathcal{M}(p+1) \subset \cdots \qquad (p \in \mathbb{Z}) \qquad (19.4)(a)$$

of quasicoherent $\mathcal{O}_Y$-submodules, satisfying

$$\mathcal{D}(p)\mathcal{M}(q) \subset \mathcal{M}(p+q), \quad \bigcup_p \mathcal{M}(p) = \mathcal{M}, \quad \bigcap_p \mathcal{M}(p) = 0 \quad (19.4)(b)$$

The first condition makes the associated graded sheaf $\operatorname{gr} \mathcal{M}$ into a sheaf of modules for $\operatorname{gr} \mathcal{D}_Y$. By (19.3)(c) and (d), we may therefore regard $\operatorname{gr} \mathcal{M}$ as a sheaf of modules on $T^*(Y)$. The filtration on $\mathcal{M}$ is called *good* if $\operatorname{gr} \mathcal{M}$ is a coherent sheaf (of $\mathcal{O}_{T^*(Y)}$-modules). (This implies in particular that $\mathcal{M}(p) = 0$ for p sufficiently negative.)

Proposition 19.5 ([12], VI.1.9, Miličić). *Suppose $\mathcal{M}$ is a coherent $\mathcal{D}_Y$-module.*

a) *$\mathcal{M}$ is locally finitely generated. More precisely, if U is any affine open subset of Y, then the restriction of $\mathcal{M}$ to U is generated by a coherent $\mathcal{O}_U$-submodule $\mathcal{N}_U$.*

b) *$\mathcal{M}$ is generated as a $\mathcal{D}_Y$-module by a coherent $\mathcal{O}_Y$-submodule $\mathcal{N}$.*

c) *$\mathcal{M}$ admits a good filtration.*

Proof. Part (a) is in [12]. The rest of the argument was provided to us by Dragan Miličić, and we are grateful for his permission to include it here. Cover Y by affine open sets $U_1, \ldots, U_n$, and choose coherent $\mathcal{O}_{U_i}$-submodules $\mathcal{N}_{U_i}$ as in (a). By [20], exercise II.5.15(d), there is for each i a coherent $\mathcal{O}_Y$-submodule $\mathcal{N}_i$ of $\mathcal{M}$, with the property that $\mathcal{N}_{U_i}$ is contained in the restriction of $\mathcal{N}_i$ to U_i. Let $\mathcal{N}$ be the $\mathcal{O}_Y$-submodule of $\mathcal{M}$ generated by the various $\mathcal{N}_i$. Then $\mathcal{N}$ is a coherent $\mathcal{O}_Y$-module. Its restriction to each U_i contains $\mathcal{N}_{U_i}$, and therefore generates the restriction of $\mathcal{M}$ to U_i. Consequently since "generating" (the surjectivity of the natural map $\mathcal{D}_Y \otimes_{\mathcal{O}_Y} \mathcal{N} \to \mathcal{M}$) is a local property, $\mathcal{N}$ must generate $\mathcal{M}$. This is (b). For (c), we simply define $\mathcal{M}(p) = \mathcal{D}_Y(p) \otimes \mathcal{N}$. Q.E.D.

We will eventually see that ${}^\vee G^{alg}$-equivariant $\mathcal{D}$-modules on $X(\mathcal{O}, {}^\vee G^\Gamma)$ admit global ${}^\vee G^{alg}$-invariant good filtrations (Proposition 21.4).

Good filtrations are certainly not unique. The next lemma allows us to compare objects defined using different filtrations, and to deduce that some such objects are independent of the filtration chosen.

Lemma 19.6 ([16], Corollary 1.3; cf. [59], Proposition 2.2). *Suppose $\mathcal{F}$ and $\mathcal{G}$ are good filtrations of the coherent $\mathcal{D}_Y$-module $\mathcal{M}$. Then there are finite filtrations*

$$\operatorname{gr}(\mathcal{M}, \mathcal{F})(0) \subset \operatorname{gr}(\mathcal{M}, \mathcal{F})(1) \subset \cdots \subset \operatorname{gr}(\mathcal{M}, \mathcal{F})(N) = \operatorname{gr}(\mathcal{M}, \mathcal{F})$$

$$\operatorname{gr}(\mathcal{M}, \mathcal{G})(0) \subset \operatorname{gr}(\mathcal{M}, \mathcal{G})(1) \subset \cdots \subset \operatorname{gr}(\mathcal{M}, \mathcal{G})(N) = \operatorname{gr}(\mathcal{M}, \mathcal{G})$$

19. Characteristic cycles and micro-packets

by graded coherent sheaves of $p_*\mathcal{O}_{T^*(Y)}$-modules), with the property that the corresponding subquotients are isomorphic:

$$\mathrm{gr}(\mathcal{M},\mathcal{F})(j)/\mathrm{gr}(\mathcal{M},\mathcal{F})(j-1) \simeq \mathrm{gr}(\mathcal{M},\mathcal{G})(j)/\mathrm{gr}(\mathcal{M},\mathcal{G})(j-1)$$
$$(0 \leq j \leq N).$$

(There is a constant N', $0 \leq N' \leq N$, with the property that the jth isomorphism shifts the grading by $j - N'$.)

In particular, the class of $\mathrm{gr}(\mathcal{M})$ in the Grothendieck group of coherent sheaves of $\mathcal{O}_{T^*(Y)}$-modules is independent of the choice of good filtration.

Definition 19.7. Suppose V is a complex algebraic variety, and $\mathcal{N}$ is a coherent sheaf of $\mathcal{O}_V$-modules. The *associated variety* (or *support*) of $\mathcal{N}$ is the set of (not necessarily closed) points $v \in V$ at which the stalk of $\mathcal{N}$ is not zero:

$$\mathcal{V}(\mathcal{N}) = \{\, v \in V \mid \mathcal{N}_v \neq 0 \,\}.$$

It is a closed subvariety of V, defined by the sheaf of ideals $\mathrm{Ann}\,\mathcal{N} \subset \mathcal{O}_V$. The set $\{W_i\}$ of irreducible components of $\mathcal{V}(\mathcal{N})$ is the set of maximal points (that is, points not in the closure of other points) of $\mathcal{V}(\mathcal{N})$. (If V is affine, these correspond to the minimal prime ideals containing the annihilator of $\mathcal{N}$.) To each such point we assign a positive integer

$$m_{W_i}(\mathcal{N}),$$

the *multiplicity of $\mathcal{N}$ along W_i*. One way to define it is as the length of $\mathcal{N}_{W_i}$ as a module for the local ring $\mathcal{O}_{V,W_i}$. The *associated cycle of $\mathcal{N}$* is the formal sum

$$\mathrm{Ch}(\mathcal{N}) = \sum_i m_{W_i}(\mathcal{N}) W_i.$$

This is a finite formal integer combination of irreducible subvarieties of V, with no containments allowed among the subvarieties. Notice that the support of $\mathcal{N}$ is the union of the subvarieties appearing in the associated cycle.

The theory of associated cycles is elementary, and well-known to experts; but it is not easy to give references with enough details for the novice. The theory is essentially local, so it suffices to discuss it for affine varieties. In that case the necessary commutative algebra is summarized in section 2 of [59].

Lemma 19.8. *Suppose $\mathcal{N}$ and $\mathcal{N}'$ are coherent sheaves of modules on the complex algebraic variety V. Assume that $\mathcal{N}$ and $\mathcal{N}'$ admit finite filtrations (as $\mathcal{O}_V$-modules) with the property that $\operatorname{gr}\mathcal{N} \simeq \operatorname{gr}\mathcal{N}'$. Then $\operatorname{Ch}\mathcal{N} = \operatorname{Ch}\mathcal{N}'$.*

Again this is well-known, and again we refer to section 2 of [59] for a proof.

Definition 19.9 Suppose $\mathcal{M}$ is a coherent $\mathcal{D}_Y$-module on the smooth algebraic variety Y. Fix a good filtration of $\mathcal{M}$ (Proposition 19.5). The *characteristic variety of $\mathcal{M}$* (or *singular support*) is the support of the $\mathcal{O}_{T^*(Y)}$-module $\operatorname{gr}\mathcal{M}$ (cf. (19.4)):

$$SS(\mathcal{M}) = \mathcal{V}(\operatorname{gr}\mathcal{M}) \subset T^*Y.$$

By Lemmas 19.6 and 19.8, this is independent of the choice of good filtration. Since $\operatorname{gr}\mathcal{M}$ is graded, the characteristic variety is a closed cone in $T^*(Y)$. The *characteristic cycle of $\mathcal{M}$* is the associated cycle of $\operatorname{gr}\mathcal{M}$:

$$\operatorname{Ch}(\mathcal{M}) = \operatorname{Ch}(\operatorname{gr}\mathcal{M}).$$

It is a formal sum of irreducible subvarieties of $T^*(Y)$, with positive integer coefficients; the union of these subvarieties is the characteristic variety of $\mathcal{M}$. Lemmas 19.6 and 19.8 guarantee that it is independent of the choice of good filtration.

Obviously the characteristic variety of $\mathcal{M}$ is contained in the inverse image under p (cf. (19.3)) of the support of $\mathcal{M}$. In fact

$$SS(\mathcal{M}) \cap T^*_y(Y) \neq \emptyset \Leftrightarrow y \in \operatorname{Supp}(\mathcal{M}). \tag{19.10}$$

In this sense the characteristic variety refines the notion of support for $\mathcal{D}$-modules (or, via the Riemann-Hilbert correspondence of Theorem 7.9, for perverse sheaves).

Recall that the dimension of the characteristic variety of $\mathcal{M}$ is (everywhere locally on $SS(\mathcal{M})$) greater than or equal to the dimension of Y ([12], Theorem VI.1.10) and that (by definition) $\mathcal{M}$ is holonomic exactly when equality holds.

Examples 19.11. If $\mathcal{M}$ is equal to $\mathcal{D}_Y$, then the usual filtration by degree is good. The associated graded module is $\mathcal{O}_{T^*(Y)}$, so $\operatorname{Ch}(\mathcal{D}_Y) = 1 \cdot T^*(Y)$.

Suppose $\mathcal{M}$ is equal to $\mathcal{O}_Y$. Then we can define

$$\mathcal{M}(p) = \mathcal{O}_Y \qquad (p = 0, 1, 2, \ldots).$$

19. Characteristic cycles and micro-packets

(This module is regular holonomic.) The associated graded module is $\mathcal{O}_Y$; that is, it is $\mathcal{O}_{T^*(Y)}$ modulo the ideal defining the zero section of the cotangent bundle. Consequently $\operatorname{Ch}(\mathcal{O}_Y) = 1 \cdot Y$

Suppose $\mathcal{M}$ is the $\mathcal{D}_Y$ module generated by a delta function supported at the point y; that is,

$$\mathcal{M} = \mathcal{D}_Y / \mathcal{D}_Y \mathcal{I}(y).$$

(Here we write $\mathcal{I}(y)$ for the ideal in $\mathcal{O}_Y$ defining the point y. Again the module is regular holonomic.) Then $\mathcal{M}$ inherits a good quotient filtration from $\mathcal{D}_Y$. The associated graded module is the sheaf of functions on the cotangent space $T_y^*(Y)$, so $\operatorname{Ch}(\mathcal{M}) = 1 \cdot T_y^*(Y)$.

We return now to the equivariant setting.

Proposition 19.12. *Suppose we are in the setting of Lemma 19.2, and that $\mathcal{M}$ is an H-equivariant coherent $\mathcal{D}_Y$-module.*

a) *For each $X \in \mathfrak{h}$, let $a(X)$ be the vector field on Y induced by the action of H, a first-order differential operator on Y. Then the action of $a(X)$ on $\mathcal{M}$ induced by the $\mathcal{D}_Y$-module structure agrees with the action of X defined by differentiating the action of H on $\mathcal{M}$.*

b) *The characteristic variety of $\mathcal{M}$ is contained in the conormal bundle to the H action (cf. (19.1)).*

c) *The H-components (Lemma 19.2) of $SS(\mathcal{M})$ are closures of conormal bundles of H-orbits on Y. Consequently*

$$\operatorname{Ch}(\mathcal{M}) = \sum_{H\text{-orbits } S} \chi_S^{mic}(\mathcal{M}) \overline{T_S^*(Y)},$$

for some non-negative integers $\chi_S^{mic}(\mathcal{M})$.

d) *The support of $\mathcal{M}$ is given by*

$$\operatorname{Supp}(\mathcal{M}) = \bigcup_{\chi_S^{mic}(\mathcal{M}) \neq 0} \overline{S}.$$

e) *The functions χ_S^{mic} are additive for short exact sequences (of H-equivariant coherent $\mathcal{D}_Y$-modules). Consequently they define $\mathbb{Z}$-linear functionals*

$$\chi_S^{mic} : K(Y, H) \to \mathbb{Z}$$

(notation (7.10)).

Proof. Part (a) is a trivial exercise in understanding the definition of equivariant $\mathcal{D}$-modules ([12], VII.12.10). (We couldn't do it ourselves, but never mind.) For (b), assume for simplicity that there is a global H-invariant good filtration. Fix $X \in \mathfrak{h}$. We may regard the vector field $a(X)$ as a first-order differential operator, and therefore

$$a(X) : \mathcal{M}(p) \to \mathcal{M}(p+1).$$

On the other hand, the action of H preserves degree in the filtration, so the differentiated action of $\mathfrak{h}$ does as well. By (a), it follows that

$$a(X) : \mathcal{M}(p) \to \mathcal{M}(p).$$

It follows that the principal symbol of $a(X)$ annihilates $\mathrm{gr}(\mathcal{M})$. This symbol is nothing but $a(X)$, regarded as a function on $T^*(Y)$. It follows that

$$\mathcal{V}(\mathrm{gr}\,\mathcal{M}) \subset \{\,(\lambda, y) \in T^*(Y) \mid \lambda(a_y(X)) = 0\,\}.$$

Comparing this with (19.1)(d) gives (b). Part (c) follows from (b) and Lemma 19.2. Part (d) follows from (c) and (19.10). Part (e) is clear from the definitions. Q.E.D.

This completes our exposition of Definition 1.30. Corollary 1.32 is an immediate consequence of Theorem 1.31 (Kashiwara's local index theorem) and Theorem 1.29. It will be convenient to introduce a bit more notation in that context, however.

Definition 19.13. In the setting of (19.1), fix $S \in \Phi(Y, H)$ (Definition 7.1); that is, an orbit of H on Y. To every complete geometric parameter $\xi' \in \Xi(Y, H)$ is associated an irreducible perverse sheaf $P(\xi')$ (cf. (7.10)(d)). The conormal bundle $T_S^*(Y)$ has a non-negative integral multiplicity $\chi_S^{mic}(\xi')$ in the characteristic cycle $\mathrm{Ch}\,P(\xi')$ (Proposition 19.12). We define the *micro-packet of geometric parameters attached to* S to be the set of ξ for which this multiplicity is non-zero:

$$\Xi(Y, H)_S^{mic} = \{\,\xi \mid \chi_S^{mic}(\xi') \neq 0\,\}.$$

Lemma 19.14. *In the setting of Definition 19.13, suppose $\xi' = (S', \mathcal{V}')$ is a geometric parameter (Definition 7.1).*
a) *If $\xi' \in \Xi(Y, H)_S^{mic}$, then $S \subset \overline{S'}$.*
b) *if $S' = S$, then $\xi' \in \Xi(Y, H)_S^{mic}$. In this case $\chi_S^{mic}(\xi')$ is equal to the rank of the local system $\mathcal{V}'$; that is, to the dimension of the corresponding representation of A_S^{loc} (Definition 7.1).*

19. Characteristic cycles and micro-packets   219

Proof. Part (a) is clear from (19.12)(d), as is the first assertion in (b). The rest of (b) (which is "well-known") can be deduced from the construction of the $\mathcal{D}_Y$-module corresponding to the perverse sheaf $P(\xi')$ ([12]; some of the main points are contained in the proof of Kashiwara's theorem VI.7.11 in [12]). Q.E.D.

Definition 19.15. In the setting of Theorem 10.4, fix an equivalence class $\phi \in \Phi^z(G/\mathbb{R})$ of Langlands parameters (Definition 5.3), and write $S = S_\phi$ for the corresponding orbit of $^\vee G$ on $X(^\vee G^\Gamma)$ (Definition 7.6). Then we can define the *micro-packet of geometric parameters attached to* ϕ as

$$\Xi^z(G/\mathbb{R})_\phi^{mic} = \Xi(X(^\vee G^\Gamma), {^\vee G}^{alg})_{S_\phi}^{mic}$$

(notation as in Definition 19.13). The *micro-packet of* ϕ is the corresponding set of irreducible representations of strong real forms of G:

$$\Pi^z(G/\mathbb{R})_\phi^{mic} = \{\, \pi(\xi') \mid \xi' \in \Xi^z(G/\mathbb{R})_\phi^{mic} \,\}.$$

Corollary 19.16. *In the setting of Theorem 10.4, fix an equivalence class* $\phi \in \Phi^z(G/\mathbb{R})$ *of Langlands parameters. Then the micro-packet* $\Pi^z(G/\mathbb{R})_\phi^{mic}$ *contains the L-packet* $\Pi^z(G/\mathbb{R})_\phi$. *If* $\pi(\xi')$ *is any irreducible representation occurring in the micro-packet, then the corresponding orbit* S' *on* $X(^\vee G^\Gamma)$ *contains* S_ϕ *in its closure. There is a strongly stable formal virtual character*

$$\eta_\phi^{mic} = \sum_{\pi' \in \Pi^z(G/\mathbb{R})_\phi^{mic}} e(\pi')(-1)^{d(\pi')-d(\phi)}\chi_\phi^{mic}(\pi')\pi'.$$

Here $e(\pi') = \pm 1$ *is Kottwitz' sign attached to the real form of which* π' *is a representation;* $d(\pi')$ *is the dimension of the orbit* $S_{\pi'}$ *on* $X(^\vee G^\Gamma)$; *and the positive integer* $\chi_\phi^{mic}(\pi')$ *is the multiplicity of the conormal bundle to* S_ϕ *in the characteristic variety of the irreducible perverse sheaf corresponding to* π'. *This is therefore a sum over the full L-packet* $\Pi^z(G/\mathbb{R})_\phi^{mic}$, *together with "correction terms" taken from L-packets corresponding to larger orbits on* $X(^\vee G^\Gamma)$.

The set $\{\,\eta_\phi^{mic}\,\}$ *(as* ϕ *varies over* $\Phi^z(G/\mathbb{R})$) *is a basis of the lattice of strongly stable formal virtual representations.*

This is just a reformulation of Corollary 1.32 (taking into account the information in Lemma 19.14). The point of the result is that under favorable circumstances the characteristic variety of an equivariant $\mathcal{D}$-module is relatively small; at best it may just be the conormal bundle of a single orbit. The sum in Corollary 19.16 therefore does not extend over too many representations. By contrast, the sum in Lemma 18.15 (when expressed in terms of irreducible representations) tends to include most of the representations for which the corresponding irreducible perverse sheaf has S_ϕ in its support.

Example 19.17. Suppose that G is adjoint. In this case strong real forms are the same as real forms, and $^\vee G$ is simply connected (so there are no coverings to consider). Take $(^\vee G^\Gamma, \mathcal{S})$ to be an L-group for G in the sense of Definition 4.14. Let $\mathcal{O} \subset {}^\vee\mathfrak{g}$ be the orbit of half sums of positive coroots; this corresponds to representations of G of infinitesimal character equal to that of the trivial representation (Lemma 15.4). The conjugacy class $\mathcal{C}(\mathcal{O})$ of (6.10) consists of the single (central) element $z(\rho)$. Since ρ is regular and integral, each canonical flat $\Lambda \subset \mathcal{F}(\mathcal{O})$ is preserved by a unique Borel subgroup ${}^dB(\Lambda)$ (notation (6.6)). We may therefore identify the geometric parameter space $X(\mathcal{O}, {}^\vee G^\Gamma)$ with the set of pairs $(y, {}^dB)$, with $y \in {}^\vee G^\Gamma - {}^\vee G$, $y^2 = z(\rho)$, and ${}^dB$ a Borel subgroup of $^\vee G$ (Definition 6.9).

Now by definition the L-group structure provides a distinguished orbit $S_0 = \mathcal{S}$ of $^\vee G$ on $X(\mathcal{O}, {}^\vee G^\Gamma)$. We write ϕ_0 for the corresponding Langlands parameter. Fix a point $(y, {}^dB) \in S_0$. Since $^\vee G$ is simply connected, the fixed point group $K(y)$ of the involution θ_y is connected. Since θ_y preserves ${}^dB$, the intersection ${}^dB \cap K(y)$ is a Borel subgroup of $K(y)$, so it is connected as well. Evidently ${}^dB \cap K(y)$ is just the isotropy group of the action of $^\vee G$; so it follows that the canonical component group $A_{\phi_0}^{loc,alg}$ is trivial. The L-packet Π_{ϕ_0} therefore consists of a single representation. Inspecting the constructions in the proof of Theorem 10.4, we find that $M(\phi_0)$ is a spherical principal series representation of the quasisplit real form of G (in our fixed inner class), and that $\pi(\phi_0)$ is the trivial representation of this group. The strongly stable standard formal virtual representation $\eta_{\phi_0}^{loc}$ of Definition 18.13 is $M(\phi)$ alone. This representation has a great many irreducible composition factors, however, occurring with relatively high multiplicity; so as a "stabilization" of the trivial representation $\pi(\phi_0)$, $M(\phi_0)$ is not very useful. (Another way to say this is that the relationship between the characters of $\pi(\phi_0)$ and $M(\phi_0)$ is quite weak.)

We consider therefore the virtual representation $\eta_{\phi_0}^{mic}$. To calculate it, we need to know the multiplicity of the conormal bundle $T_{S_0}^*(X(\mathcal{O}, {}^\vee G))$ in the characteristic variety of any irreducible perverse sheaf. We will consider some techniques for making such calculations in Chapters 21 and 27 (see for example Theorem 27.18). For the moment we simply state the result: the micro-packet of ϕ_0 consists of one trivial representation of each real form of G in our inner class. The multiplic-

19. Characteristic cycles and micro-packets   221

ities $\chi_{S_0}^{mic}$ are equal to 1, and the differences in orbit dimensions have the same parity as Kottwitz' invariant. The formula in Corollary 19.16 therefore reduces to

$$\eta_{\phi_0}^{mic} = \sum_{\text{real forms } \delta} (\text{ trivial representation of } G(\mathbb{R},\delta)).$$

This is a charming (if rather obvious) stabilization of the trivial representation. Notice that all real forms (in the inner class) appear, even though ϕ_0 is relevant only for the quasisplit one.

20. Characteristic cycles and Harish-Chandra modules

Recall that Theorem 8.5 related the category of equivariant perverse sheaves (or, equivalently, equivariant $\mathcal{D}$-modules) on the geometric parameter space to certain categories of Harish-Chandra modules. Our goal in this chapter is to see what this relationship has to say about the characteristic cycles. Theorem 8.5 implies that the characteristic cycles must somehow be encoded in the Harish-Chandra modules. We are not able to break that code, but we get some useful information about it (Theorem 20.18). This will later be the key to relating our definition of Arthur's unipotent representations to the original one of Barbasch and Vogan. Roughly speaking, Arthur's representations will be characterized on the E-group side by the occurrence of certain "regular" components in a characteristic cycle (Definition 20.7). We therefore seek to understand that occurrence on the level of Harish-Chandra modules.

We begin with analogues of Propositions 7.14 and 7.15.

Proposition 20.1. *Suppose X and Y are smooth algebraic varieties, and*

$$f : X \to Y$$

is a smooth surjective morphism having connected fibers of dimension d. Suppose $\mathcal{M}$ is a $\mathcal{D}_Y$-module.
a) There is a natural inclusion

$$f^*(T^*Y) \hookrightarrow T^*X$$

identifying the pullback of the cotangent bundle of Y with a subbundle of T^*X. This induces a bundle map

$$\tilde{f} : f^*(T^*Y) \to T^*Y$$

which is again a smooth surjective morphism having connected geometric fibers of dimension d.
b) The $\mathcal{O}_X$-module $f^*\mathcal{M}$ carries a natural $\mathcal{D}_X$-module structure.
c) The functor f^* is a fully faithful exact functor from coherent $\mathcal{D}_Y$ modules to coherent $\mathcal{D}_X$ modules.

d) *Suppose $\mathcal{M}(p)$ is a good filtration of $\mathcal{M}$. Then*

$$(f^*\mathcal{M})(p) = f^*(\mathcal{M}(p))$$

is a good filtration of $f^*\mathcal{M}$. The associated graded module $\mathrm{gr}(f^*\mathcal{M})$ is defined on the subvariety $f^*(T^*Y)$ of T^*X, where it may be computed by

$$\mathrm{gr}(f^*\mathcal{M}) \simeq \tilde{f}^*\mathrm{gr}(\mathcal{M}).$$

The characteristic cycle of $f^*\mathcal{M}$ is therefore

$$\mathrm{Ch}(f^*\mathcal{M}) = \tilde{f}^{-1}(\mathrm{Ch}(\mathcal{M})).$$

e) *Suppose we are in the setting of Proposition 7.15 (that is, that a pro-algebraic group H acts on everything). Suppose $\mathcal{M}$ is an H-equivariant coherent $\mathcal{D}_Y$-module, and S is an H-orbit on Y. Define f^*S to be the unique dense H-orbit in $f^{-1}S$ (cf. (7.16)). Then the multiplicities of Proposition 19.12 are given by*

$$\chi^{mic}_{f^*S}(f^*\mathcal{M}) = \chi^{mic}_S(\mathcal{M}).$$

Proof. Part (a) is formal. Part (b) may be found in [12], VI.4.1. Part (c) is [12], Proposition VI.4.8. For (d), we know that f^* is an exact functor on $\mathcal{O}_Y$-modules (since f is smooth), and that it preserves coherence; the only thing to check is the degree condition in (19.4)(b). This is immediate from the definition of the $\mathcal{D}_X$-module structure. The rest of (d) is a straightforward consequence. (Notice that the inverse image under a smooth map with connected fibers of an irreducible variety is irreducible, so the last assertion makes sense.) The formula of (e) just interprets the last relation in (d). Q.E.D.

For an analogue of Proposition 7.14, we will need to recall (in that setting) that the induction functor $G \times_H$ is an equivalence of categories from H-equivariant (quasi)-coherent $\mathcal{O}_Y$-modules to G-equivariant (quasi)-coherent $\mathcal{O}_X$-modules. We have already used this for local systems, but we will need it now more generally. (The definitions of the functors may be found for example in the appendix to [61].)

Proposition 20.2. *Suppose we are in the setting of Proposition 7.14. Regard $\mathfrak{g} \times TY$ as a vector bundle over Y, containing the bundle $\mathfrak{h} \times Y$ of (19.1). We make H act on the bundle by the adjoint action on*

$\mathfrak{g}$. Write $i : \mathfrak{h} \times Y \to \mathfrak{g} \times Y$ for the inclusion, and consider the bundle map

$$i \times a : \mathfrak{h} \times Y \to \mathfrak{g} \times TY.$$

Write Q for the quotient bundle: the fiber at y is

$$Q_y = (\mathfrak{g} \times T_y Y) / \{(X, a_y(X)) \mid X \in \mathfrak{h}\}.$$

a) The tangent bundle TX of $X = G \times_H Y$ is naturally isomorphic to the bundle on X induced by Q:

$$TX \simeq G \times_H Q.$$

b) The action mapping $a_x : \mathfrak{g} \to T_x X$ (cf. (19.1)) may be computed as follows. Fix a representative (g, y) for the point x of $G \times_H Y$, and an element $Z \in \mathfrak{g}$. Then

$$a_x(Z) = \text{class of } (g, (\mathrm{Ad}(g^{-1})Z, (y, 0))).$$

Here $(y, 0)$ is the zero element of $T_y Y$, so the term paired with g on the right side represents a class in Q_y.

c) The conormal bundle to the G action on $G \times_H Y$ is naturally induced by the conormal bundle to the H action on Y:

$$T^*_G(G \times_H Y) \simeq G \times_H T^*_H(Y).$$

Suppose $\mathcal{M}$ is an H-equivariant coherent $\mathcal{D}_Y$-module admitting an H-invariant good filtration $\mathcal{M}(p)$. Let $\mathcal{N}$ be the corresponding G-equivariant $\mathcal{D}_X$-module (Proposition 7.14(d)).

d) As a G-equivariant $\mathcal{O}_X$-module, $\mathcal{N}$ is isomorphic to $G \times_H \mathcal{M}$.

e) The filtration

$$\mathcal{N}(p) = G \times_H \mathcal{M}(p)$$

is a G-equivariant good filtration of $\mathcal{N}$. In particular,

$$\mathrm{gr}(\mathcal{N}) \simeq G \times_H \mathrm{gr}(\mathcal{M}).$$

f) The characteristic cycle of $\mathcal{N}$ may be identified as a cycle in $T^*_G X$ (see (c) above) as

$$\mathrm{Ch}(\mathcal{N}) = G \times_H \mathrm{Ch}(\mathcal{M}).$$

20. Characteristic cycles and Harish-Chandra modules

In particular, the multiplicities are given by

$$\chi^{mic}_{G \times_H S}(\mathcal{N}) = \chi^{mic}_S(\mathcal{M}).$$

Proof. Parts (a) and (b) are elementary. Part (c) is a consequence of (b). Parts (d) and (e) of course involve the explicit equivalence of categories in Proposition 7.14(d), which we have not written down. The idea, as in Proposition 7.14, is to apply Proposition 20.1 to the two smooth morphisms $f_G : G \times Y \to Y$ (projection on the second factor) and $f_H : G \times Y \to G \times_H Y$ (dividing by the H action). Then $f_H^* \mathcal{M} \simeq f_G^* \mathcal{N}$; this is more or less the definition of $\mathcal{N}$. Part (f) follows from (e). Q.E.D.

We work now in the setting of Theorem 8.3. Fix a complex connected reductive algebraic group G, and a complete homogeneous space Y for G. If y is any point of Y, then the isotropy group G_y is a parabolic subgroup of G; in this way Y may be identified with a conjugacy class $\mathcal{P}$ of parabolic subgroups in G (or parabolic subalgebras in $\mathfrak{g}$). As in (19.1) or (8.1) there is a bundle map

$$a : \mathfrak{g} \times Y \twoheadrightarrow TY. \qquad (20.3)(a)$$

It is surjective because Y is a homogeneous space. The transpose of a is a bundle map

$$a^* : T^*Y \hookrightarrow \mathfrak{g}^* \times Y. \qquad (20.3)(b)$$

It is a closed immersion (and therefore projective) since a is surjective. Composing a^* with projection on the first factor gives the *moment map*

$$\mu : T^*Y \to \mathfrak{g}^* \qquad (20.3)(c)$$

([13], 2.3). The first projection is a projective morphism because Y is a projective variety; so μ is also projective. Clearly

$$\mu(T^*_y(Y)) = (\mathfrak{g}/\mathfrak{g}_y)^*, \qquad (20.3)(d)$$

so that the image of μ consists of all linear functionals annihilating some parabolic subalgebra in the class $\mathcal{P}$ ([13], Proposition 2.4). We write

$$\mathcal{N}^*_Y = \mu(T^*Y), \qquad (20.3)(e)$$

an irreducible closed cone in $\mathfrak{g}^*$.

Proposition 20.4 (Richardson — see [43] or [13], 2.6). *In the setting of (20.3), the cone $\mathcal{N}_Y^*$ is contained in the cone of nilpotent elements of $\mathfrak{g}^*$. It is the union of finitely many orbits under G. In particular there is a unique open orbit*

$$\mathcal{Z}_Y \subset \mathcal{N}_Y^*,$$

the Richardson orbit attached to Y. The dimension of $\mathcal{Z}_Y$ is equal to the dimension of T^*Y; so μ is generically finite. The preimage $\mu^{-1}(\mathcal{Z}_Y)$ is an open orbit of G on T^*Y.

The degree of μ may be computed as follows. Fix $z \in \mathcal{Z}_Y$, and a point $\zeta \in T_y^*(Y)$ with $\mu(\zeta) = z$. Write G_z for the isotropy group of z, G_y for the isotropy group of y (a parabolic subgroup in $\mathcal{P}$), and G_ζ for the isotropy group of ζ. Then

$$G_\zeta = G_y \cap G_z, \quad G_z \supset G_\zeta \supset (G_z)_0.$$

The degree of μ is equal to the index of G_ζ in G_z. Equivalently, it is the number of distinct parabolic subalgebras in $\mathcal{P}$ annihilated by the linear functional $z \in \mathfrak{g}^*$.

We should say a word about the meaning of "nilpotent elements" of $\mathfrak{g}^*$. The Lie algebra $\mathfrak{g}$ is reductive, and so may be identified with its dual using an invariant bilinear form. Using this identification, we may therefore speak of nilpotent or semisimple elements in $\mathfrak{g}^*$. A more intrinsic discussion may be found in section 5 of [59].

Finally, suppose M is a finitely generated $U(\mathfrak{g})$-module. By an argument parallel to (but technically much simpler than) Definition 19.9, we can define the *associated variety of M*,

$$\mathcal{V}(M) \subset \mathfrak{g}^* \tag{20.5}$$

(see [59] or [13], 4.1.) Again the irreducible components of the associated variety have well-defined multiplicities, so that we can define the *associated cycle* Ch(M) ([59], Definition 2.4 or [14], 1.2).

Theorem 20.6 (Borho-Brylinski — see [14], Theorem 1.9). *In the setting (20.3), suppose M is a finitely generated module for $U(\mathfrak{g})/I_Y$ (cf. (8.1)(c)) and $\mathcal{M}$ is the corresponding $\mathcal{D}_Y$-module (Theorem 8.3). Then*

$$\mathcal{V}(M) = \mu(\mathrm{SS}(\mathcal{M})).$$

20. Characteristic cycles and Harish-Chandra modules

Keeping track of multiplicities is a more delicate task, because the fibers of μ can be quite complicated.

Theorem 20.7 (Borho-Brylinski). *In the setting of Theorem 20.6, fix good filtrations of $\mathcal{M}$ and M. Consider the higher direct images*

$$R^i \mu_*(\operatorname{gr} \mathcal{M}) = N^i,$$

which we regard as finitely generated graded $S(\mathfrak{g})$-modules. Then N^i is supported on

$$\{z \in \mathcal{N}_Y^* \mid \dim(\mu^{-1}(z) \cap \operatorname{SS} \mathcal{M}) \geq i\} \subset \mathcal{V}(M).$$

In the Grothendieck group of coherent modules supported on $\mathcal{V}(M)$, we have

$$\operatorname{gr} M = \sum_i (-1)^i N^i.$$

Proof. That N^i is coherent follows from the fact that μ is projective ([20], Theorem III.8.8). The statement about support follows from [20], Theorem III.11.1. A version of the last assertion is proved in [14], Corollary A.6 after twisting by a sufficiently positive line bundle on Y. (This has the effect of killing the higher direct images.) Their argument proves this result in general. Q.E.D.

There are several ways to describe the modules N^i of Theorem 20.6. One is

$$N^i \simeq H^i(T^*Y, \operatorname{gr} \mathcal{M}) \qquad (20.8)(a)$$

([20], Proposition III.8.5). Here we are using (19.3) to regard $\operatorname{gr} \mathcal{M}$ as a sheaf on T^*Y. Elements of $S(\mathfrak{g})$ define global functions on T^*Y, and so act on this cohomology. Another is

$$N^i \simeq H^i(Y, \operatorname{gr} \mathcal{M}) \qquad (20.8)(b)$$

([20], Exercise III.8.2). Here we are regarding $\operatorname{gr} \mathcal{M}$ as a quasicoherent sheaf on Y that happens to be also a module for $p_* \mathcal{O}_{T^*Y}$. The extra structure is needed to make N^i an $S(\mathfrak{g})$-module, but not to compute it as a vector space or as a module for some group.

Definition 20.9. In the setting of Proposition 20.4, we say that an element $\zeta \in T^*Y$ is *regular* if it belongs to the inverse image of the G-orbit $\mathcal{Z}_Y$. Similarly, an element $z \in \mathcal{N}_Y^*$ is Y-regular (or simply regular) if it belongs to $\mathcal{Z}_Y$. An irreducible subvariety W of T^*Y or $\mathcal{N}_Y^*$ is regular if it contains a regular element; equivalently, if its subset of regular elements is open and dense.

Corollary 20.10. *In the setting of Proposition 20.4, suppose that $W \subset T^*Y$ is a regular irreducible subvariety (Definition 20.9). Then $\mu(W)$ is a Y-regular irreducible subvariety of $\mathcal{N}_Y^*$, and every Y-regular irreducible subvariety arises in this way. If in addition μ has degree 1, then this correspondence of regular irreducible subvarieties is bijective.*

In the setting of Theorem 20.7, assume that μ has degree 1. Then μ defines a multiplicity-preserving bijection from the regular irreducible components of $\mathrm{SS}(\mathcal{M})$ *onto the regular irreducible components of* $\mathcal{V}(M)$.

Definition 20.11. A *symmetric pair* consists of a complex reductive Lie algebra $\mathfrak{g}$, an involutive automorphism θ of $\mathfrak{g}$, and a reductive pro-algebraic group K, with the following properties:

(a) $(\mathfrak{g}, K)$ is a compatible pair (Definition 8.2);
(b) $\mathfrak{k} = \mathfrak{g}^\theta$; and
(c) θ commutes with $\mathrm{Ad}(K)$.

We will sometimes write $\mathfrak{s}$ for the -1-eigenspace of θ on $\mathfrak{g}$, so that

$$\mathfrak{g} = \mathfrak{k} + \mathfrak{s},$$

the *Cartan decomposition*.

In the setting of (20.3), we assume in addition now that

$$(\mathfrak{g}, K) \text{ is a symmetric pair, and } K \text{ acts compatibly on } Y \quad (20.12)(a)$$

Since Y may be identified with a conjugacy class $\mathcal{P}$ of parabolic subalgebras, the action of K may be identified with

$$k \cdot \mathfrak{p} = \mathrm{Ad}(k)(\mathfrak{p}). \quad (20.12)(b)$$

(The only reason we need to assume in (a) that the action exists is that if K is disconnected, it could fail to preserve $\mathcal{P}$.) Recall from (19.1) the conormal bundle $T_K^*(Y)$ to the K action on Y (the union of the conormal bundles of the orbits). The fiber of this bundle at the point y may be identified with

$$T_{K,y}^* = (\mathfrak{g}/(\mathfrak{g}_y + \mathfrak{k}))^*. \quad (20.12)(c)$$

Set

$$\mathcal{N}_{Y,K}^* = \mu(T_K^*(Y)) = \bigcup_{\mathfrak{p}\in\mathcal{P}} (\mathfrak{g}/(\mathfrak{p}+\mathfrak{k}))^*; \qquad (20.12)(d)$$

the second description is immediate from (20.12)(c). This is a closed cone in $\mathcal{N}_Y^*$.

Lemma 20.13. *In the setting (20.12), suppose $M \in \mathcal{F}(\mathfrak{g}, K, I_Y)$ (Definition 8.2); that is, that M is a finite-length $(\mathfrak{g}, K)$-module annihilated by the kernel of the operator representation on Y. Let $\mathcal{M}$ be the associated K-equivariant $\mathcal{D}_Y$-module (Theorem 8.3)(c)), so that M may be identified with the space of global sections of $\mathcal{M}$.*

a) The associated variety $\mathcal{V}(M)$ is a closed K-invariant cone in $\mathcal{N}_Y^*$.
b) The module $\mathcal{M}$ admits a K-invariant good filtration.

Proof. Part (a) follows from Theorem 20.6, Proposition 19.12(c), and (20.12)(d). For (b), choose a finite-dimensional K-invariant generating subspace $M(0)$ of $M \simeq \Gamma\mathcal{M}$, and let $\mathcal{M}(0)$ be the (coherent) $\mathcal{O}_Y$-submodule of $\mathcal{M}$ generated by $M(0)$. Then $\mathcal{M}(0)$ generates $\mathcal{M}$ as a $\mathcal{D}_Y$ module (Theorem 8.3). It follows that

$$\mathcal{M}(p) = \mathcal{D}_Y(p)\mathcal{M}(0)$$

is a K-invariant good filtration of $\mathcal{M}$. Q.E.D.

Lemma 20.14 (Kostant-Rallis; see [31] or [59], Corollary 5.20). *In the setting of (20.12), the group K has a finite number of orbits on $\mathcal{N}_{Y,K}^*$. If $z \in \mathcal{N}_{Y,K}^*$, then*

$$\dim K \cdot z = 1/2 \dim G \cdot z.$$

In particular, z is regular (Definition 20.9) if and only if $\dim K \cdot z = \dim Y$.

Theorem 20.15 (Borho-Brylinski) *In the setting (20.12), suppose $M \in \mathcal{F}(\mathfrak{g}, K, I_Y)$ (Definition 8.2); that is, that M is a finite-length $(\mathfrak{g}, K)$-module annihilated by the kernel of the operator representation on Y. Let $\mathcal{M}$ be the associated K-equivariant $\mathcal{D}_Y$-module (Theorem 8.3)(c)), so that M may be identified with the space of global sections of $\mathcal{M}$. Then the following four conditions are equivalent:*

1) the Gelfand-Kirillov dimension $\operatorname{Dim} M$ is equal to the dimension of Y;
2) the annihilator $\operatorname{Ann} M$ is equal to I_Y;

3) the associated variety $\mathcal{V}(M)$ contains Y-regular elements (Definition 20.9); and

4) the singular support $SS\,M$ contains regular elements (Definition 20.9).

Proof. We know that the Gelfand-Kirillov dimension of M is half that of $U(\mathfrak{g})/\mathrm{Ann}\,M$ ([52], Theorem 1.1); so condition (1) is equivalent to

$$\mathrm{Dim}\,U(\mathfrak{g})/\mathrm{Ann}\,M = 2\dim Y.$$

It is well-known that the right side is equal to the Gelfand-Kirillov dimension of $U(\mathfrak{g})/I_Y$. (This follows from Theorem 20.6, for example.) On the other hand, I_Y is a prime ideal (since it is the kernel of a map into a ring without zero divisors), and $\mathrm{Ann}\,M \supset I_Y$. It follows that

$$\mathrm{Dim}\,U(\mathfrak{g})/\mathrm{Ann}\,M \geq \mathrm{Dim}\,U(\mathfrak{g})/I_Y,$$

with equality only if $\mathrm{Ann}\,M = I_Y$. This gives the equivalence of (1) and (2). Since the Gelfand-Kirillov dimension of M is the dimension of the largest component of $\mathcal{V}(M)$, the equivalence of (1) and (3) follows from Lemma 20.14. That of (3) and (4) follows from Corollary 20.10. Q.E.D.

In light of this result and Proposition 19.12, we now seek to understand the K-orbits S on Y with the property that T_S^*Y is regular.

Lemma 20.16. *In the setting (20.12), suppose S is an orbit of K on Y. Then $\mu(\overline{T_S^*Y})$ is the closure of a single orbit $\mathcal{Z}_S$ of K on $\mathcal{N}_{Y,K}^*$. We have*

$$\dim \mathcal{Z}_S = \dim Y - \text{fiber dimension of } \mu|_{T_S^*Y}.$$

Proof. Since T_S^*Y is a vector bundle over a homogeneous space for K, it is "K-irreducible;" that is, K permutes its irreducible components transitively. Consequently $\mu(T_S^*Y)$ is also K-irreducible. By Lemma 20.14, it must be contained in the closure of a single orbit. Since the dimension of T_S^*Y is equal to the dimension of Y, the last statement follows. Q.E.D.

Definition 20.17. In the setting (20.12), an orbit S of K on Y is called *regular* if any of the following equivalent conditions is satisfied:

(a) T_S^*Y is regular;
(b) $\mu(T_S^*Y)$ is Y-regular;

(c) the restriction of μ to T_S^*Y has finite degree; or
(d) $\dim \mu(T_S^*Y) = \dim Y$.

The equivalence of the conditions follows from Corollary 20.10, Lemma 20.14, and Lemma 20.16.

Theorem 20.18. *In the setting (20.12), suppose that the moment map μ has degree 1 on T^*Y. Then the correspondence of Lemma 20.16 defines a bijection from the set of regular orbits S of K on Y onto the set of regular orbits $\mathcal{Z}_S$ of K on $\mathcal{N}_{Y,K}^*$. This bijection is multiplicity-preserving in the following sense. Suppose the K-equivariant $\mathcal{D}_Y$-module $\mathcal{M}$ corresponds to the finite-length $(\mathfrak{g}, K)$-module M (Theorem 8.3). Then for every regular K-orbit S on Y, the multiplicity $\chi_S^{mic}(\mathcal{M})$ (Proposition 19.12) is equal to the multiplicity of $\overline{\mathcal{Z}_S}$ in the associated cycle $\mathrm{Ch}(M)$.*

This is a special case of Corollary 20.10. It is not very difficult to analyze the case when μ has degree greater than one. For example, if we fix a regular orbit $\mathcal{Z}_0$ of K on $\mathcal{N}_{Y,K}^*$, then

$$\sum_{\mathcal{Z}_S = \mathcal{Z}_0} \text{degree of } \mu \text{ on } T_S^*Y = \text{degree of } \mu \text{ on } T^*Y. \tag{20.19}$$

We conclude this chapter with some examples illustrating the notion of regular.

Example 20.20. Suppose that Y is the variety of Borel subgroups of G. Then $\mathcal{N}_Y^*$ is the nilpotent cone in $\mathfrak{g}^*$, and $\mathcal{Z}_Y$ is the open orbit of principal nilpotent elements. Suppose S is a regular orbit. We claim that S consists of θ-stable Borel subalgebras having no compact simple roots. (These are the Borel subalgebras of large type for θ in the sense of Definition 4.9.) To see this, suppose $z \in \mu(T_{S,y}^*Y)$ is a regular element. If y corresponds to the Borel subalgebra $\mathfrak{b}$, this means (by (20.12)(c)) that $z \in (\mathfrak{g}/(\mathfrak{b} + \mathfrak{k}))^*$; that is, that z annihilates $\mathfrak{b}$. Since $\theta\mathfrak{b} \subset \mathfrak{b} + \mathfrak{k}$, z also annihilates $\theta\mathfrak{b}$. The linear span of any two distinct Borel subalgebras must contain a larger parabolic subalgebra; so if $\mathfrak{b} \neq \theta\mathfrak{b}$, then there is a parabolic $\mathfrak{p}$ properly containing $\mathfrak{b}$ and annihilated by z. It follows from Proposition 20.4 that the G orbit of z has dimension at most twice the dimension of G/P; so z is not principal, a contradiction. Consequently $\mathfrak{b} = \theta\mathfrak{b}$. If some simple root α is compact, we find that z must also annihilate $s_\alpha\mathfrak{b}$, and reach a contradiction again. Conversely, if $(\theta, \mathfrak{b})$ is large, then it is easy to find a principal nilpotent element z annihilating $\mathfrak{b}$ with $\theta z = -z$ ([1], Proposition A.7). Then z annihilates $\mathfrak{k}$, and so is a regular element of $\mu(T_{S,y}^*Y)$. We have therefore completely characterized the regular orbits in this case. (The moment map has degree one, since the stabilizer of a principal nilpotent element is contained in a Borel subgroup.)

Example 20.21. Suppose $G = Sp(4)$, and $Y = \mathbb{P}^3$, the variety of (isotropic) lines in $\mathbb{C}^4$; the Levi subgroup for the corresponding parabolic is $GL(1) \times Sp(2)$. It turns out that $\mathcal{Z}_Y$ consists of all short root vectors in the Lie algebra, and that the moment map has degree 2. We take $K = GL(2)$, acting on $\mathbb{C}^4 = \mathbb{C}^2 + \mathbb{C}^2$ by block diagonal matrices with blocks g and $(g^t)^{-1}$. The two copies of $\mathbb{C}^2$ are in natural duality by the symplectic form. We therefore think of $\mathbb{C}^4$ as consisting of pairs (v, η), with $v \in \mathbb{C}^2$ and η a linear functional on $\mathbb{C}^2$. Obviously there are four orbits of K on Y:

$$S_1^+ = \text{lines with } v = 0, \qquad S_1^- = \text{lines with } \eta = 0 \qquad (20.21)(a)$$

$$S_2 = \text{lines with } \eta(v) = 0 \text{ not in } S_1^\pm, \qquad S_3 = \text{lines with } \eta(v) \neq 0. \qquad (20.21)(b)$$

The subscripts indicate the dimensions of the orbits. The orbit S_3 is obviously not regular, since its conormal bundle is zero. It turns out that the other three orbits are all regular, and that the corresponding orbits on $\mathcal{N}^*_{K,Y}$ are precisely the three regular orbits there. Consequently (cf. (20.19)) the moment map has degree 2 on the conormal bundle of each regular orbit.

Example 20.22. Take $G = Sp(2n)$ and Y the variety of Lagrangian subspaces of $\mathbb{C}^{2n}$; the corresponding parabolic has Levi factor $GL(n)$, and the moment map has degree 1. We take $K = GL(n)$, acting on $\mathbb{C}^n + (\mathbb{C}^n)^*$ by diagonal blocks g and $(g^t)^{-1}$. Using a little linear algebra, we can identify Y with the set of triples (V_0, V_1, Q). Here $V_0 \subset V_1 \subset \mathbb{C}^n$ are subspaces, and Q is a non-degenerate quadratic form on V_1/V_0. We will not recall in detail why this is so, but here are the definitions. If L is a Lagrangian subspace, define

$$V_0 = L \cap \mathbb{C}^n, \qquad V_1 = \{ v \in \mathbb{C}^n \mid (v, \lambda) \in L \text{ for some } \lambda \in (\mathbb{C}^n)^* \}. \qquad (20.22)(a)$$

Suppose then that (v, λ) and (v', λ') belong to L. The quadratic form is defined by

$$Q(v, v') = \lambda(v') = \lambda'(v); \qquad (20.22)(b)$$

these are equal because their difference is the symplectic pairing of (v, λ) and (v', λ'). From this description of Y, it is clear that the orbits of K are parametrized by integers p and q with $p+q \leq n$: we can take $p = \dim V_0$, $q = \dim(\mathbb{C}^n/V_1)$. The orbit $S_{p,q}$ is open if and only if $p = q = 0$, and closed if and only if $p+q = n$. The $n+1$ closed orbits are also precisely the regular orbits; they correspond by Theorem 20.18 to the $n+1$ regular orbits of K on $\mathcal{N}_{Y,K}^*$.

21. The classification theorem and Harish-Chandra modules for the dual group

We assemble here the results we have established linking the representation theory of real forms of G to Harish-Chandra modules for certain subgroups of $^\vee G$. This is for the most part a reformulation of the results of [56]. We work therefore in the setting of Proposition 6.24. Specifically, we fix always a weak E-group $^\vee G^\Gamma$, and a semisimple orbit

$$\mathcal{O} \subset {}^\vee\mathfrak{g} \qquad (21.1)(a)$$

We fix a canonical flat $\Lambda \subset \mathcal{O}$, or sometimes just a point $\lambda \in \mathcal{O}$. In either case we get as in (6.6) a well-defined semisimple element

$$e(\Lambda) = \exp(2\pi i \lambda) \in {}^\vee G \qquad (\lambda \in \Lambda) \qquad (21.1)(b)$$

and subgroups

$$P(\Lambda) \subset {}^\vee G(\Lambda)_0 \qquad (21.1)(c)$$

The reductive group $^\vee G(\Lambda)_0$ is the identity component of the centralizer of $e(\Lambda)$. Its parabolic subgroup $P(\Lambda)$ is the stabilizer of Λ in the action of $^\vee G$ on $\mathcal{F}(\mathcal{O})$ (cf. (6.10)). Define $\mathcal{P}(\Lambda)^0$ as in Proposition 6.24 to be the $^\vee G(\Lambda)_0$-orbit of Λ. Then

$$\mathcal{P}(\Lambda)^0 \simeq {}^\vee G(\Lambda)_0/P(\Lambda), \qquad (21.1)(d)$$

a flag manifold for $^\vee G(\Lambda)_0$. (This space will play the rôle of Y in Chapter 20.) Recall from (8.1) the ideal

$$I_{\mathcal{P}(\Lambda)^0} \subset U({}^\vee\mathfrak{g}(\Lambda)), \qquad (21.1)(e)$$

and from Proposition 20.4 the nilpotent orbit

$$\mathcal{Z}_{\mathcal{P}(\Lambda)^0} \subset {}^\vee\mathfrak{g}(\Lambda)^*. \qquad (21.1)(f)$$

Recall from (6.10)(f) the set $\mathcal{I}(\Lambda)$, on which $^\vee G(\Lambda)_0$ acts with finitely many orbits $\mathcal{I}_1^0(\Lambda), \ldots, \mathcal{I}_s^0(\Lambda)$; choose representatives

$$y_j \in \mathcal{I}_j^0(\Lambda). \qquad (21.1)(g)$$

21. Harish-Chandra modules for the dual group

Write θ_j for the involutive automorphism of $^\vee G(\Lambda)_0$ defined by y_j, K_j^0 for its group of fixed points, and

$$^\vee \mathfrak{g}(\Lambda) = \mathfrak{k}_j + \mathfrak{s}_j \qquad (21.1)(h)$$

for the corresponding Cartan decomposition. Finally, recall from Proposition 6.24 that the geometric parameter space $X(\mathcal{O}, {}^\vee G^\Gamma)$ is the disjoint union of s smooth connected subvarieties $X_j(\mathcal{O}, {}^\vee G^\Gamma)$, with

$$X_j(\mathcal{O}, {}^\vee G^\Gamma) \simeq {}^\vee G \times_{K_j^0} ({}^\vee G(\Lambda)_0/P(\Lambda)) \qquad (21.1)(i)$$

In particular, the orbits of $^\vee G$ on $X_j(\mathcal{O}, {}^\vee G^\Gamma)$ are naturally in one-to-one correspondence with the orbits of K_j^0 on $^\vee G(\Lambda)_0/P(\Lambda)$.

We will sometimes abuse this notation in the following way. The element $e(\Lambda)$, the group $^\vee G(\Lambda)_0$, and the sets $\mathcal{I}_j^0(\Lambda)$ depend only on the $^\vee G(\Lambda)_0$ orbit $\mathcal{P}(\Lambda)^0$, and not on Λ itself. We may therefore allow Λ to vary over $\mathcal{P}(\Lambda)^0$, rather than keep it fixed.

Here is a reformulation of Theorem 8.5.

Theorem 21.2. *Suppose we are in the setting (21.1). Then there are natural bijections among the following sets:*

1) *irreducible $(^\vee \mathfrak{g}(\Lambda), (K_j^0)^{alg})$-modules annihilated by $I_{\mathcal{P}(\Lambda)^0}$;*
2) *irreducible $(K_j^0)^{alg}$-equivariant $\mathcal{D}_{\mathcal{P}(\Lambda)^0}$-modules; and*
3) *irreducible $^\vee G^{alg}$-equivariant $\mathcal{D}$-modules on $X_j(\mathcal{O}, {}^\vee G^\Gamma)$.*

These bijections arise from natural equivalences among the three corresponding categories. If in addition we are in the setting of Theorem 10.4, then the union over j of these sets is in natural bijection with

4) *equivalence classes of irreducible canonical projective representations of type z and infinitesimal character $\mathcal{O}$ of strong real forms of G.*

Proof. The bijection between (1) and (2) is the Beilinson-Bernstein localization theorem 8.3. That between (2) and (3) is the smooth base change of Proposition 7.14 (for which the hypotheses are established in Proposition 6.24). That between (3) and (4) is Theorem 10.4. Q.E.D.

We want to apply the results of Chapter 20 on characteristic varieties to these bijections.

Proposition 21.3. *Suppose we are in the setting (21.1). Regard $x = (y_j, \Lambda)$ as a point of $X(\mathcal{O}, {}^\vee G^\Gamma)$ (Definition 6.9), so that*

$$S = {}^\vee G \cdot x \subset X(\mathcal{O}, {}^\vee G^\Gamma), \qquad S_K = K_j^0 \cdot \Lambda \subset \mathcal{P}(\Lambda)^0$$

are corresponding orbits.

a) *The isotropy group of the action of $^\vee G$ at x (or of the action of K_j^0 at Λ) is $P(\Lambda) \cap K_j^0$.*
b) *The conormal space to the orbit S at x (or to the orbit S_K at Λ) may be identified naturally (as a module for $P(\Lambda) \cap K_j^0$) with*

$$T^*_{S,x}(X(\mathcal{O},{}^\vee G^\Gamma)) = T^*_{{}^\vee G,x}(X(\mathcal{O},{}^\vee G^\Gamma)) = ({}^\vee\mathfrak{g}(\Lambda)/(\mathfrak{p}(\Lambda)+\mathfrak{k}_j))^*.$$

c) *Fix a non-degenerate $^\vee G^\Gamma$-invariant symmetric bilinear form on $^\vee\mathfrak{g}$, and use it to identify $^\vee\mathfrak{g}(\Lambda)^*$ with $^\vee\mathfrak{g}(\Lambda)$. Then the subspace appearing in (b) corresponds to*

$$({}^\vee\mathfrak{g}(\Lambda)/(\mathfrak{p}(\Lambda)+\mathfrak{k}_j))^* \simeq \mathfrak{n}(\Lambda) \cap \mathfrak{s}_j.$$

Proof. Part (a) is obvious (and contained in Proposition 6.16). For (b), combine the last description of $X(\mathcal{O},{}^\vee G^\Gamma)$ in Proposition 6.24 with Proposition 20.2(a) and (20.12)(d). For (c), the left side is clearly identified with $\mathfrak{p}(\Lambda)^\perp \cap \mathfrak{k}_j^\perp$; here the orthogonal complements are to be taken with respect to the bilinear form. It is well known (and easy to check) that the orthogonal complement of a parabolic subalgebra is its nil radical; and the Cartan decomposition of (20.11) is orthogonal since (by assumption) $\mathrm{Ad}(y_j)$ respects the form. Q.E.D.

Proposition 21.4. *Suppose we are in the setting (21.1).*
a) *Every $^\vee G^{alg}$-equivariant $\mathcal{D}$-module on $X(\mathcal{O},{}^\vee G^\Gamma)$ admits a $^\vee G^{alg}$-invariant good filtration.*
b) *Suppose that the $^\vee G^{alg}$-equivariant $\mathcal{D}$-module $\mathcal{N}$ corresponds to the $K_j^{0,alg}$-equivariant $\mathcal{D}$-module $\mathcal{M}$ in the equivalence of Theorem 21.2. Suppose S is any orbit of $^\vee G$ on $X(\mathcal{O},{}^\vee G^\Gamma)$; write S_K for the corresponding orbit of $K_j^{0,alg}$ on $\mathcal{P}^0(\Lambda)$. Then the multiplicities in the characteristic cycle are related by*

$$\chi_S^{mic}(\mathcal{N}) = \chi_{S_K}^{mic}(\mathcal{M}).$$

Proof. Part (a) follows from Proposition 20.2(e) and Lemma 20.13. Part (b) is Proposition 20.2(f). Q.E.D.

Recall now from (20.3) and (20.12) the moment map

$$\mu : T^*(\mathcal{P}(\Lambda)^0) \to \mathcal{N}^*_{\mathcal{P}(\Lambda)^0} \subset {}^\vee\mathfrak{g}(\Lambda)^* \qquad (21.5)(a)$$

and its restriction

$$\mu_{K_j^0} : T^*_{K_j^0}(\mathcal{P}(\Lambda)^0) \to \mathcal{N}^*_{\mathcal{P}(\Lambda)^0, K_j^0} \subset ({}^\vee\mathfrak{g}(\Lambda)/\mathfrak{k}_j)^*. \qquad (21.5)(b)$$

Write

$$\mathcal{Z}_{\mathcal{P}(\Lambda)^0, K_j^0} = \mathcal{Z}_{\mathcal{P}(\Lambda)^0} \cap (^\vee\mathfrak{g}(\Lambda)/\mathfrak{k}_j)^*, \qquad (21.5)(c)$$

a finite union of orbits of K_j^0 of dimension equal to $\dim \mathcal{P}(\Lambda)^0$ (Lemma 20.14). Recall (Definition 20.17) that an orbit S_K of K_j^0 on $\mathcal{P}(\Lambda)^0$ is called regular if and only if $\mu(T^*_{S_K}(\mathcal{P}(\Lambda)^0))$ meets $\mathcal{Z}_{\mathcal{P}(\Lambda)^0}$. We say that an orbit S of $^\vee G$ on $X_j(\mathcal{O}, {^\vee G^\Gamma})$ is *regular* if the corresponding orbit of K_j^0 is regular (cf. (21.1)(i)).

Theorem 21.6. *Suppose we are in the setting (21.1); use the notation (21.5). Suppose that the irreducible $(^\vee\mathfrak{g}(\Lambda), (K_j^0)^{alg})$-module M, the irreducible $(K_j^0)^{alg}$-equivariant $\mathcal{D}_{\mathcal{P}(\Lambda)^0}$-module $\mathcal{M}$, and the irreducible $^\vee G^{alg}$-equivariant $\mathcal{D}$-module $\mathcal{N}$ correspond in the bijections of Theorem 21.2. Then the following conditions are equivalent.*

a) *The characteristic variety $SS(\mathcal{N})$ contains the conormal bundle of a regular orbit S (Definition 19.9 and (21.5)).*
b) *The characteristic variety $SS(\mathcal{M})$ contains the conormal bundle of a regular orbit S_K.*
c) *The associated variety $\mathcal{V}(M)$ meets $\mathcal{Z}_{\mathcal{P}(\Lambda)^0, K_j^0}$ ((20.5) and (21.5)).*
d) *The annihilator $\operatorname{Ann} M$ is equal to $I_{\mathcal{P}(\Lambda)^0}$ ((21.1)(e)).*
e) *The Gelfand-Kirillov dimension $\operatorname{Dim} M$ is equal to $\dim \mathcal{P}(\Lambda)^0$.*

Proof. The equivalence of (a) and (b) follows from the definition of regular and Proposition 21.4(b). The rest of the result is Theorem 20.15. Q.E.D.

Theorem 21.7. *In the setting of (21.1), assume that the moment map μ of (21.5)(a) has degree one. Then there are natural bijections among the following three sets:*

1) *regular orbits S of $^\vee G$ on $X_j(\mathcal{O}, {^\vee G^\Gamma})$;*
2) *regular orbits S_K of K_j^0 on $\mathcal{P}(\Lambda)^0$; and*
3) *orbits $\mathcal{Z}_S$ of K_j^0 on $\mathcal{Z}_{\mathcal{P}(\Lambda)^0, K_j^0}$ ((21.5)(c)).*

This bijection is multiplicity-preserving in the following sense. Suppose S, S_K, and $\mathcal{Z}_S$ correspond as above, and $\mathcal{N}$, $\mathcal{M}$, and M correspond as in Theorem 21.2. Then $\chi_S^{mic}(\mathcal{N}) = \chi_{S_K}^{mic}(\mathcal{M})$, and these are equal to the multiplicity of $\overline{\mathcal{Z}_S}$ in the associated cycle $\operatorname{Ch}(M)$ (cf. (20.5)).

Proof. The first bijection is a definition, and the equality $\chi_S^{mic}(\mathcal{N}) = \chi_{S_K}^{mic}(\mathcal{M})$ is Proposition 21.4(b). The rest of the result is Theorem 20.18.
Q.E.D.

Theorem 21.8 (cf. [7], Proposition 3.24). *Suppose we are in the setting (21.1). Then there is a natural order-reversing bijection*

$$I \to {}^\vee I$$

from the set of primitive ideals in $U(\mathfrak{g})$ of infinitesimal character λ, onto the set of primitive ideals in $U({}^\vee\mathfrak{g}(\Lambda))$ containing $I_{\mathcal{P}(\Lambda)^o}$. In particular, the maximal primitive ideal $J_{max}(\lambda)$ of infinitesimal character λ corresponds to $I_{\mathcal{P}(\Lambda)^o}$:

$$^\vee J_{max}(\lambda) = I_{\mathcal{P}(\Lambda)^o}.$$

In the setting of Theorem 21.2 and Theorem 10.4, suppose that the irreducible $({}^\vee\mathfrak{g}(\Lambda), K_j^0)$-module M corresponds to the irreducible canonical projective representation π of $G(\mathbb{R}, \delta)^{can}$. Then the annihilators correspond:

$$^\vee(\operatorname{Ann} \pi) = \operatorname{Ann} M.$$

In particular, the annihilator of π is a maximal primitive ideal if and only if the annihilator of M is equal to $I_{\mathcal{P}(\Lambda)^o}$.

Recall that Theorem 21.6 provides several characterizations of the modules M with annihilator equal to $I_{\mathcal{P}(\Lambda)^o}$.

22. Arthur parameters

In this chapter we recall from [3] the parameters appearing in Arthur's conjectures. We begin with some simple facts about conjugacy classes in complex algebraic groups, which will be applied to E-groups.

Definition 22.1. Suppose H is a complex algebraic group. An element $h \in H$ is called *elliptic* if the closure in the analytic topology of the group $\langle h \rangle$ generated by h is compact. Similarly, an element $X \in \mathfrak{h}$ is called *elliptic* if the closure in the analytic topology of the one-parameter subgroup $\{\exp(tX) \mid t \in \mathbb{R}\}$ is compact.

Lemma 22.2. *Suppose H is a complex algebraic group, $h \in H$, and $X \in \mathfrak{h}$.*

a) *If h is elliptic, then h is semisimple; and if X is elliptic, then X is semisimple.*
b) *The element h is elliptic if and only if h^n is elliptic for every non-zero integer n.*

Suppose $T \subset H$ is an algebraic torus. Define

$$\mathfrak{t}_\mathbb{R} = X_*(T) \otimes_\mathbb{Z} \mathbb{R} \subset \mathfrak{t}$$

(cf. (9.1)(d) and Lemma 9.9), so that $X_*(T)$ is a lattice in the real vector space $\mathfrak{t}_\mathbb{R}$. Finally, define

$$T_c = e(\mathfrak{t}_\mathbb{R}) \simeq \mathfrak{t}_\mathbb{R}/X_*(T),$$

a compact analytic torus in T (cf. Lemma 9.9).

c) *Suppose $h \in T$. Then h is elliptic if and only if $h \in T_c$.*
d) *Suppose $X \in \mathfrak{t}$. Then X is elliptic if and only if $X \in i\mathfrak{t}_\mathbb{R}$.*

We leave the elementary argument to the reader. The point of (a) and (b) is that they reduce the problem of testing for ellipticity (even if H is disconnected) to the cases treated in (c) and (d)).

Definition 22.3. Suppose $^\vee G^\Gamma$ is a weak E-group. A Langlands parameter $\phi \in P(^\vee G^\Gamma)$ (Definition 5.2) is said to be *tempered* if it satisfies any of the following equivalent conditions.

(a) The closure of $\phi(W_\mathbb{R})$ in the analytic topology is compact.

240   The Langlands Classification and Irreducible Characters

(b) In the notation of Proposition 5.6, the Lie algebra element $\lambda + \mathrm{Ad}(y)\lambda$ is elliptic.
(c) Suppose ${}^dT^\Gamma$ is a Cartan subgroup of ${}^\vee G^\Gamma$, and that $\phi(W_\mathbb{R}) \subset {}^dT^\Gamma$. Fix an E-group structure on ${}^dT^\Gamma$, and let T be an algebraic torus defined over $\mathbb{R}$ admitting ${}^dT^\Gamma$ as an E-group. Then the canonical projective character $\pi(\phi)$ of $T(\mathbb{R})^{can}$ associated to ϕ (Proposition 10.6) is unitary.
(d) Fix an E-group structure on ${}^\vee G^\Gamma$, and let $(G^\Gamma, \mathcal{W})$ be an extended group admitting ${}^\vee G^\Gamma$ as an E-group. Then at least one of the irreducible canonical projective representations of strong real forms of G parametrized by ϕ (Theorem 10.4) is tempered.
e) In the setting of (d), all the representations parametrized by ϕ are tempered.

We write $P_{temp}({}^\vee G^\Gamma)$ for the set of tempered Langlands parameters, and $\Phi_{temp}({}^\vee G^\Gamma)$ for the corresponding set of equivalence classes.

To see that the conditions in the definition are equivalent, we can argue as follows. First, the Weil group is the direct product of the positive real numbers $\mathbb{R}^+$ and a compact group; so (a) is equivalent to ellipticity of the generator of $\phi(\mathbb{R}^+)$. This generator is precisely $\lambda + \mathrm{Ad}(y)\lambda$ (cf. (5.5)(a)). Hence (a) and (b) are equivalent. In the setting of (c), $T(\mathbb{R})$ is the direct product of a compact group and a vector group ([1], Corollary 3.16), so $\pi(\phi)$ is unitary if and only if its restriction to the vector group is. Now $\lambda + \mathrm{Ad}(y)\lambda$ is identified with the differential of that restriction, so the equivalence of (b) and (c) follows ([1], Proposition 4.11). From the definitions underlying Theorem 10.4 (Proposition 13.12), it follows that (c) is equivalent to versions of (d) and (e) in which "tempered representation" is replaced by "unitary limit character" (Definitions 11.2 and 12.1). To finish, we need to observe that in the classification of Theorem 11.4, tempered representations correspond precisely to unitary final limit characters. This is a fairly serious result, but it is well-known (see [34]).

Definition 22.4 (cf. [3], section 6). Suppose ${}^\vee G^\Gamma$ is a weak E-group (Definition 4.3). An *Arthur parameter* is a homomorphism

$$\psi : W_\mathbb{R} \times SL(2, \mathbb{C}) \to {}^\vee G^\Gamma$$

(see Definition 5.2) satisfying

(a) the restriction of ψ to $W_\mathbb{R}$ is a tempered Langlands parameter; and
(b) the restriction of ψ to $SL(2, \mathbb{C})$ is holomorphic.

Two such parameters are called *equivalent* if they are conjugate by the action of ${}^\vee G$. The set of Arthur parameters is written $Q({}^\vee G^\Gamma)$; the set

22. Arthur parameters

of equivalence classes is written $\Psi({}^\vee G^\Gamma)$. In the setting of Definition 5.3, we may also write $\Psi(G/\mathbb{R})$.

Suppose ψ is an Arthur parameter. In analogy with Definition 5.11, we define

$${}^\vee G_\psi = \text{centralizer in } {}^\vee G \text{ of } \psi(W_\mathbb{R} \times SL(2,\mathbb{C})),$$

and

$$A_\psi = {}^\vee G_\psi / ({}^\vee G_\psi)_0,$$

the *Arthur component group for* ψ. Similarly,

$$A_\psi^{alg} = {}^\vee G_\psi^{alg} / ({}^\vee G_\psi^{alg})_0,$$

the *universal component group for* ψ.

The *associated Langlands parameter for* ψ is the homomorphism

$$\phi_\psi : W_\mathbb{R} \to {}^\vee G, \qquad \phi_\psi(w) = \psi\left(w, \begin{pmatrix} |w|^{1/2} & 0 \\ 0 & |w|^{-1/2} \end{pmatrix}\right) \quad (w \in W_\mathbb{R}).$$

Attached to ϕ_ψ is an orbit S_{ϕ_ψ} of ${}^\vee G$ on $X({}^\vee G^\Gamma)$ (Proposition 6.17); define

$$S_\psi = S_{\phi_\psi}.$$

Write

$$P_{Arthur}({}^\vee G^\Gamma) = \{\phi_\psi \mid \psi \in Q({}^\vee G^\Gamma)\} \subset P({}^\vee G^\Gamma)$$

for the image of the map on parameters, and

$$\Phi_{Arthur}({}^\vee G^\Gamma) \subset \Phi({}^\vee G^\Gamma)$$

for the corresponding set of equivalence classes.

Proposition 22.5 (Arthur [2], p.10). *Suppose ${}^\vee G^\Gamma$ is a weak E-group. The map $\psi \mapsto \phi_\psi$ from Arthur parameters to Langlands parameters defines a bijection on the level of equivalence classes*

$$\Psi({}^\vee G^\Gamma) \simeq \Phi_{Arthur}({}^\vee G^\Gamma).$$

We have
$$\Phi_{temp}({}^\vee G^\Gamma) \subset \Phi_{Arthur}({}^\vee G^\Gamma) \subset \Phi({}^\vee G^\Gamma).$$

If ψ is an Arthur parameter, there is an inclusion
$$ {}^\vee G_\psi^\Gamma \subset {}^\vee G_{\phi_\psi}^\Gamma.$$

The induced maps on component groups
$$A_\psi \to A_{\phi_\psi}^{loc}, \qquad A_\psi^{alg} \to A_{\phi_\psi}^{loc,alg}$$

are surjective.

Here is the definition of the "Arthur packets" discussed in the introduction.

Definition 22.6. In the setting of Theorem 10.4, suppose $\psi \in \Psi({}^\vee G^\Gamma)$ is an Arthur parameter (Definition 22.4). The *Arthur packet* Π_ψ^z of ψ is by definition the micro-packet of the Langlands parameter ϕ_ψ attached to ψ (Definition 19.15, Definition 22.4):
$$\Pi^z(G/\mathbb{R})_\psi = \Pi^z(G/\mathbb{R})_{\phi_\psi}^{mic}.$$

A little more explicitly, write $S = S_\psi$ as in Definition 22.4. Then Π_ψ^z consists of those irreducible representations with the property that the corresponding irreducible perverse sheaf (Theorem 1.24 or Theorem 15.12) contains the conormal bundle $T_S^*(X({}^\vee G^\Gamma))$ in its characteristic cycle (Proposition 19.12).

Theorem 22.7. *Suppose $(G^\Gamma, \mathcal{W})$ is an extended group for G (Definition 1.12), and $({}^\vee G^\Gamma, \mathcal{D})$ is an E-group for the corresponding inner class of real forms, with second invariant z (Definition 4.6). Fix an Arthur parameter $\psi \in \Psi^z(G/\mathbb{R})$ (Definition 22.4). Define the Arthur packet Π_ψ as in Definition 22.6.*

a) Π_ψ contains the L-packet Π_{ϕ_ψ}, and at most finitely many additional representations of each strong real form of G.

b) There is a strongly stable formal virtual character
$$\eta_\psi = \sum_{\pi' \in \Pi_\psi} e(\pi')(-1)^{d(\pi')-d(\phi_\psi)} \chi_\psi(\pi') \pi'.$$

Here the terms are as explained in Corollary 19.16. In particular, $e(\pi') = \pm 1$ is Kottwitz' sign attached to the real form of which π' is a representation, and $\chi_\psi(\pi')$ is a positive integer.

22. Arthur parameters

This is immediate from Corollary 19.16 and the definitions. Essentially it solves Problems A, C, and E from the introduction. For Problem B, we must attach to each $\pi' \in \Pi_\psi$ a representation of A_ψ^{alg}, of dimension equal to $\chi_\psi(\pi')$. This will turn out to be standard micro-local geometry. To see that, however, we need a micro-local interpretation of A_ψ^{alg}.

We turn therefore to an analysis of the geometry of an Arthur parameter ψ. Write ψ_0 for the restriction of ψ to $W_\mathbb{R}$, a tempered Langlands parameter. According to Proposition 5.6, ψ_0 is determined by a pair

$$(y_0, \lambda_0), \qquad y_0 \in {}^\vee G^\Gamma - {}^\vee G, \qquad \lambda_0 \in {}^\vee\mathfrak{g}. \qquad (22.8)(a)$$

Write ψ_1 for the restriction of ψ to $SL(2,\mathbb{C})$, an algebraic homomorphism into ${}^\vee G$. We can define

$$y_1 = \psi_1 \begin{pmatrix} i & 0 \\ 0 & -i \end{pmatrix} \in {}^\vee G, \qquad \lambda_1 = d\psi_1 \begin{pmatrix} 1/2 & 0 \\ 0 & -1/2 \end{pmatrix} \in {}^\vee\mathfrak{g}. \qquad (22.8)(b)$$

An elementary calculation from Definition 22.4 and Proposition 5.6 shows that the Langlands parameter ϕ_ψ is associated to the pair (y, λ), with

$$y = y_0 y_1, \qquad \lambda = \lambda_0 + \lambda_1. \qquad (22.8)(c)$$

(Here and elsewhere in the following discussion, one should keep in mind that elements of $\psi_0(W_\mathbb{R})$ commute with elements of $\psi_1(SL(2,\mathbb{C}))$.) Using λ and y, we can introduce all of the structure considered in Chapter 6 (cf. (6.1), (6.10)). We also write θ_y for the conjugation action of y on ${}^\vee G(\lambda)$, and

$$ {}^\vee\mathfrak{g}(\lambda) = \mathfrak{k}(y) + \mathfrak{s}(y) \qquad (22.8)(c)$$

for the corresponding Cartan decomposition. Define

$$E_\psi = d\psi_1 \begin{pmatrix} 0 & 1 \\ 0 & 0 \end{pmatrix} \in {}^\vee\mathfrak{g}, \qquad F_\psi = d\psi_1 \begin{pmatrix} 0 & 0 \\ 1 & 0 \end{pmatrix} \in {}^\vee\mathfrak{g}; \qquad (22.8)(d)$$

these are nilpotent elements. By calculation in $\mathfrak{sl}(2)$, E_ψ lies in the $+1$-eigenspace of $\operatorname{ad}(\lambda_1)$. Since λ_0 and E_ψ commute, it follows that

$$E_\psi \in {}^\vee\mathfrak{g}(\lambda)_1 \subset \mathfrak{n}(\lambda). \qquad (22.8)(e)$$

Similarly, we check that $\operatorname{Ad}(y_1)(E_\psi) = -E_\psi$, and deduce

$$E_\psi \in \mathfrak{s}(y). \qquad (22.8)(f)$$

Here is a micro-local description of the Arthur component group. The result is a slight generalization of work of Kostant and Rallis in [31]; the proof follows their ideas very closely.

Proposition 22.9. *Suppose* $^\vee G^\Gamma$ *is an E-group, and* $\psi \in Q(^\vee G^\Gamma)$ *is an Arthur parameter. Let* $x = p(\phi_\psi) \in X(^\vee G^\Gamma)$ *(Proposition 6.17) be the corresponding point, and* $S = {^\vee G} \cdot x$ *the corresponding orbit. Fix a non-degenerate* $^\vee G^\Gamma$-*invariant symmetric bilinear form on* $^\vee \mathfrak{g}$, *and use it as in Proposition 21.3 to identify the conormal space* $T^*_{S,x}(X(^\vee G^\Gamma))$ *with* $\mathfrak{n}(\lambda) \cap \mathfrak{s}(y)$. *Accordingly regard* E_ψ *as an element of this conormal space (cf. (22.8)(e),(f)):*

$$E_\psi \in \mathfrak{n}(\lambda) \cap \mathfrak{s}(y) \simeq T^*_{S,x}(X(^\vee G^\Gamma)).$$

a) *We have*

$$[\mathfrak{p}(\lambda) \cap \mathfrak{k}(y), E_\psi] = \mathfrak{n}(\lambda) \cap \mathfrak{s}(y).$$

b) *The orbit* $(P(\lambda) \cap K(y)) \cdot E_\psi$ *(with the natural action of the isotropy group on the conormal space) is Zariski dense in* $T^*_{S,x}(X(^\vee G^\Gamma))$.
c) *The orbit* $^\vee G \cdot E_\psi$ *is Zariski dense in the full conormal bundle* $T^*_S(X(^\vee G^\Gamma))$.
d) *The isotropy groups of* E_ψ *in the actions of (b) and (c) coincide; write* $^\vee G_{E_\psi}$ *for this subgroup of* $^\vee G_x$. *Then* $^\vee G_\psi$ *is a Levi subgroup of* $^\vee G_{E_\psi}$.
e) *The inclusion in (d) identifies the Arthur component group (Definition 22.4) naturally with the component group of* $^\vee G_{E_\psi}$:

$$A_\psi \simeq {^\vee G_{E_\psi}}/({^\vee G_{E_\psi}})_0.$$

Similarly,

$$A_\psi^{alg} \simeq {^\vee G_{E_\psi}^{alg}}/({^\vee G_{E_\psi}^{alg}})_0.$$

Proof. By the representation theory of $\mathfrak{sl}(2)$, $\mathrm{ad}(\lambda_1)$ has half-integral eigenvalues. (For this discussion, we call any element of $1/2\mathbb{Z}$ a half-integer.) By definition of $\mathfrak{g}(\lambda)$, $\mathrm{ad}(\lambda)$ has integral eigenvalues on $\mathfrak{g}(\lambda)$. It follows that $\lambda_0 = \lambda - \lambda_1$ has half-integral eigenvalues on $\mathfrak{g}(\lambda)$. In analogy with (6.1)(b), we define for half-integers r and s

$$^\vee \mathfrak{g}_{r,s} = \{\mu \in {^\vee \mathfrak{g}} \mid [\lambda_0, \mu] = r\mu, [\lambda_1, \mu] = s\mu\}. \qquad (22.10)(a)$$

Then
$$^\vee\mathfrak{g}(\lambda)_n = \sum_{r+s=n} {}^\vee\mathfrak{g}_{r,s}. \qquad (22.10)(b)$$

Consulting the definitions (6.1), we deduce that

$$\mathfrak{n}(\lambda) = \sum_{r+s \in \mathbb{N}-\{0\}} {}^\vee\mathfrak{g}_{r,s}, \quad \mathfrak{p}(\lambda) = \sum_{r+s \in \mathbb{N}} {}^\vee\mathfrak{g}_{r,s}. \qquad (22.10)(c)$$

Now $\theta_y(\lambda_0) = \theta_{y_0}(\lambda_0)$ commutes with λ_0 and with λ_1 (Proposition 5.6(c)) and so acts diagonalizably on each ${}^\vee\mathfrak{g}_{r,s}$. Since ψ_0 is tempered, $\lambda_0 + \theta_{y_0}(\lambda_0)$ has purely imaginary eigenvalues (Definition 22.3(b)). It follows that $\theta_y(\lambda_0)$ acts by $-r$ on ${}^\vee\mathfrak{g}_{r,s}$. On the other hand,

$$\theta_y(\lambda_1) = \mathrm{Ad}(y_1)(\lambda_1) = \lambda_1;$$

so $\theta_y(\lambda_1)$ acts by s on ${}^\vee\mathfrak{g}_{r,s}$. Combining these two facts, we get

$$\theta_y : {}^\vee\mathfrak{g}_{r,s} \to {}^\vee\mathfrak{g}_{-r,s}. \qquad (22.10)(d)$$

Combining the descriptions of θ_y and $\mathfrak{p}(\lambda)$ in terms of the eigenspaces ${}^\vee\mathfrak{g}_{r,s}$, we get

$$\mathfrak{k}(y) \cap \mathfrak{p}(\lambda) = \{\, X + \theta_y X \mid X \in {}^\vee\mathfrak{g}_{r,s}, s \pm r \in \mathbb{N} \,\} \qquad (22.10)(e)$$

$$\mathfrak{s}(y) \cap \mathfrak{n}(\lambda) = \{\, X - \theta_y X \mid X \in {}^\vee\mathfrak{g}_{r,s}, s \pm r \in \mathbb{N} - \{0\} \,\} \qquad (22.10)(f).$$

To study the adjoint action of E_ψ, we use the representation theory of $\mathfrak{sl}(2)$. We deduce that

$$\mathrm{ad}(E_\psi) : {}^\vee\mathfrak{g}_{r,s-1} \to {}^\vee\mathfrak{g}_{r,s}; \qquad (22.10)(g)$$

this map is surjective for $s \geq 1/2$ and injective for $s \leq 1/2$.

We can now prove (a). Since both sides are vector spaces, it is enough by (22.10)(f) to show that every element of the form $X - \theta_y X$ belongs to the left side, with $X \in {}^\vee\mathfrak{g}_{r,s}$ and $s \pm r$ positive integers. By (22.10)(g), we can find $Y \in {}^\vee\mathfrak{g}_{r,s-1}$ with $[Y, E_\psi] = X$. Since $\theta_y E_\psi = -E_\psi$, it follows that

$$[Y + \theta_y Y, E_\psi] = X - \theta_y X.$$

Furthermore $(s-1)\pm r$ must be non-negative integers, so (by (22.10)(e)) $Y + \theta_y Y \in \mathfrak{k}(y) \cap \mathfrak{p}(\lambda)$. This is (a).

For (b), recall that the conormal space $T^*_{S,x}(X(^\vee G^\Gamma))$ has been identified with $\mathfrak{s}(y) \cap \mathfrak{n}(\lambda)$. This identification sends the isotropy action of $^\vee G_x = K(y) \cap P(\lambda)$ (Proposition 6.16) to the restriction of the adjoint action. It follows that the tangent space to the orbit $(P(\lambda) \cap K(y)) \cdot E_\psi$ is just $[\mathfrak{k}(y) \cap \mathfrak{p}(\lambda), E_\psi]$. (Here we identify the tangent space to the vector space $\mathfrak{s}(y) \cap \mathfrak{n}(\lambda)$ at the point E_ψ with the vector space.) By (a), this tangent space is the full tangent space at E_ψ; so the orbit is open, as we wished to show.

For (c), the conormal bundle $T^*_S(X(^\vee G^\Gamma))$ is an equivariant bundle over the homogeneous space $S = {^\vee G}/{^\vee G_x}$. The orbits of such an action are in one-to-one correspondence with those of $^\vee G_x$ on the fiber over x (cf. Lemma 6.15); this bijection preserves isotropy groups and codimension of orbits. Since $(P(\lambda) \cap K(y)) \cdot E_\psi$ has codimension 0 in $T^*_{S,x}(X(^\vee G^\Gamma))$, it follows that $^\vee G \cdot E_\psi$ has codimension 0 in $T^*_S(X(^\vee G^\Gamma))$, as we wished to show.

We have already explained the first claim of (d). For the rest, we begin with the Levi decomposition

$$P(\lambda) \cap K(y) = (L(\lambda) \cap K(y))(N_{\theta_y} \cap K(y)) \qquad (22.11)(a)$$

of Proposition 6.17 and Lemma 6.18. We claim that this decomposition is inherited by the subgroup $^\vee G_{E_\psi}$:

$$^\vee G_{E_\psi} = (L(\lambda) \cap K(y))_{E_\psi} (N_{\theta_y} \cap K(y))_{E_\psi}. \qquad (22.11)(b)$$

So suppose that ln is the decomposition of an element of $^\vee G_{E_\psi}$; it is enough to show that $l \in {^\vee G_{E_\psi}}$. Using (22.10), one calculates easily that

$$\mathfrak{n}_{\theta_y} = \sum_{s \geq |r|, s > 0, s+r \in \mathbb{Z}} {^\vee \mathfrak{g}_{r,s}}. \qquad (22.11)(c)$$

It follows that

$$\mathrm{Ad}(n)(E_\psi) = E_\psi + \text{(terms in } ^\vee \mathfrak{g}_{r,s} \text{ with } s > 1). \qquad (22.11)(d)$$

(Recall that $E_\psi \in {^\vee \mathfrak{g}_{0,1}}$.) Applying $\mathrm{Ad}(l)$ to this, we get

$$E_\psi = \mathrm{Ad}(ln)(E_\psi) = \mathrm{Ad}(l)(E_\psi) + \text{(terms in } ^\vee \mathfrak{g}_{r,s} \text{ with } s > 1).$$
$$(22.11)(e)$$

Projecting on the 1-weight space of λ_1, we find that $\mathrm{Ad}(l)(E_\psi) = E_\psi$, proving (22.11)(b).

22. Arthur parameters

To complete the proof of (d), it suffices to show that

$$(L(\lambda) \cap K(y))_{E_\psi} = {}^\vee G_\psi. \qquad (22.12)(a)$$

That the left side contains the right is trivial from the definitions; so suppose $l \in (L(\lambda) \cap K(y))_{E_\psi}$. We must show that l centralizes the image of ψ. It is enough to show that it fixes the images under $d\psi_1$ of the basis $\{E_\psi, 2\lambda_1, F_\psi\}$ of $\mathfrak{sl}(2)$ (see (22.8)). Since l commutes with λ and y, it commutes also with $\theta_y \lambda$. Now it follows from (22.10)(c) that the only eigenvalue of $\lambda_0 + \theta_y \lambda_0$ on ${}^\vee\mathfrak{g}(\lambda)$ is 0; so it is central in ${}^\vee\mathfrak{g}(\lambda)$, and in particular is fixed by l. Hence

$$2\lambda_1 = (\lambda + \theta_y \lambda) - (\lambda_0 + \theta_y \lambda_0) \qquad (22.12)(b)$$

is also fixed by l. By the injectivity statement in (22.10)(g), there is at most one element $F \in {}^\vee\mathfrak{g}_{0,-1}$ with the property that

$$[E_\psi, F] = 2\lambda_1. \qquad (22.12)(c)$$

By calculation in $\mathfrak{sl}(2)$, F_ψ satisfies (22.12)(c). Since l fixes E_ψ and $2\lambda_1$, $\mathrm{Ad}(l)(F_\psi)$ is a second solution of (22.12)(d). By the uniqueness, l fixes F_ψ. Consequently l commutes with E_ψ, $2\lambda_1$, and F_ψ, as we wished to show.

Part (e) follows from (d) exactly as in Lemma 7.5. Q.E.D.

23. Local geometry of constructible sheaves

As preparation for the serious geometry of the next chapter, we present here some trivial reformulations of material from Chapter 7. The goal is to find local results analogous to the microlocal ones we want. Constructible sheaves have a local behavior analogous to the microlocal behavior of perverse sheaves, so we will concentrate on them.

We work in the setting of Definition 7.7, so that the pro-algebraic group H acts on the smooth algebraic variety Y with finitely many orbits:

$$Y = \bigcup_{H\text{-orbits } S} S. \qquad (23.1)(a)$$

Each orbit S is smooth and locally closed. If $y \in S$, then

$$S \simeq H/H_y \qquad (23.1)(b)$$

Proposition 23.2. *In the setting (23.1), suppose C is an H-equivariant constructible sheaf on Y. Then the restriction of C to an H-orbit S is an H-equivariant local system $Q_S^{loc}(C)$ on S.*

a) *The rank of $Q_S^{loc}(C)$ is equal to the dimension of the stalks of C along S.*

b) *Suppose that $\xi = (S, \mathcal{V})$ is a complete geometric parameter for H acting on Y (Definition 7.1), and that $\mu(\xi)$ (the extension of ξ by zero — see (7.10(c))) is the corresponding irreducible H-equivariant constructible sheaf. Then $Q_S^{loc}(\mu(\xi)) = \xi$, and $Q_{S'}^{loc}(\mu(\xi)) = 0$ for $S' \neq S$.*

c) *Q_S^{loc} is an exact functor from $\mathcal{C}(Y, H)$ to H-equivariant local systems on S.*

This is entirely trivial. Using the description of equivariant local systems in Lemma 7.3, we deduce

Corollary 23.3. *In the setting (23.1), suppose C is an H-equivariant constructible sheaf on Y, and $S \subset Y$ is an H-orbit. Attached to C there is a representation $\tau_S^{loc}(C)$ of the equivariant fundamental group A_S^{loc} (Definition 7.1).*

a) The dimension of $\tau_S^{loc}(C)$ is equal to the dimension of the stalks of C on S.

b) Suppose that $\xi = (S, \tau)$ is a local complete geometric parameter for H acting on Y (with τ an irreducible representation of A_S^{loc}), and $\mu(\xi)$ is the corresponding irreducible constructible sheaf (see (7.10)(c)). Then $\tau_S^{loc}(\mu(\xi)) = \tau$, and $\tau_{S'}^{loc}(\mu(\xi)) = 0$ for $S \neq S'$.

c) The functor τ_S^{loc} from $\mathcal{C}(Y, H)$ to representations of A_S^{loc} is exact. In particular, it gives a well-defined map (also denoted τ_S^{loc}) from the Grothendieck group $K(Y, H)$ (cf. (7.10)) to virtual representations of A_S^{loc}.

Definition 23.4. Suppose $P \in \mathcal{P}(Y, H)$ is an H-equivariant perverse sheaf on Y. (In fact we could take for P any H-equivariant constructible complex on Y.) Then the cohomology sheaves $H^i P$ are H-equivariant constructible sheaves on Y (Lemma 7.8), so we can attach to each orbit S an H-equivariant local system

$$(Q_S^{loc})^i(P) = Q_S^{loc}(H^i P)$$

on S (Proposition 23.2), and a representation

$$(\tau_S^{loc})^i(P) = \tau_S^{loc}(H^i P)$$

of the equivariant fundamental group A_S^{loc} (Corollary 23.3). We write

$$\tau_S^{loc}(P) = \sum (-1)^i (\tau_S^{loc})^i(P),$$

a virtual representation of A_S^{loc}.

Because of Lemma 7.8, this notation is consistent with that of Corollary 23.3(c). We have already seen the virtual representation $\tau_S^{loc}(P)$, at least for P irreducible, as the geometric character matrix. More precisely, suppose $\gamma = (S_\gamma, \tau_\gamma)$ and $\xi = (S_\xi, \tau_\xi)$ are complete geometric parameters (Definition 7.1), with τ_γ an irreducible representation of $A_{S_\gamma}^{loc}$ and τ_ξ an irreducible representation of $A_{S_\xi}^{loc}$. Write $P(\gamma)$ for the irreducible perverse sheaf parametrized by γ (cf. (7.10)). Then

$$\text{multiplicity of } \tau_\xi \text{ in } \tau_{S_\xi}^{loc}(P(\gamma)) = (-1)^{\dim S_\xi} c_g(\xi, \gamma) \qquad (23.5)$$

(notation (7.11)).

Definition 23.6. Suppose $F \in K(Y, H)$ (notation (7.10)), S is an H-orbit on Y, and $\sigma \in A_S^{loc}$. The *local trace of σ on F* is

$$\chi_S^{loc}(F)(\sigma) = \text{tr}(\tau_S^{loc}(F)(\sigma)). \qquad (23.6)(a)$$

(The trace of an element in a finite-dimensional virtual representation of a group is well-defined.) If C is an H-equivariant constructible sheaf, or perverse sheaf, or constructible complex representing F (cf. (7.10)(a)), then we write $\chi_S^{loc}(C)(\sigma)$ instead of $\chi_S^{loc}(F)(\sigma)$. Explicitly, fix $y \in S$, so that A_S^{loc} is the group of connected components of H_y, and choose a representative $s \in H_y$ for σ. Then

$$\chi_S^{loc}(C)(\sigma) = \sum (-1)^i \operatorname{tr}(s \text{ on } (H^i C)_y), \qquad (23.6)(b)$$

the alternating sum of the traces of s on the stalks of the cohomology sheaves of C.

The *local multiplicity of F along S* is

$$\chi_S^{loc}(F) = \chi_S^{loc}(F)(1) \qquad (23.6)(c)$$

In the setting of (23.6)(b), this is the alternating sum of the dimensions of the stalks of the cohomology sheaves. (In particular, it is non-negative for constructible sheaves.)

We can use this notation to reformulate the result of Proposition 7.18.

Proposition 23.7. *Suppose we are in the setting (7.17).*

a) *Suppose $C \in \operatorname{Ob} \mathcal{C}(X, G)$ is a G-equivariant constructible sheaf on X, so that $\epsilon^* C$ is an H-equivariant constructible sheaf on Y (Proposition 7.18). Fix an orbit S of H on Y. Then the representation $\tau_S^{loc}(\epsilon^* C)$ (Corollary 23.3) is given by*

$$\tau_S^{loc}(\epsilon^* C) = A^{loc}(\epsilon) \circ \tau_{\Phi(\epsilon)S}^{loc}(C)$$

(notation (7.17)).

b) *Suppose $\xi' = (S', \tau') \in \Xi(X, G)$ is a local complete geometric parameter for G acting on X, with τ' an irreducible representation of $A_{S'}^{loc}$ (Definition 7.1). Then*

$$\epsilon^* \mu(\xi') = \sum_{\substack{\xi=(S,\tau)\in\Xi(Y,H) \\ S'=\Phi(\epsilon)S}} (\text{multiplicity of } \tau \text{ in } A^{loc}(\epsilon) \circ \tau') \mu(\xi).$$

c) *Suppose $F \in K(X, G)$ (cf. (7.10)(a)). Fix an H-orbit $S \subset Y$, and an element $\sigma \in A_S^{loc}$ (Definition 7.1). Define*

$$S' = \Phi(\epsilon) S \subset X, \qquad \sigma' = A^{loc}(\epsilon)(\sigma) \in A_{S'}^{loc}.$$

23. Local geometry of constructible sheaves

Then the local trace of σ on $\epsilon^* F$ (Definition 23.6) is

$$\chi_S^{loc}(\epsilon^* F)(\sigma) = \chi_{S'}^{loc}(F)(\sigma').$$

In particular, the local multiplicity of $\epsilon^* F$ along S is equal to the local multiplicity of F along S'.

Proof. Part (a) is a reformulation of Proposition 7.18. Part (b) follows from (a) and Corollary 23.3(b). If F is represented by a constructible sheaf C, then (c) follows from (a) by taking the trace of the action of σ. The general case follows from the fact that both sides are homomorphisms from $K(X, G)$ to $\mathbb{C}$. Q.E.D.

24. Microlocal geometry of perverse sheaves

Up until now we have described the integers that are to be the dimensions of the representations whose existence is predicted by Arthur's conjectures (Theorem 22.7). Here the key idea was that of the multiplicity of a component of a characteristic cycle. To produce the representations themselves, we must exhibit such a multiplicity as the dimension of a natural vector space, on which something like the Arthur component group can act. Now it is relatively easy to find a vector space whose dimension is the multiplicity, and in this way to construct (in the notation of Definition 22.4) a representation of something like ${}^\vee G_\psi^{alg}$. (Something very similar is done in [59], Theorem 2.13.) The difficulty arises in showing that this representation is trivial on the identity component. (The analogous result in [59] is Theorem 8.7.) Before going into further detail, we introduce some notation.

We work in the setting of Definition 7.7, so that the pro-algebraic group H acts on the smooth algebraic variety Y with finitely many orbits. Recall from (19.1) the conormal bundle

$$T_H^*(Y) = \bigcup_{H\text{-orbits } S} T_S^*(Y). \tag{24.1}(a)$$

A covector $(\lambda, y) \in T_S^*(Y)$ is called *degenerate* if it belongs to the closure in T^*Y of some other conormal bundle $T_{S'}^*(Y)$ (cf. [17], Definition I.1.8). (In this case necessarily $S \subset \overline{S'}$.) We indicate non-degenerate conormal covectors with the subscript reg, so that

$$T_H^*(Y)_{reg} = \bigcup_{H\text{-orbits } S} T_S^*(Y)_{reg}. \tag{24.1}(b)$$

The set of degenerate covectors at y is evidently a closed H_y-invariant cone in the conormal space, so the complementary set is an open cone. It follows that

$$T_S^*(Y)_{reg} \simeq H \times_{H_y} T_{S,y}^*(Y)_{reg}, \tag{24.1}(c)$$

a smooth cone bundle over $S \simeq H/H_y$.

24. Microlocal geometry of perverse sheaves

At this point things become much simpler if H has an open orbit on $T_S^*(Y)$ (or, equivalently, if H_y has an open orbit on the vector space $T_{S,y}^*(Y)$). This is automatic in the case of orbits attached to Arthur parameters (Proposition 22.9) but not in general. Here are two examples.

Example 24.2. Suppose Y is the projective line $\mathbb{CP}^1$, and H is the additive group $\mathbb{C}$ (acting by translation on the affine line, and fixing the point at infinity). Then the conormal bundle to the fixed point y is just the cotangent space at y: a one-dimensional vector space. The only degenerate covector is 0. The action of $H = H_y$ on this cotangent space is unipotent algebraic, and therefore trivial; so there are infinitely many orbits of H_y, and none is open. (This example cannot be realized as a geometric parameter space.)

For a more interesting example, take Y to be the complete flag variety of Borel subgroups in $SL(4)$, and $H = B$ the standard Borel subgroup of upper triangular matrices. (This example — or rather the essentially equivalent (Proposition 20.2) $SL(4) \times_H Y$ — does occur as a geometric parameter space, in connection with representations of $PGL(4, \mathbb{C})$.) According to the Bruhat decomposition, B has 24 orbits on Y, corresponding to the various B-conjugacy classes of Borel subgroups. Suppose B' is such a subgroup, corresponding to a point $y \in Y$ and an orbit $S \subset Y$. Write N and N' for the corresponding unipotent radicals, and T for a maximal torus in $B \cap B'$. Then

$$B_y = B \cap B' = T(N \cap N'), \qquad T_{S,y}^*(Y) \simeq \mathfrak{n} \cap \mathfrak{n}' \qquad (24.2)(a)$$

(compare Proposition 20.14). Suppose now B' is chosen so that $N \cap N'$ is abelian. Then the unipotent radical $N \cap N'$ of $B \cap B'$ acts trivially on $\mathfrak{n} \cap \mathfrak{n}'$, so

$$\text{orbits of } B_y \text{ on } T_{S,y}^*(Y) \simeq \text{orbits of } T \text{ on } \mathfrak{n} \cap \mathfrak{n}'. \qquad (24.2)(b)$$

Now let B' be obtained from B by permuting the first two and the last two coordinates. Then

$$\mathfrak{n} \cap \mathfrak{n}' = \begin{pmatrix} 0 & A \\ 0 & 0 \end{pmatrix}; \qquad (24.2)(c)$$

here we use two by two block matrices. Consequently $\mathfrak{n} \cap \mathfrak{n}'$ is abelian and four-dimensional. Since T is three-dimensional, there can be no open orbits of T on $\mathfrak{n} \cap \mathfrak{n}'$. By (24.2)(b), there are no open orbits of B_y on $T_{S,y}^*(Y)$.

Lemma 24.3. *Suppose the complex algebraic group G acts on the smooth irreducible variety X. Write G_x for the stabilizer in G of the point $x \in X$.*

a) *There is a variety G_X and a morphism $f : G_X \to X$ (a group scheme over X) with the property that the fiber over x is G_x.*

b) *Put $m = \dim G_X - \dim X$. There is an open G-invariant subvariety U_0 of X with the property that $\dim G_x = m$ for all $x \in U_0$.*

c) *There is an open G-invariant subvariety $U_1 \subset U_0$ with the property that $G_{U_1} = f^{-1}(U_1)$ is smooth, and the restriction of f to G_{U_1} is a smooth morphism .*

d) *There is an open G-invariant subvariety $U_2 \subset U_1$ with the following property. Put $G_{U_2} = f^{-1}(U_2)$, and let $(G_{U_2})_0$ be the union of the identity components $(G_x)_0$ (for $x \in U_2$). Then $(G_{U_2})_0$ is (the set of closed points of) an open and closed subgroup scheme of G_{U_2}.*

e) *There is a an étale group scheme A_{U_2} over U_2, with the property that the fiber A_x over $x \in U_2$ is isomorphic to the component group $G_x/(G_x)_0$.*

f) *There is an open G-invariant subvariety $U_3 \subset U_2$ over which the étale group scheme of (e) is finite. The family of component groups $A_x = G_x/(G_x)_0$ is in a natural way a local system of finite groups over U_3.*

g) *For each $x \in U_3$, the (algebraic) fundamental group $\pi_1(U_3; x)$ acts by automorphisms on A_x. If y is any other point of U_3, then A_y is isomorphic to A_x by an isomorphism that is canonical up to an automorphism coming from $\pi_1(U_3; x)$.*

Proof. The definition of the action provides a morphism $a : G \times X \to X$. We define G_X to be the subvariety of $G \times X$

$$G_X = \{\, (g, x) \mid a(g, x) = x \,\}.$$

Obviously this is a closed group subscheme of $G \times X$. The morphism f is just the restriction to G_X of projection on the second factor. Part (b) is a general property of morphisms of varieties ([20], Exercise II.3.23). (More precisely, we choose U' as in the general property, then let $U_0 = G \cdot U'$.)

To prove (c), we work in the analytic topology; this does not affect the notion of smooth points of a variety or a morphism. The strategy we use for showing that f is smooth near some point (g_0, x_0) in G_X is this. We look for a neighborhood X_0 of x_0, a neighborhood Z_0 of 0 in $\mathbb{C}^m$, and a holomorphic immersion $\gamma : Z_0 \times X_0 \to G \times X$, with the following properties: $\gamma(z, x) \in G_X$, $f \circ \gamma(z, x) = x$, and $\gamma(0, x_0) = (g_0, x_0)$. If such a map exists, then f is smooth of relative dimension m at (g_0, x_0), and the image of γ is a neighborhood of (g_0, x_0) in G_X. (We will call γ

a *local trivialization*, even though it is not really trivializing the whole bundle G_X over X_0.) Conversely, if G_X and f are smooth of relative dimension m at (g_0, x_0), then the implicit function theorem guarantees the existence of a local trivialization.

For (c), we may as well assume $X = U_0$, so that G_x has dimension m for all x. We will first show that f is smooth near $e \times X \subset G_X$. The family of Lie algebras $\mathfrak{g}_x$ may be regarded as a map from the smooth variety X to the Grassmanian variety of m-dimensional subspaces of $\mathfrak{g}$. In an analytic neighborhood X_0 of any point x_0 on X, we can find m holomorphic functions

$$s_i : X_0 \to \mathfrak{g} \qquad (i = 1, \ldots, m)$$

so that $\{s_i(x)\}$ is a basis of $\mathfrak{g}_x$. For a sufficiently small neighborhood of zero Z_0 in $\mathbb{C}^m$, the map

$$\gamma_e : X_0 \times Z_0 \to G_X, \qquad \gamma_e(x, z) = (\exp(\sum z_i s_i(x)), x) \in G \times X$$

is a local trivialization at (e, x_0); so f is smooth near $e \times X$.

Next, fix an irreducible component C of G_X, and suppose that the restriction of f to C is dominant (that is, that $f(C)$ contains an open set in X). Fix a smooth point (g_0, x_0) of C at which the restriction of f to C is smooth; this is possible by [20], Lemma III.10.5. Then on an appropriate analytic neighborhood X_0 of x_0, we can choose a holomorphic section

$$\sigma : X_0 \to C, \qquad \sigma(x_0) = (g_0, x_0).$$

Now the group scheme structure μ (that is, the multiplication in G in the first factor) gives a holomorphic map

$$\gamma : X_0 \times Z_0 \to C, \qquad \gamma(x, z) = \mu(\gamma_e(x, z), \sigma(x)).$$

The map γ is a local trivialization at (g_0, x_0), and it follows that the restriction of f to C is smooth of relative dimension m.

If C is an irreducible component of G_X on which f is not dominant, then $f(C)$ is contained in a Zariski-closed G-invariant proper subvariety of X. After removing such subvarieties, we may as well assume that no such components exist. Write V for the (Zariski open and dense) set of points of G_X at which f is smooth of relative dimension m. The argument of the preceding paragraph now shows that V is a subgroup scheme in G_X. (We multiply a local section at one smooth point (g_0, x_0)

by a local trivialization at another (g_0', x_0) to get a local trivialization at the product point $(g_0 g_0', x_0)$.) We also know that V contains the identity component of G_x for every x. Write Z for the complement of V. Every fiber Z_x is a union of cosets of $(G_x)_0$, so has dimension equal to m (if it is non-empty). Since Z has dimension less than the dimension of G_X, $f(Z)$ must be contained in a proper G-invariant subvariety of X. After removing this subvariety, we are left with f smooth everywhere.

For (d), we replace X by U_1 as in (c). Corollary VI-B 4.4 (page 349) of [15] guarantees the existence of an open subgroup scheme $(G_X)_0$ with the property that the fibers of f on $(G_X)_0$ are the identity components $(G_x)_0$. Let Z be the complement of $(G_X)_0$ in its closure. Then $\dim Z < m + \dim X$. On the other hand, the fibers of f on Z are unions of cosets of $(G_x)_0$, and therefore have dimension m. It follows that $f(Z)$ is contained in a proper G-invariant closed subvariety of X. After removing such a subvariety, we find that $(G_X)_0$ is closed, as we wished to show.

We will not prove (e) in detail (but see [4], Lemma 1.17, where a related result is also not proved in detail). Of course A_X is the quotient of G_X by $(G_X)_0$. If we had not arranged for $(G_X)_0$ to be closed as well as open, this quotient would exist only as an algebraic space, and not as a scheme.

For (f), an étale morphism must be finite over an open set ([20], exercise II.3.7). Part (g) is a general statement about local systems of finite groups. Q.E.D.

Example 24.4. Let G be the group of upper triangular two-by-two matrices of determinant one, and X the Lie algebra of G (with the adjoint action). Then the stabilizer of the origin is all of G, and so has dimension 2. Every other point has a one-dimensional stabilizer. The stabilizers of the semisimple points (those with non-zero diagonal entries) are the (connected) maximal tori of G. The stabilizer of a non-zero nilpotent element is the subgroup of upper triangular matrices with diagonal entries ± 1; it has two connected components. Explicitly,

$$G_X = \left\{ \left(\begin{pmatrix} a & b \\ 0 & a^{-1} \end{pmatrix}, \begin{pmatrix} x & y \\ 0 & -x \end{pmatrix} \right) \mid 2xab - y(a^2 - 1) = 0 \right\},$$
(24.4)(a)

a three-dimensional irreducible variety. The morphism f is projection on the second factor. By the Jacobian criterion (that is, by inspection of the differential of the defining equation) G_X is smooth except at the two points $(\pm I, 0)$. At smooth points of G_X, we can test for smoothness of f by determining whether df is surjective. The conclusion is that f is smooth except at $x = y = 0$. In the notation of Lemma 24.3, we can therefore take $U_0 = U_1 = X - 0$. The open subgroup scheme discussed in the proof of (d) is $G_{U_1} - \{x = 0, a = -1\}$. This is dense in G_{U_1}, and the complementary closed set is the two-dimensional variety

$$Z = \{a = -1, x = 0, y \neq 0\}.$$

Its image in X is the one-dimensional cone of non-zero nilpotent elements; so we take

$$U_2 = \left\{ \begin{pmatrix} x & y \\ 0 & -x \end{pmatrix} \mid x \neq 0 \right\}.$$

The group scheme A_{U_2} is trivial, so we can take $U_3 = U_2$.

Definition 24.5. In the setting of Lemma 24.3, the finite group A_x attached to any $x \in U_3$ is called the *generic component group* for the action of G on X. If G is only assumed to be pro-algebraic, then A_x is pro-finite.

The hypotheses of Lemma 24.3 may be weakened slightly. For example, we need not assume that X is smooth; there will in any case be a G-invariant open smooth subvariety. We can allow X to be reducible, as long as G permutes the irreducible components transitively. It would be nice to have a well-defined "generic isotropy group," but this is not possible. The difficulty is that the unipotent radical U_x of G_x may vary continuously over an open set. Nevertheless the dimension of U_x and the isomorphism class of the (reductive) quotient $L_x = G_x/U_x$ will be constant on an open set. We will make no use of these facts, however.

Lemma 24.6. *Suppose the complex pro-algebraic group G acts on the smooth irreducible variety X. Write $A = G/G_0$ for the pro-finite component group of G. Choose an open G-invariant subvariety $U \subset X$ as in Lemma 24.3(f), so that the pro-finite component groups $A_x = G_x/(G_x)_0$ form a local system over U.*

a) There is a locally constant family of natural homomorphisms

$$i_x : A_x \to A \qquad (x \in U).$$

The image of these homomorphisms is independent of x.

b) If G has an open orbit on X, then the homomorphisms of (a) are surjective.

c) Suppose $\mathcal{V}$ is a G-equivariant local system of complex r-dimensional vector spaces on X. Then $\mathcal{V}$ defines a local system of r-dimensional representations of A_x.

Proof. The maps in (a) are just

$$i_x : G_x/(G_x)_0 \to G/G_0.$$

That they are locally constant is clear from the construction of the local system A_U in Lemma 24.3. Since U is connected, the image must be constant. Clearly the image of i_x consists of the classes of elements in G that preserve the orbit $G_0 \cdot x$. If X has an open orbit S, then we may take x in S. Since X is irreducible, S must be connected; so $S = G \cdot x = G_0 \cdot x$, and i_x is surjective. For (c), we apply Lemma 7.3 to the restriction of $\mathcal{V}$ to the orbit $G \cdot x$. That is, we take the isotropy action of G_x on the fiber $\mathcal{V}_x$. Q.E.D.

Definition 24.7. Suppose Y is a smooth complex algebraic variety on which the pro-algebraic group H acts with finitely many orbits. Fix an orbit S of H on Y, and a point $y \in S$. Consider the action of H_y on the regular part $T^*_{S,y}(Y)_{reg}$ of the conormal space (cf. (24.1)), and choose an open set $U_{S,y} \subset T^*_{S,y}(Y)_{reg}$ as in Lemma 24.3(f). Notice that

$$U_S = H \times_{H_y} U_{S,y}$$

is then an open set in $T^*_S(Y)_{reg}$. Fix a point $\nu \in U_{S,y}$. The *(local) equivariant micro-fundamental group at* (y, ν) is the pro-finite group

$$A^{mic}_{y,\nu} = H_{y,\nu}/(H_{y,\nu})_0.$$

Notice that this is just the generic component group for the action of H_y on $T^*_{S,y}(Y)$, or for the action of H on $T^*_S(Y)$ (Definition 24.5). By Definition 7.1 and Lemma 24.6, it comes equipped with a natural homomorphism

$$i_{y,\nu} : A^{mic}_{y,\nu} \to A^{loc}_y,$$

which is surjective if H_y has an open orbit on $T^*_{S,y}(Y)$ (or, equivalently, if H has an open orbit on $T^*_S(Y)$).

The *equivariant micro-fundamental group for S* is

$$A^{mic}_S = A^{mic}_{y,\nu} \qquad ((y,\nu) \in U_S \subset T^*_S(Y)).$$

By Lemma 24.3, it is independent of the choice of (y, ν) up to automorphism; if H has an open orbit on $T^*_S(Y)$, then it is independent of choices up to *inner* automorphism. There is a natural homomorphism

$$i_S : A^{mic}_S \to A^{loc}_S,$$

24. Microlocal geometry of perverse sheaves

which is surjective if H has an open orbit on $T_S^*(Y)$.

Theorem 24.8 ([25], [17]). *In the setting (24.1), suppose P is an H-equivariant perverse sheaf on Y. Attached to P there is an H-equivariant local system $Q^{mic}(P)$ of complex vector spaces on $T_H^*(Y)_{reg}$ (notation (24.1)(b)).*

a) *The rank of $Q^{mic}(P)$ at any point (y,ν) of $T_S^*(Y)_{reg}$ is equal to the multiplicity $\chi_S^{mic}(P)$ of P in the characteristic cycle of P (Proposition 19.12).*
b) *Suppose that P is supported on the closure of the H-orbit S. Then the restriction of $Q^{mic}(P)$ to $T_S^*(Y)_{reg}$ is the pullback of $(Q_S^{loc})^{-\dim S}(P)$ (Definition 23.4) by the projection $T_S^*(Y)_{reg} \to S$.*
c) *Q^{mic} is an exact functor from $\mathcal{P}(Y, H)$ to H-equivariant local systems on $T_H^*(Y)_{reg}$.*

Corollary 24.9. *In the setting (24.1), suppose P is an H-equivariant perverse sheaf on Y, and $S \subset Y$ is an H-orbit. Attached to P there is a representation $\tau_S^{mic}(P)$ of the equivariant micro-fundamental group A_S^{mic} (Definition 24.7).*

a) *The dimension of $\tau_S^{mic}(P)$ is equal to the multiplicity $\chi_S^{mic}(P)$.*
b) *Suppose that $\xi = (S, \tau)$ is a local complete geometric parameter for H acting on Y (with τ an irreducible representation of A_S^{loc}), and $P(\xi)$ is the corresponding irreducible perverse sheaf. Then $\tau_S^{mic}(P(\xi)) = \tau \circ i_S$ (Definition 24.7).*
c) *The functor τ_S^{mic} from H-equivariant perverse sheaves on Y to representations of A_S^{mic} is exact. In particular, it gives rise to a map (also denoted τ_S^{mic}) from the Grothendieck group $K(Y, H)$ (cf. (7.10)) to virtual representations of A_S^{mic}.*

This theorem is absolutely fundamental to "microlocal geometry," and it is well-known to all mathematicians working in the field. Nevertheless it appears to be very difficult to find a complete proof in the literature. We are certainly not qualified to fill this gap, but we will try to indicate what is involved.

There are several ways to define the local system $Q^{mic}(P)$. Kashiwara and Kawai begin with the H-equivariant $\mathcal{D}$-module $\mathcal{M}$ corresponding to P under Theorem 7.9. By extension of scalars, they get a module $\mathcal{M}^{mic}$ for the sheaf $\mathcal{E}$ of micro-differential operators. The restriction of $\mathcal{M}^{mic}$ to $T_S^*(Y)_{reg}$ can be understood completely: it gives rise to the local system that we want ([25], Theorem 1.3.1, and [24], Theorem 3.2.1). Unfortunately, even to formulate their definition of the local system requires not only $\mathcal{E}$, but also the much larger sheaf of algebras $\mathcal{E}^{\mathbb{R}}$. Some of the necessary proofs appear in [24], but several key steps (like the flatness results that make the extensions of scalars exact) may apparently be found only in [44].

A closely related possibility is to work with the $\mathcal{O}_{T^*Y}$-module $\mathrm{gr}(\mathcal{M})$ (cf. (19.4)). We first consider what can be done easily. If the filtration of $\mathcal{M}$ is chosen to be H-invariant (as is possible on the geometric parameter space by Proposition 20.15) then it follows easily that $\mathrm{gr}(\mathcal{M})$ is annihilated by the defining ideal of $T_H^*(Y)$ (see the proof of Proposition 19.12(b)). We may therefore regard $\mathrm{gr}(\mathcal{M})$ as a sheaf of modules on $T_H^*(Y)$, which we may restrict to a single conormal space $T_S^*(Y)$. Because it is H-equivariant, this restriction is of the form $H \times_{H_y} \mathcal{N}$ for some H_y-equivariant coherent $\mathcal{O}_{T_{S,y}^*(Y)}$-module $\mathcal{N}$ (cf. (24.1)(c)). There is an open H_y-invariant subvariety U of $T_{S,y}^*(Y)$ over which $\mathcal{N}$ is (the sheaf of sections of) an H_y-equivariant vector bundle $\mathcal{V}$ (compare [46], Proposition VI.3.1). Essentially by definition, the rank of $\mathcal{V}$ is the multiplicity of $T_S^*(Y)$ in the characteristic cycle of $\mathcal{M}$. If $\nu \in U$ is arbitrary, we get an isotropy representation $\tau_{y,\nu}$ of $H_{y,\nu}$ on the stalk of $\mathcal{V}$ at ν. It is a consequence of Lemma 19.6 that the class of $\tau_{y,\nu}$ in the Grothendieck group of virtual representations of $H_{y,\nu}$ is independent of the choice of good filtration on $\mathcal{M}$. In order to prove at least Corollary 24.9 (which is all we will use) the main difficulty is to prove that $\tau_{y,\nu}$ must be trivial on the identity component of $H_{y,\nu}$. It seems likely that this can be done in a direct and elementary way, as in the proof of Theorem 8.7 of [59]. We have not done this, however. It turns out that the "canonical" good filtration constructed in [25], Corollary 5.1.11 has the property that $\mathrm{gr}(\mathcal{M})$ is actually a local system on $T_H^*(Y)_{reg}$; but this uses again the machinery of [44].

Another possibility is to use the vanishing cycles functor to construct Q^{mic}. One basic result is then [16], Proposition 7.7.1. (The proof given there appears to use analytic microlocal methods extensively.) This statement falls short of Theorem 24.8, however, and it is not easy for a non-expert to evaluate the difficulty of extending it.

The approach we will adopt is the Morse-theoretic method of [17]. The results actually proved in [17] are not quite as general as we need, but at least one can find there most of the statements and techniques needed in general (see in particular sections II.6.3 and II.6.A). We outline the construction. Fix $(y,\nu) \in T_S^*(Y)_{reg}$. First, we choose a (smooth) complex analytic submanifold N of Y that meets S transversally at y. Second, we choose a (real-valued smooth) function f on Y with $f(y) = 0$ and $df(y) = \nu$. (The assumptions on f and ν guarantee that f is Morse on N near y, with respect to the stratification by H-orbits intersected with N ([17], I.2.1.) Finally, we choose a Riemannian metric on Y, and sufficiently small positive numbers ϵ and δ ([17], I.3.6). Write $B_\delta(y)$ for the closed ball of radius δ about y in Y, and define a pair of compact

24. Microlocal geometry of perverse sheaves

spaces

$$J = N \cap B_\delta(y) \cap f^{-1}[-\epsilon, \epsilon] \supset K = N \cap B_\delta(y) \cap f^{-1}(-\epsilon) \quad (24.10)(a)$$

Then the stalk of the local system $Q^{mic}(P)$ at the point (y, ν) is by definition

$$Q^{mic}(P)_{y,\nu} = H^{-\dim S}(J, K; P), \quad (24.10)(b)$$

the hypercohomology of the pair (J, K) with coefficients in the constructible complex P. That this is a local system on $T_H^*(Y)_{reg}$ is [17], Proposition II.6.A.1.

We have not been able to find a good reference for Theorem 24.8(a) (which is stated for example in [17], II.6.A.4). One reasonable approach is to drop our old definition of $\chi_S^{mic}(P)$ completely, replacing it by Theorem 24.8(a) as a definition. We then need a proof of Theorem 1.31 using the new definition of χ_S^{mic}. We will outline such a proof at the end of this chapter.

Let us verify (b). By the transversality of the intersection of S and N, $N \cap B_\delta(y)$ meets $\overline{S}$ (the support of P) only in the point y. In particular, K does not meet the support of P at all. Consequently

$$Q^{mic}(P)_{y,\nu} = H^{-\dim S}(\{y\}, \emptyset; P) = \text{stalk at } y \text{ of } H^{-\dim S}(P).$$

By Definition 23.4, this is the stalk of $(Q_S^{loc})^{-\dim S}(P)$ at y, as we wished to show.

The exactness of $Q^{mic}(P)$ in P (part (c)) is an immediate consequence of the "purity" theorem of Kashiwara-Schapira and Goresky-MacPherson ([26], section 7.2 and Theorem 9.5.2, or [17], section II.6.A and Theorem II.6.4), which says simply that

$$H^i(J, K; P) = 0 \quad (P \text{ perverse and } i \neq -\dim S). \quad (24.10)(c)$$

(The proof in [26] uses [44], and the definition of pure is formulated a little differently.)

This concludes our discussion of Theorem 24.8. The construction allows us to formulate an analogue of Definition 23.4.

Definition 24.11. Suppose C is any H-equivariant constructible sheaf on Y. (In fact we could take for C any H-equivariant constructible complex on Y.) Then there is attached to C a family $(Q^{mic})^i(C)$ of H-equivariant local systems on $T_H^*(Y)_{reg}$, as follows. Fix a point $(y, \nu) \in$

$T^*_S(Y)_{reg})$ as in (24.10), and define $J \supset K$ as in (24.10)(a). Then the stalk of $(Q^{mic})^i(C)$ at (y, ν) is

$$H^{i-\dim S}(J, K; C).$$

That this is a local system is [17], Proposition II.6.A.1. By Lemma 24.6, it carries a representation $(\tau^{mic}_S)^i(C)$ of A^{mic}_S. (When we wish to emphasize the base point, we will write $(\tau^{mic}_{y,\nu})^i(C)$.) The virtual representation of A^{mic}_S in Corollary 24.9(c) is

$$\tau^{mic}_S(C) = \sum_i (-1)^i (\tau^{mic}_S)^i(C).$$

Definition 24.12. Suppose $F \in K(Y, H)$ (notation (7.10)), S is an H-orbit on Y, and $\sigma \in A^{mic}_S$. The *microlocal trace of σ on F* is

$$\chi^{mic}_S(F)(\sigma) = \text{tr}(\tau^{mic}_S(F)(\sigma)). \quad (24.12)(a)$$

If C is an H-equivariant constructible sheaf, or perverse sheaf, or constructible complex representing F (cf. (7.10)(a)), then we write $\chi^{mic}_S(C)(\sigma)$ instead of $\chi^{mic}_S(F)(\sigma)$. Explicitly, fix $(y, \nu) \in T^*_S(Y)_{reg}$, so that A^{mic}_S is the group of connected components of $H_{y,\nu}$, and choose a representative $s \in H_{y,\nu}$ for σ. Then

$$\chi^{mic}_S(C)(\sigma) = (-1)^{\dim S} \sum (-1)^i \text{tr}(s \text{ on } H^i(J, K; C)).$$

(If s preserves the pair (J, K), then its action on the cohomology is the natural one. In general it will carry (J, K) to another pair (J', K') of the same type, and we need to use the canonical isomorphism $H^i(J, K; C) \simeq H^i(J', K'; C)$ explained in [17] to define the action of s.)

Definition 24.13. In the setting of Theorem 10.4, fix an equivalence class $\phi \in \Phi^z(G/\mathbb{R})$ of Langlands parameters (Definition 5.3), and write $S = S_\phi$ for the corresponding orbit of ${}^\vee G$ on $X({}^\vee G^\Gamma)$ (Definition 7.6). The *micro-component group for ϕ* is by definition the ${}^\vee G^{alg}$-equivariant micro-fundamental group A^{mic}_S (Definition 24.7):

$$A^{mic,alg}_\phi = A^{mic}_S.$$

By Definition 24.7, there is a natural homomorphism

$$i_\phi : A^{mic,alg}_\phi \to A^{loc,alg}_\phi,$$

24. Microlocal geometry of perverse sheaves

which is surjective if $^\vee G$ has an open orbit on $T_S^*(X(^\vee G^\Gamma))$. To each irreducible representation $\pi \in \Pi^z(G/\mathbb{R})_\phi^{mic}$ (Definition 19.15), we associate a (possibly reducible) representation $\tau_\phi^{mic}(\pi)$, as follows. Let $P(\pi)$ be the irreducible perverse sheaf corresponding to π. Then

$$\tau_\phi^{mic}(\pi) = \tau_S^{mic}(P(\pi))$$

(Corollary 24.9). This definition makes sense for any irreducible representation π (not necessarily in the micro-packet of ψ). Corollary 24.9(a) guarantees that $\tau_\phi^{mic}(\pi) \neq 0$ if and only if $\pi \in \Pi^z(G/\mathbb{R})_\phi^{mic}$. Following Definition 24.12, we write for $\sigma \in A_\phi^{mic,alg}$

$$\chi_\phi^{mic}(\pi)(\sigma) = \operatorname{tr}(\tau_\phi^{mic}(\pi)(\sigma)).$$

Theorem 24.14. *Suppose $(G^\Gamma, \mathcal{W})$ is an extended group for G, and $(^\vee G^\Gamma, \mathcal{D})$ is an E-group for the corresponding inner class of real forms, with second invariant z (Definition 4.6). Fix an equivalence class $\phi \in \Phi^z(G/\mathbb{R})$ of Langlands parameters, and define the map τ_ψ (from irreducible representations in the micro-packet $\Pi^z(G/\mathbb{R})_\phi^{mic}$ to representations of $A_\phi^{mic,alg}$) as in Definition 24.13.*

a) Suppose π belongs to the L-packet Π_ϕ^z, and $\tau_\phi^{loc}(\pi)$ is the corresponding irreducible representation of $A_\phi^{loc,alg}$ (Definition 5.11 and Theorem 10.4). Then

$$\tau_\phi^{mic}(\pi) = \tau_\phi^{loc}(\pi) \circ i_\phi.$$

b) The dimension of $\tau_\phi^{mic}(\pi)$ is the multiplicity $\chi_\phi^{mic}(\pi)$ (Corollary 19.16).

This is immediate from Corollary 24.9 and the definitions. In the special case of Arthur parameters, we get a solution to Problem B of the introduction. Here it is.

Definition 24.15. In the setting of Theorem 10.4, suppose $\psi \in \Psi(^\vee G^\Gamma)$ is an Arthur parameter (Definition 22.4), and $\pi \in \Pi_\psi^z$ is an irreducible representation in the corresponding Arthur packet (Definition 22.6). We define a (possibly reducible) representation $\tau_\psi(\pi)$ of the universal component group A_ψ^{alg} as follows. Let $x = p(\phi_\psi) \in X(^\vee G^\Gamma)$ be the geometric parameter corresponding to the Langlands parameter ϕ_ψ, and $S \subset X(^\vee G^\Gamma)$ the corresponding orbit. Define $E_\psi \in T_{S,x}^*(^\vee G^\Gamma)$ as in Proposition 22.9, so that the orbit of E_ψ is open and dense in $T_S^*(^\vee G^\Gamma)$.

By Proposition 22.9 and Definition 24.7, A_ψ^{alg} is naturally identified with the equivariant micro-fundamental group A_S^{mic} for the action of ${}^\vee G^{alg}$ on $X({}^\vee G^\Gamma)$; that is, with $A_{\phi_\psi}^{mic,alg}$. We can therefore define

$$\tau_\psi(\pi) = \tau_{\phi_\psi}^{mic,alg}(\pi), \qquad \chi_\psi(\pi)(\sigma) = \operatorname{tr} \tau_\psi(\pi)(\sigma)$$

(Definition 24.13).

We leave to the reader the formulation of a special case of Theorem 24.14 for Arthur packets, in analogy with the deduction of Theorem 22.7 from Corollary 19.16.

We conclude this chapter with a proof of Theorem 1.31 using Theorem 24.8(a) as the definition of χ_S^{mic}. The main point is the following observation of MacPherson taken from [41].

Proposition 24.16. *Suppose X is a compact space with a finite Whitney stratification $\mathcal{S} = \{S_j\}$ having connected strata, and C is a complex of sheaves on X constructible with respect to $\mathcal{S}$. Define F_{ij} to be the dimension of the stalks of the ith cohomology sheaf $H^i C$ at points of S_j, and set*

$$F_j = \sum_i (-1)^i F_{ij}.$$

Then the Euler characteristic of the hypercohomology

$$\chi(X; C) = \sum (-1)^i \dim H^i(X; C)$$

depends only on the integers F_j. More precisely, define

$$c_j = \chi_c(S_j) = \sum_i (-1)^i \dim H_c^i(S_j; \mathbb{C}),$$

the Euler characteristic with compact supports of the stratum S_j. Then

$$\chi(X; C) = \sum_j c_j F_j.$$

The elementary proof is sketched in [18], 11.3. (Proofs that all the cohomology groups are finite-dimensional may be found in [11], Lemma V.10.13.) Connectedness of the strata is needed only to make F_{ij} well-defined; we may replace the S_j by arbitrary unions of strata, as long as the stalks of $H^i C$ have constant dimension on each S_j.

24. Microlocal geometry of perverse sheaves

To prove Theorem 1.31, fix S and any point $(y, \nu) \in T_S^*(Y)_{reg}$. Define $J \supset K$ as in (24.10). These compact sets have Whitney stratifications so that the various $J \cap S'$ and $K \cap S'$ (with S' an H-orbit containing S in its closure) are unions of strata. The cohomology sheaves $H^i(P)$ are locally constant on these sets, of constant rank $\dim(\tau_{S'}^{loc})^i(P)$ (Definition 23.4). The alternating sum of these ranks is $\chi_{S'}^{loc}(P)$ (Definition 23.6):

$$\sum_i (-1)^i \dim H^i(P)_x = \chi_{S'}^{loc}(P) \qquad (x \in J \cap S'). \qquad (24.17)(a)$$

Define

$$c(S, S') = (-1)^{\dim S} \chi_c(J \cap S', K \cap S') \qquad (24.17)(b)$$
$$= (-1)^{\dim S} \sum_i (-1)^i \dim H_c^i(J \cap S', K \cap S'; \mathbb{C}).$$

Finally, recall that we are defining

$$\chi_S^{mic}(P) = (-1)^{\dim S} \sum_i \dim H^i(J, K; P). \qquad (24.17)(c)$$

Then Theorem 1.31 is immediate from Proposition 24.16 and (24.17).
Q.E.D.

25. A fixed point formula

We alluded in the introduction to a fixed point formula relevant to the theory of endoscopic lifting. In this chapter we will explain the formula; the connection with endoscopy will appear in the next chapter.

We work in the setting of Definition 7.7, so that the pro-algebraic group G acts on the smooth variety X with a finite number of orbits. We wish to study the action of an automorphism of finite order on this situation. That is, we assume that we are given compatible automorphisms of finite order

$$\sigma : G \to G, \quad \sigma : X \to X \qquad (25.1)(a)$$

(A little more formally, we could consider an action of some finite cyclic group $\mathbb{Z}/n\mathbb{Z}$ on everything, with σ the action of the distinguished generator $1 + n\mathbb{Z}$.) The compatibility means for example that

$$(\sigma \cdot g) \cdot (\sigma \cdot x) = \sigma \cdot (g \cdot x). \qquad (25.1)(b)$$

We fix a subgroup H of G, with the property that

$$(G^\sigma)_0 \subset H \subset G^\sigma. \qquad (25.1)(c)$$

We fix also a subvariety Y of X^σ, with the property that

$$H \cdot Y = Y, \text{ and } Y \text{ is open and closed in } X^\sigma. \qquad (25.1)(d)$$

We write

$$\epsilon : Y \to X, \quad \epsilon : H \to G \qquad (25.1)(e)$$

for the inclusions. Typically we will also have an G-equivariant constructible complex C on X. We will assume that C is also endowed with an automorphism σ of finite order, compatible with those of (25.1)(a):

$$\sigma : C \to C \qquad (25.1)(f)$$

25. A fixed point formula

According to Definition 24.11, we can attach to C a family of G-equivariant local systems $(Q^{mic})^i(C)$ on $T_G^*(X)_{reg}$; the stalks at a point (x,ν) are given by

$$(Q^{mic})^i(C)_{x,\nu} = H^{i-\dim G \cdot x}(J, K; C), \qquad (25.1)(g)$$

with $J \supset K$ as in (24.10). Accordingly we get an action of σ on $(Q^{mic})^i(C)$:

$$\sigma : (Q^{mic})^i(C)_{x,\nu} \to (Q^{mic})^i(C)_{\sigma \cdot (x,\nu)}. \qquad (25.1)(h)$$

Assume finally that we are given a point $x_0 \in Y \subset X$, and a conormal covector

$$(x_0, \nu_0) \in T_G^*(X)_{reg}, \qquad \sigma \cdot (x_0, \nu_0) = (x_0, \nu_0), \qquad (25.1)(i)$$

so that σ defines an automorphism (denoted $(\tau_{x_0,\nu_0}^{mic})^i(C)(\sigma)$, in analogy with Definition 24.11) of finite order of each of the finite-dimensional vector spaces $(Q^{mic})^i(C)_{x_0,\nu_0}$. In analogy with Definition 24.12, we write

$$\chi_{x_0,\nu_0}^{mic}(C)(\sigma) = \sum (-1)^i \mathrm{tr}(\tau_{x_0,\nu_0}^{mic})^i(C)(\sigma), \qquad (25.1)(j)$$

the *microlocal trace of σ on C at (x_0, ν_0)*. The problem we consider is the calculation of this microlocal trace. Because of (25.1)(g), it may be regarded as a Lefschetz number — that is, as an alternating sum of traces on cohomology groups — so we may hope to compute it on the fixed points of σ. That is what we will do in Theorem 25.8.

For our purposes the most important examples of (25.1) arise in the following way. Fix an element $s \in G$ of finite order, and put

$$\sigma \cdot g = sgs^{-1} \qquad (g \in G). \qquad (25.2)(a)$$

We make σ act on X and C as s does. In this case $(\tau_{x_0,\nu_0}^{mic})^i(C)(\sigma)$ is just the action of s on the stalk of the local system $(Q^{mic})^i(C)$ at the point (x_0, ν_0). Suppose in addition that (x_0, ν_0) is sufficiently generic that

$$G_{x_0,\nu_0}/(G_{x_0,\nu_0})_0 = A_{x_0,\nu_0}^{mic} \qquad (25.2)(b)$$

(Definition 24.7). Then the element s (which by assumption (25.1)(i) belongs to G_{x_0,ν_0}) defines a coset

$$\overline{s} = s(G_{x_0,\nu_0})_0 \in A_{x_0,\nu_0}^{mic} \qquad (25.2)(c)$$

(Definition 24.7). It follows that

$$(\tau^{mic}_{x_0,\nu_0})^i(C)(\sigma) = (\tau^{mic}_{x_0,\nu_0})^i(C)(\bar{s}) \qquad (25.2)(d)$$

(Definition 24.11). Taking the alternating sum of traces, we get

$$\chi^{mic}_{x_0,\nu_0}(C)(\sigma) = \chi^{mic}_{x_0,\nu_0}(C)(\bar{s}). \qquad (25.2)(e)$$

Here the term on the left is defined in (25.1)(j), and that on the right by Definition 24.12.

In order to formulate the fixed point theorem, we need to know that the pair (Y, H) of (25.1) satisfies the requirements of Definition 7.7.

Lemma 25.3. *Suppose Z is a smooth algebraic variety, and σ is an algebraic automorphism of Z of finite order. Write Z^σ for the subvariety of fixed points.*

a) Z^σ is a closed smooth subvariety of Z.

b) The tangent space to Z^σ at a fixed point of σ may be identified with the fixed points of the differential of σ on the tangent space to Z:

$$T_z(Z^\sigma) = T_z(Z)^{d\sigma}.$$

Proof. (This is well-known, and we have omitted many much more difficult arguments; the reader may take it as an opportunity to relax a moment.) Every point z of Z^σ belongs to an open affine set U, and therefore to a σ-invariant affine open set $\bigcap_{n\in\mathbb{Z}} \sigma^n \cdot U$. (The intersection is finite since σ has finite order.) We may therefore assume $Z = \operatorname{Spec} A$ is affine. Then σ arises from an automorphism (also called σ) of A, also of finite order. The subvariety Z^σ is defined by the ideal I generated by elements $f - \sigma \cdot f$, with $f \in A$. Since σ acts trivially on A/I and has finite order, we have

$$A/I \simeq A^\sigma/(A^\sigma \cap I). \qquad (25.4)(a)$$

Finally, write $\mathfrak{m}$ for the maximal ideal of the point z; necessarily $I \subset \mathfrak{m}$. The Zariski cotangent spaces to Z and Z^σ at z are

$$T_z^*(Z) = \mathfrak{m}/\mathfrak{m}^2, \qquad T_z^*(Z^\sigma) = \mathfrak{m}/(\mathfrak{m}^2 + I) \simeq \mathfrak{m}^\sigma/(\mathfrak{m}^\sigma \cap (\mathfrak{m}^2 + I)). \qquad (25.4)(b)$$

(The last isomorphism uses (25.4)(a).) The (equivalent) cotangent version of (b) in the lemma therefore amounts to

$$\mathfrak{m}^\sigma/(\mathfrak{m}^\sigma \cap \mathfrak{m}^2) \simeq \mathfrak{m}^\sigma/(\mathfrak{m}^\sigma \cap (\mathfrak{m}^2 + I)). \qquad (25.4)(c)$$

25. A fixed point formula

Here the first space evidently maps surjectively to the second, so the isomorphism is equivalent to

$$(\mathfrak{m}^2 + I)^\sigma \subset \mathfrak{m}^2.$$

Since σ preserves both $\mathfrak{m}^2$ and I, this in turn is equivalent to

$$I^\sigma \subset \mathfrak{m}^2 \qquad (25.4)(d)$$

(This says that a σ-invariant function vanishing on Z^σ must actually vanish to first order there.) To prove (25.4)(d), suppose $f \in I^\sigma$. By the definition of I, we can write $f = \sum f_i(g_i - \sigma \cdot g_i)$. We can expand each f_i and g_i in eigenfunctions of σ. This gives

$$f = \sum f_j(g_j - \sigma \cdot g_j), \qquad \sigma \cdot f_j = \lambda_j f_j, \qquad \sigma \cdot g_j = \mu_j g_j$$

for some roots of unity λ_j and μ_j. Consequently

$$f = \sum (1 - \mu_j) f_j g_j, \qquad \sigma \cdot f = \sum (1 - \mu_j) \lambda_j \mu_j (f_j g_j).$$

It follows that the terms in the expansion of f in eigenfunctions of σ may be found by restricting to the summands with $\lambda_j \mu_j$ fixed. Since f is fixed by σ, we may throw away all the terms with $\lambda_j \neq (\mu_j)^{-1}$. Of course we may also throw away the terms with $\mu_j = 1$. This leaves

$$f = \sum (1 - \mu_j) f_j g_j, \qquad \sigma \cdot f_j = (\mu_j)^{-1} f_j, \qquad \sigma \cdot g_j = \mu_j g_j, \qquad \mu_j \neq 1.$$

Therefore

$$f = \sum (1 - \mu_j^{-1})^{-1} (f_j - \sigma \cdot f_j)(g_j - \sigma \cdot g_j),$$

which exhibits f as an element of $I^2 \subset \mathfrak{m}^2$. This proves (25.4)(d), and hence (b) of the lemma.

For (a), we must show that the dimension of the Zariski cotangent space is locally constant on Z^σ; for the dimension must jump up at a singular point. If $z \in Z^\sigma$, choose functions $f_1, \cdots, f_n$ on Z so that $(df_1, \cdots, df_n)$ is a basis of $T_z^* Z$. Clearly we may assume that the f_i are eigenfunctions for σ; say $\sigma \cdot f_i = \lambda_i f_i$. There is a Zariski open set $U \subset Z$ with the property that the differentials of the f_i are a basis of the tangent spaces at each point of U. By (b), it follows that

$$\dim T_y^*(Z^\sigma) = \text{number of } i \text{ such that } \lambda_i = 1$$

for all $y \in U \cap Z^\sigma$. Thus this dimension is constant on a neighborhood of z in Z^σ, as we wished to show. Q.E.D.

This argument can be extended to the case of a reductive group action on a smooth variety without essential change.

Lemma 25.5. *Suppose S is a homogeneous space for an algebraic group G, and σ is an automorphism of finite order of the pair (S, G). Then the orbits of the group of fixed points G^σ on S^σ are both open and closed; they are finite in number.*

Proof. The tangent space to S at a point x may be identified (using the action mapping of (19.1)) with the quotient of Lie algebras

$$T_x S \simeq \mathfrak{g}/\mathfrak{g}_x.$$

Here G_x is the isotropy group of the action at x. If $\sigma \cdot x = x$, then σ must preserve G_x, and so act on $\mathfrak{g}_x$. By Lemma 25.3(b),

$$T_x(S^\sigma) \simeq \mathfrak{g}^\sigma/\mathfrak{g}_x^\sigma.$$

This is also the tangent space to the orbit $G^\sigma \cdot x$. We conclude that the (smooth) orbit $G^\sigma \cdot x$ has the same dimension at x as the (smooth) fixed point set S^σ, and therefore that the orbit contains a neighborhood of x in S^σ. It follows that every orbit of G^σ on S^σ is open. As S^σ is the disjoint union of orbits of G^σ, every orbit must also be closed. Since an algebraic variety has only finitely many connected components, the lemma follows. Q.E.D.

Theorem 25.6. *Suppose we are in the setting (25.1)(a)–(e). Then Y is a smooth algebraic variety on which H acts with finitely many orbits. In particular, the map $\epsilon : (Y, H) \to (X, G)$ of (25.1) has the properties considered in (7.17). Fix a point $y \in Y$.*

a) *The conormal bundle to the H action on Y at y (cf. (19.1)) may be identified naturally as*

$$T^*_{H,y}(Y) = \left(T^*_{G,y}(X)\right)^\sigma.$$

b) *In the identification of (a), we have*

$$\left(T^*_{G,y}(X)_{reg}\right)^\sigma \subset T^*_{H,y}(Y)_{reg}.$$

c) *Suppose $y \in Y$. Then there is a natural homomorphism of equivariant fundamental groups (Definition 7.1)*

$$A^{loc}(\epsilon) : A^{loc}_y(Y, H) \to A^{loc}_y(X, G).$$

d) Suppose $(y,\nu) \in T^*_{H,y}(Y)$; use (a) to identify (y,ν) with a point of $T^*_{G,y}(X)$. Assume that these points belong to the open sets $U_{H \cdot y}$ and $U_{G \cdot y}$ of Definition 24.7. Then there is a natural homomorphism of equivariant micro-fundamental groups

$$A^{mic}(\epsilon) : A^{mic}_{y,\nu}(Y, H) \to A^{mic}_{y,\nu}(X, G).$$

Proof. Through (a), this is a straightforward consequence of Lemmas 25.3 and 25.5 and the definitions. For (b), suppose that the normal covector (y,ν) to $H \cdot y$ fails to be regular. By (24.1), (y,ν) is a limit of normal covectors to orbits in Y distinct from $H \cdot y$. By (a), these may be identified with normal covectors to orbits in X. By Lemma 25.5, the corresponding points of X must (except for finitely many) belong to orbits of G distinct from $G \cdot y$. Consequently (y,ν) fails to be regular for X, as we wished to show. The homomorphisms in (c) and (d) arise from the corresponding inclusions of isotropy groups; part (c) is just (7.17)(d), and part (d) is similar. (The regularity hypothesis on ν is included only so that we can call the component group a micro-fundamental group; it is not needed for the proof.) Q.E.D.

In the setting (25.1), we can introduce categories

$$\mathcal{C}(X, G; \sigma), \qquad \mathcal{P}(X, G; \sigma), \qquad \mathcal{D}(X, G; \sigma) \qquad (25.7)(a)$$

in analogy with Definition 7.7; we include in the objects an automorphism σ of finite order, compatible with σ on (X, G). Thus for example an object of $\mathcal{C}(X, G; \sigma)$ is an G-equivariant constructible sheaf on X, endowed with an automorphism σ of finite order, and satisfying some obvious compatibility conditions. In the special case (25.2), we simply use the categories of Definition 7.7. There are obvious analogues of Lemma 7.8 and Theorem 7.9 in this setting. In particular, the three categories of (25.7)(a) have a common Grothendieck group

$$K(X, G; \sigma). \qquad (25.7)(b)$$

By Theorem 25.6 and Proposition 7.18, there are corresponding categories

$$\mathcal{C}(Y, H; \sigma), \qquad \mathcal{P}(Y, H; \sigma), \qquad \mathcal{D}(Y, H; \sigma) \qquad (25.7)(c)$$

and a homomorphism

$$\epsilon^* : K(X, G; \sigma) \to K(Y, H; \sigma). \qquad (25.7)(d)$$

Now the action of σ on (Y, H) is trivial. It follows that any object in one of these categories is a direct sum of "eigenobjects" on which σ acts by scalar multiplication (necessarily by a root of unity). For example, a constructible sheaf $C \in \mathrm{Ob}\,\mathcal{C}(Y, H; \sigma)$ can be written

$$C = \sum_{\lambda \in \mathbb{C}^\times} C_\lambda \qquad (25.7)(e)$$

using the eigenspace decomposition of the action of σ on each stalk. This decomposition is inherited by $K(Y, H; \sigma)$. (An unnecessarily fancy way to say this is that $K(Y, H; \sigma)$ is the tensor product of $K(Y, H)$ with the group algebra of the group of roots of unity.)

Theorem 25.8. *Suppose we are in the setting (25.1). (Thus σ is an automorphism of finite order of a triple (X, G, C); here G has finitely many orbits on the smooth variety X, and C is an G-equivariant constructible complex on X.) Then the microlocal trace of σ at the σ-fixed point $(x_0, \nu_0) \in T_G^*(X)_{reg}$ (cf. (25.1)(j)) may be computed along the fixed points of σ. More precisely, use Theorem 25.6(b) to identify (x_0, ν_0) as a point of $T_H^*(Y)_{reg}$. Write $\epsilon^* C$ for the restriction to Y of C, with its inherited action of σ (Proposition 7.18). Then*

$$\chi^{mic}_{x_0,\nu_0}(C)(\sigma) = (-1)^{\dim G \cdot x_0 - \dim H \cdot x_0} \chi^{mic}_{x_0,\nu_0}(\epsilon^* C)(\sigma) \qquad (25.9)(a)$$

(notation (25.1)(j)). (The trace on the left is for X, and on the right for Y.)

Suppose in particular that we are in the setting (25.2). Then

$$\chi^{mic}_{x_0,\nu_0}(C)(\bar{s}) = (-1)^{\dim G \cdot x_0 - \dim H \cdot x_0} \chi^{mic}_{x_0,\nu_0}(\epsilon^* C)(A^{mic}(\epsilon)(\bar{s})) \qquad (25.9)(b)$$

(notation as in Theorem 25.6(d)).

We will be most interested in the last case, with C an irreducible perverse sheaf. In that case the left side of (25.9)(b) is a value of a character of a representation. The complex $\epsilon^* C$ need not be perverse, however, so the right side is only the value of a character of a virtual representation.

Proof. Recall from (24.10) the construction of the pair of spaces $J \supset K$ attached to the regular conormal covector (x_0, ν_0). The normal slice N may be taken (near x_0) to be the set of common zeros of any set of holomorphic functions vanishing at x_0, whose differentials form a basis of the conormal space $T^*_{G,x_0}(X)$. Obviously we may choose these functions to be eigenfunctions of σ; and in this case N will be preserved by σ. Of course we can choose the Riemannian metric to be preserved

25. A fixed point formula

by σ. Since (x_0, ν_0) is assumed to be fixed by σ, we may replace the smooth function f by the average of its (finitely many) translates by σ without affecting the requirements $f(x_0) = 0$ and $df(x_0) = \nu_0$. That is, we may assume that $\sigma \cdot f = f$. When the choices are made in this way, the automorphism σ preserves $J \supset K$, and so we may interpret the microlocal trace directly as a Lefschetz number:

$$\chi^{mic}_{x_0,\nu_0}(C)(\sigma) = (-1)^{\dim G \cdot x_0} \sum (-1)^i \mathrm{tr}\,(\sigma \text{ on } H^i(J, K; C)). \quad (25.9)(a)$$

By the long exact sequence for the pair, this is also the difference of the Lefschetz numbers of σ on J and K with coefficients in C.

By the analysis of tangent spaces in Lemma 25.3, and the choice of N made above, we see that N^σ is a smooth submanifold meeting $H \cdot x_0$ transversally in the single point x_0. The restriction of f to Y still has differential ν_0; so we see that the pair of spaces $J^\sigma \supset K^\sigma$ may be taken as the ones constructed in Y using (x_0, ν_0) (cf. (24.10)). Consequently

$$\chi^{mic}_{x_0,\nu_0}(\epsilon^* C)(\sigma) = (-1)^{\dim H \cdot x_0} \sum (-1)^i \mathrm{tr}\,(\sigma \text{ on } H^i(J^\sigma, K^\sigma; \epsilon^* C)). \quad (25.9)(b)$$

Again this is a difference of Lefschetz numbers for J^σ and K^σ separately.

We now apply to each of J and K the Lefschetz fixed point formula in the form established in [18]. We take for the "indicator map" needed in [18] the function

$$t(y) = (0, \text{ distance to } Y);$$

this is σ-invariant (since the Riemannian metric is) and therefore satisfies Goresky and MacPherson's requirements

$$t^{-1}(0,0) = Y, \qquad t_1(\sigma \cdot y) \geq t_1(y), \qquad t_2(\sigma \cdot y) \leq t_1(y).$$

Their local group A^i_3 ([18], Definition 4.4) for J is precisely

$$H^i(J^\sigma; \epsilon^* C),$$

and similarly for K. Their Theorem 1 ([18], section 4.7) now says

$$\sum (-1)^i \mathrm{tr}\,(\sigma \text{ on } H^i(J; C)) = \sum (-1)^i \mathrm{tr}\,(\sigma \text{ on } H^i(J^\sigma; \epsilon^* C)), \quad (25.9)(c)$$

and similarly for K. Combining (25.9)(a), (b), and (c) gives the formula we want. Q.E.D.

Most fixed point theorems (including the one in [18]) are concerned with the possibility of pathological behavior of the map whose fixed points are considered, and this is what makes them difficult. For automorphisms of finite order things are much simpler, and one should expect the most naïve results — for example, that the Lefschetz number of the automorphism is the Euler characteristic of the fixed point set — to hold under very mild hypotheses. To prove such a statement for finite polyhedra is an easy exercise. Our topological skills were insufficient to produce a direct elementary proof of Theorem 25.8, but we still believe that one exists.

26. Endoscopic lifting

The theory of endoscopic lifting created by Langlands and Shelstad concerns correspondences from stable characters of various smaller reductive groups into the (unstable virtual) characters of real forms of G. We begin with a fairly general setting for Langlands functoriality. Suppose $(G^\Gamma, \mathcal{W})$ and $(H^\Gamma, \mathcal{W}_\mathcal{H})$ are extended groups (Definition 1.12). (Recall that this essentially means that G and H are complex connected reductive algebraic groups endowed with inner classes of real forms.) Suppose $(^\vee G^\Gamma, \mathcal{D})$ and $(^\vee H^\Gamma, \mathcal{D}_H)$ are E-groups for these extended groups (Definition 4.6), say with second invariants $z \in Z(^\vee G)^{\theta_z}$ and $z_H \in Z(^\vee H)^{\theta_z}$ respectively. Suppose we are given an L-homomorphism

$$\epsilon : {}^\vee H^\Gamma \to {}^\vee G^\Gamma \qquad (26.1)(a)$$

(Definition 5.1). As in Definition 10.10, fix a quotient Q of $\pi_1(^\vee G)^{alg}$, and form the corresponding quotient

$$1 \to Q \to {}^\vee G^Q \to {}^\vee G \to 1 \qquad (26.1)(b)$$

of $^\vee G^{alg}$. As in (5.13)(b), we can pull this extension back by ϵ to

$$1 \to Q \to {}^\vee H^Q \to {}^\vee H \to 1$$

Define Q_H to be the intersection of Q with the identity component $(^\vee H^Q)_0$, so that

$$1 \to Q_H \to {}^\vee H^{Q_H} \to {}^\vee H \to 1 \qquad (26.1)(c)$$

is a connected pro-finite covering of $^\vee H$. It is easy to see that this is a quotient of the canonical cover of $^\vee H$, so Q_H is naturally a quotient of $\pi_1(^\vee H)^{alg}$. Proposition 7.18 provides a natural homomorphism

$$\epsilon^* : KX(^\vee G^\Gamma)^Q \to KX(^\vee H^\Gamma)^{Q_H} \qquad (26.1)(d)$$

(cf. (7.19)(d)).

As in Theorem 10.11, we now define a subgroup $J \subset Z(G)^{\sigma_z, fin}$ so that

$$\widehat{Q} \simeq J, \qquad (26.1)(e)$$

and similarly for J_H. Since Q_H is by its definition a subgroup of Q, J_H may be regarded as a quotient of J. The Langlands classification theorem provides bijections

$$\Pi^z(G/\mathbb{R})_J \leftrightarrow \Xi^z(G/\mathbb{R})^Q, \qquad \Pi^z(H/\mathbb{R})_{J_H} \leftrightarrow \Xi^z(H/\mathbb{R})^{Q_H}. \quad (26.1)(f)$$

To formulate Langlands functoriality in this setting, we need to consider formal complex-linear combinations of representations. Here is an appropriate modification of Corollary 1.26.

Theorem 26.2. *Suppose we are in the setting of Theorem 10.11. Write*

$$\overline{K}_{\mathbb{C}}\Pi^z(G/\mathbb{R})_J$$

for the set of (possibly infinite) formal complex combinations of irreducible canonical projective representations of type z (Definition 10.3) of strong real forms of G of type J (Definition 10.10). Then $\overline{K}_{\mathbb{C}}\Pi^z(G/\mathbb{R})_J$ may be identified with the space of complex-valued linear functionals on the Grothendieck group $KX({}^\vee G^\Gamma)^Q$:

$$\overline{K}_{\mathbb{C}}\Pi^z(G/\mathbb{R})_J \simeq \mathrm{Hom}_{\mathbb{Z}}(KX({}^\vee G^\Gamma)^Q, \mathbb{C}).$$

We will call elements of $\overline{K}_{\mathbb{C}}\Pi^z(G/\mathbb{R})_J$ *formal complex virtual representations*, or simply *virtual representations* if no confusion can arise. As in Definition 18.6, we write $\overline{K}_{\mathbb{C},f}\Pi^z(G/\mathbb{R})_J$ for the virtual representations involving only finitely many irreducible representations of each strong real form; these will have well-defined characters (Definition 18.6).

Definition 26.3. Suppose we are in the setting (26.1). *Langlands functoriality* is a linear map from formal complex virtual (canonical projective of type z_H) representations of (strong real forms of type J_H of) H to formal complex virtual representations of G,

$$\epsilon_* : \overline{K}_{\mathbb{C}}\Pi^{z_H}(H/\mathbb{R})^{Q_H} \to \overline{K}_{\mathbb{C}}\Pi^z(G/\mathbb{R})^Q. \qquad (26.3)(a)$$

It is by definition the transpose of ϵ^* (cf. (26.1)(d)) with respect to the isomorphisms of Proposition 26.2 for H and G. In terms of the pairing of Definition 15.11, this means that the defining property is

$$\langle \epsilon_* \eta_H, F_G \rangle_G = \langle \eta_H, \epsilon^* F_G \rangle_H. \tag{26.3}(b)$$

Here η_H is a formal complex virtual representation for H, and F_G belongs to the Grothendieck group of equivariant constructible sheaves on $X({}^\vee G^\Gamma)$.

We can calculate the Langlands functoriality map ϵ_* using Definition 15.11 and Theorem 15.12. To see what it does to standard representations for H, we need to see what the restriction map ϵ^* does to irreducible constructible sheaves on $X({}^\vee G)$. This is easy (Proposition 23.7). To see what it does to irreducible representations for H, we need to see what ϵ^* does to irreducible perverse sheaves. This is much harder. Here is a precise statement.

Proposition 26.4. *Suppose we are in the setting (26.1). Fix complete geometric parameters $\xi_H = (S_H, \tau_H) \in \Xi^{z_H}(H/\mathbb{R})^{Q_H}$ and $\xi_G = (S_G, \tau_G) \in \Xi^z(G/\mathbb{R})^Q$. (Here $S_H \in \Phi(H/\mathbb{R})$ is an orbit of ${}^\vee H$ on $X({}^\vee H^\Gamma)$, τ_H is an irreducible representation of the corresponding Q_H-component group $A_{S_H}^{loc, Q_H}$, and similarly for G.)*

a) *If $\Phi(\epsilon)(S_H) \neq S_G$ (cf. (7.19)(b)) then the standard representation $M(\xi_G)$ does not occur in the expression of $\epsilon_* M(\xi_H)$ in terms of standard representations.*

b) *Suppose $\Phi(\epsilon)(S_H) = S_G$; recall from (7.19)(c) the induced map $A^{loc}(\epsilon) : A_{S_H}^{loc, Q_H} \to A_{S_G}^{loc, Q}$. Then the multiplicity of $M(\xi_G)$ in the expression of $\epsilon_* M(\xi_H)$ in terms of standard representations is equal to the multiplicity of τ_H in the representation $A^{loc}(\epsilon) \circ \tau_G$ of $A_{S_H}^{loc, Q_H}$, multiplied by the quotient $e(\xi_G)/e(\xi_H)$ of the Kottwitz invariants (Definition 15.8).*

c) *If $\Phi(\epsilon)(S_H)$ is not contained in the closure of S_G, then the irreducible representation $\pi(\xi_G)$ does not occur in the expression of $\epsilon_* \pi(\xi_H)$ in terms of irreducible representations.*

d) *Suppose $\Phi(\epsilon)(S_H) \subset \overline{S_G}$. Recall from (7.10) the irreducible equivariant perverse sheaf $P(\xi_G)$. Its restriction $\epsilon^* P(\xi_G)$ is a ${}^\vee H^{Q_H}$-equivariant constructible complex on $X({}^\vee H^\Gamma)$, and may therefore be written (in the Grothendieck group $KX({}^\vee H^\Gamma)$) as a sum of irreducible equivariant perverse sheaves, with integer coefficents. Write m for the coefficient of $P(\xi_H)$ in this sum. Then the multiplicity of $\pi(\xi_G)$ in the expression of $\epsilon_* \pi(\xi_H)$ in terms of irreducible representations is equal to*

$$m(e(\xi_G)(-1)^{\dim S_G})/(e(\xi_H)(-1)^{\dim S_H}).$$

Proof. According to Definition 15.11, the multiplicity of $M(\xi_G)$ in any formal virtual representation η_G is equal to $e(\xi_G)\langle \eta_G, \mu(\xi_G)\rangle_G$. Applying Definition 26.3(b), we find that the multiplicity of $M(\xi_G)$ in $\epsilon_* M(\xi_H)$ is equal to

$$e(\xi_G)\langle \epsilon_* M(\xi_H), \mu(\xi_G)\rangle_G = e(\xi_G)\langle M(\xi_H), \epsilon^* \mu(\xi_G)\rangle_H.$$

Applying Definition 15.11 for H, we find that the right side is $e(\xi_G)/e(\xi_H)$ times the multiplicity of $\mu(\xi_H)$ in $\epsilon^* \mu(\xi_G)$. This last number is computed by Proposition 23.7(b), which gives (a) and (b). Parts (c) and (d) are proved in exactly the same way, using Theorem 15.12 in place of Definition 15.11. Q.E.D.

In part (d) of the proposition, we could use the alternating sum of the perverse cohomology groups ${}^p H^i(\epsilon^* P(\xi_G))$ (see [9]) to represent the image of $\epsilon^* P(\xi_G)$ in the Grothendieck group of equivariant perverse sheaves. These cohomology groups are equivariant perverse sheaves, so m is the alternating sum of the non-negative integers

$$m_i = \text{multiplicity of } P(\xi_H) \text{ in } {}^p H^i(\epsilon^* P(\xi_G)).$$

This perhaps sounds a little more concrete, but seems to add nothing in the way of computability to the formulation in the proposition.

The reader may wonder why we define Langlands functoriality in such generality, when our applications will be to the very special case of endoscopy. One reason is that it seems to us to be easier to understand the definition in a setting including only what is needed. Another is that we expect to find applications more general than endoscopy. For example, the cohomological induction functors used to construct discrete series representations from characters of compact Cartan subgroups implement some non-endoscopic examples of functoriality; this was critical to the proof of the Langlands classification theorem. In any case, the reader is of course welcome to think of H as an endoscopic group.

We now introduce some virtual representations that behave particularly well under functoriality.

Definition 26.5. In the setting of Theorem 10.11, suppose $\phi \in P({}^\vee G^\Gamma)$ is a Langlands parameter, and $x \in X({}^\vee G^\Gamma)$ is the corresponding point of the geometric parameter space. Write S for the orbit of x, so that the ${}^\vee G^Q$-equivariant fundamental group of S is isomorphic to the Q-component group for ϕ (Definition 7.6):

$$A_S^{loc,Q} = A_x^{loc,Q} \simeq A_\phi^{loc,Q}$$

(Definition 7.6). Fix $\sigma \in A_\phi^{loc,Q}$. The *standard formal complex virtual representation attached to ϕ and σ* is a sum of the standard final limit representations in the L-packet $\Pi^z(G/\mathbb{R})_\phi^Q$, with coefficients given by character values for the corresponding representation of $A_\phi^{loc,Q}$. Specifically,

$$\eta_\phi^{loc,Q}(\sigma) = \sum_{\xi=(\phi,\tau)\in\Xi^z(G/\mathbb{R})^Q} e(\xi)\operatorname{tr}\tau(\sigma)M(\xi);$$

the sum is over complete Langlands parameters of type Q with ϕ fixed (Definition 5.11 and (5.13)). We may also write this as $\eta_S^{loc,Q}(\sigma)$ or $\eta_x^{loc,Q}(\sigma)$. When we wish to emphasize the group G, we will write $\eta_{\phi,G}^{loc,Q}(\sigma)$. Incorporating the definition of $e(\xi)$ (Definition 15.8), we may rewrite the definition as

$$\eta_\phi^{loc,Q}(\sigma) = \sum_{\tau \in (A_\phi^{loc,Q})\widehat{}} \operatorname{tr}\tau(\sigma\overline{z(\rho)})M(\phi,\tau),$$

a sum over the irreducible representations of the Q-component group.

This virtual representation is locally finite (Definition 18.6). If δ is a strong real form of G of type J (Definition 10.10) we can write

$$\eta_\phi^{loc,Q}(\sigma)(\delta) = e(G(\mathbb{R},\delta)) \sum_{\substack{\xi=(\phi,\tau)\in\Xi^z(G/\mathbb{R}) \\ \delta(\xi)=\delta}} \operatorname{tr}\tau(\sigma)M(\xi)$$

for the part of $\eta_\phi^{loc}(\sigma)$ living on $G(\mathbb{R},\delta)$. This is a finite combination of irreducible representations with complex coefficients, so it has a well-defined character

$$\Theta(\eta_\phi^{loc,Q}(\sigma),\delta),$$

a generalized function on $G(\mathbb{R},\delta)^{can}$.

The term $\overline{z(\rho)}$ appears unnatural here; it is justified by Shelstad's Theorem 18.12.

Lemma 26.6. *In the setting of Definition 26.5, the isomorphism of Theorem 26.2 identifies the formal complex virtual representation $\eta_\phi^{loc,Q}(\sigma)$ with the map $\chi_S^{loc,Q}(\cdot)(\sigma)$ (Definition 23.6) that assigns to a constructible equivariant sheaf C the trace of σ acting on the stalk C_x.*

In terms of irreducible representations, we therefore have

$$\eta_\phi^{loc,Q}(\sigma) = \sum_{\xi \in \Xi^z(G/\mathbb{R})^Q} e(\xi)(-1)^{d(\xi)} \chi_S^{loc,Q}(P(\xi))(\sigma)\pi(\xi).$$

Here as usual $P(\xi)$ is the irreducible perverse sheaf and $\pi(\xi)$ the irreducible representation corresponding to the complete geometric parameter ξ.

This is proved in exactly the same way as Lemma 18.15; the $\overline{z(\rho)}$ has disappeared because of its occurrence in Definition 15.11.

Proposition 26.7. *Suppose we are in the setting (26.1). Fix a Langlands parameter $\phi_H \in P({}^\vee H^\Gamma)$, so that $\phi_G = \epsilon \circ \phi_H$ is a Langlands parameter for ${}^\vee G^\Gamma$. Fix an element $\sigma_H \in A_{\phi_H}^{loc,Q_H}$, and write*

$$\sigma_G = A^{loc}(\epsilon)(\sigma_H) \in A_{\phi_G}^{loc,Q}$$

(cf. (5.14)(d)). Then Langlands functoriality sends the virtual representation attached to ϕ_H and σ_H to the one attached to ϕ_G and σ_G:

$$\epsilon_* \eta_{\phi_H}^{loc,Q_H}(\sigma_H) = \eta_{\phi_G}^{loc,Q}(\sigma_G).$$

Proof. This is a reformulation of Proposition 23.7(c) in a special case. It may also be deduced from Proposition 26.4(a) and (b), using the formulas in Definition 26.5 to express the standard virtual representations in terms of standard representations. Q.E.D.

The reason this result is so easy is that it has so little content. We have shown that Langlands functoriality is computable, but not that it has any nice properties.

Using the results of Chapter 23, we can now give an unstable generalization of the microlocal stable characters of Chapter 19.

Definition 26.8. In the setting of Theorem 10.11, suppose $\phi \in P({}^\vee G^\Gamma)$ is a Langlands parameter, and $x \in X({}^\vee G^\Gamma)$ is the corresponding point of the geometric parameter space. Write S for the orbit of x, so that the equivariant micro-fundamental group of S is isomorphic to the micro-component group for ϕ (Definition 24.13):

$$A_S^{mic,Q} \simeq A_\phi^{mic,Q}.$$

Fix $\sigma \in A_\phi^{mic,Q}$. The *formal complex virtual representation attached to ϕ and σ* is a sum of the irreducible representations in the micropacket $\Pi^z(G/\mathbb{R})_{J,\phi}^{mic}$, with coefficients given by character values for the corresponding representation of $A_\phi^{mic,Q}$. Specifically,

$$\eta_\phi^{mic,Q}(\sigma) = \sum_{(\pi,\delta)\in\Pi^z(G/\mathbb{R})_{J,\phi}^{mic}} e(\delta)(-1)^{d(\pi)-d(\phi)} \left(\chi_\phi^{mic}(\pi)(\sigma)\right)\pi$$

(notation 24.13). As in Definition 26.5, we may replace the Kottwitz invariant $e(\delta)$ with an extra factor $z(\rho)$ next to σ. We may also separate the terms $\eta_\phi^{mic,Q}(\sigma)(\delta)$ corresponding to each strong real form, and attach a generalized function to each of these.

Suppose $\psi \in Q({}^\vee G^\Gamma)$ is an Arthur parameter (Definition 22.4). We identify the Q-component group A_ψ^Q with $A_{\phi_\psi}^{mic,Q}$ (Definition 24.15). This allows us to define

$$\eta_\psi^Q(\sigma) = \eta_{\phi_\psi}^{mic,Q}(\sigma)$$

for any $\sigma \in A_\psi^Q$. Using the notation of Definitions 22.6 and 24.15, this amounts to

$$\eta_\psi^Q(\sigma) = \sum_{(\pi,\delta)\in\Pi^z(G/\mathbb{R})_{J,\psi}} e(\delta)(-1)^{d(\pi)-d(\psi)} \left(\chi_\psi(\pi)(\sigma)\right)\pi.$$

Lemma 26.9. *In the setting of Definition 26.8, fix a sufficiently generic point $(x,\nu) \in T_S^*(X({}^\vee G^\Gamma))_{reg}$ (Definition 24.7). The formal complex virtual representation $\eta_\phi^{mic,Q}(\sigma)$ corresponds in the isomorphism of Theorem 26.2 to the map $(-1)^{d(\phi)}\chi_{(x,\nu)}^{mic,Q}(\cdot)(\sigma)$ (Definition 24.12) that assigns to a ${}^\vee G^Q$-equivariant perverse sheaf P the trace of σ acting on the stalk of $Q^{mic}(P)$ at (x,ν), multiplied by $(-1)^{d(\phi)}$. In terms of standard representations, we therefore have*

$$\eta_\phi^{mic,Q}(\sigma) = (-1)^{d(\phi)} \sum_{\xi\in\Xi^z(G/\mathbb{R})} e(\xi)\chi_S^{mic}(\mu(\xi))(\sigma)M(\xi).$$

Proof. We use Theorem 15.12 to compute the pairing of $\eta_\phi^{mic,Q}(\sigma)$ with the irreducible perverse sheaf $P(\pi)$. The result is

$$e(\delta)(-1)^{d(\pi)-d(\phi)} \left(\chi_\phi^{mic}(\pi)(\sigma)\right)\langle\pi,P(\pi)\rangle = (-1)^{d(\phi)}\chi_\phi^{mic}(\pi)(\sigma).$$

The first assertion follows. The second is a consequence, because of Definition 15.11. Q.E.D

We should make here a few remarks about the computability of these virtual characters. The parametrization of standard final limit representations in Theorem 10.11 is fairly concrete; there is no difficulty in listing all the induced from limits of discrete series representations corresponding to a given Langlands parameter ϕ, or in describing the associated characters of $A_\phi^{loc,Q}$. In this sense the virtual representations $\eta_\phi^{loc,Q}(\sigma)$ are very well understood. The characters of standard limit representations are computable as well, although it is not quite so easy to get answers in closed form. Nevertheless the generalized functions $\Theta(\eta_\phi^{loc,Q}(\sigma),\delta)$ may be regarded as known. To describe $\eta_\phi^{loc,Q}(\sigma)$ in terms of irreducible representations is not very different from describing the composition series of the standard representations. This is accomplished in principle by the Kazhdan-Lusztig algorithms (see Chapters 15 – 17), although one can get useful answers in closed form only under very special circumstances.

The situation for $\eta_\phi^{mic,Q}(\sigma)$ is less satisfactory. We do not know even in principle how to compute characteristic cycles of perverse sheaves, so we cannot compute even the size of the micro-packets of Definition 19.15 in general. Nevertheless, we expect them to be moderately small — larger than L-packets, but not too much. It should actually be simpler to compute characteristic cycles for irreducible perverse sheaves than for irreducible constructible sheaves (or at least the answer should be simpler). The expression of $\eta_\phi^{mic,Q}(\sigma)$ in terms of standard representations (which would provide explicit formulas for its character) therefore promises to be very complicated.

We would like a version of Proposition 26.7 for the virtual representations $\eta_\phi^{mic,Q}(\sigma)$. It is immediately clear that some additional hypotheses are required: there is no homomorphism of equivariant microfundamental groups $A^{mic}(\epsilon)$ attached to a general L-homomorphism, so we cannot even formulate an analogous statement. But we found in Chapter 24 geometric conditions under which such a homomorphism does exist, and even leads to a result like Proposition 23.7 (which was the geometric part of Proposition 26.7). In the present context, these conditions lead directly to the Langlands-Shelstad theory of endoscopic groups. Before we discuss that theory, it is convenient to develop a small extension of the results of Chapter 18.

Definition 26.10. Suppose we are in the setting (5.14). Recall from Proposition 4.4 the group $Z(^\vee G)^{\theta_z}$; write $Z(^\vee G)^{\theta_z,Q}$ for its preimage in $^\vee G^Q$, so that we have a short exact sequence

$$1 \to Q \to Z(^\vee G)^{\theta_z,Q} \to Z(^\vee G)^{\theta_z} \to 1.$$

If ϕ is any Langlands parameter for ${}^\vee G^\Gamma$, then this group centralizes the image of ϕ:

$$Z({}^\vee G)^{\theta_z, Q} \subset {}^\vee G_\phi^Q \qquad (26.10)(a)$$

(notation as in (5.14)). The *universal Q-component group for ${}^\vee G^\Gamma$* is

$$A_{univ}^Q({}^\vee G^\Gamma) = Z({}^\vee G)^{\theta_z, Q}/(Z({}^\vee G)^{\theta_z, Q})_0. \qquad (26.10)(b)$$

The map of (26.10)(a) provides a natural homomorphism

$$i_{univ}^{loc} : A_{univ}^Q({}^\vee G^\Gamma) \to A_\phi^{loc, Q} \qquad (26.10)(c)$$

for every Langlands parameter ϕ; similarly we get maps into the various geometrically defined groups of Chapter 7, into the Q-component group for an Arthur parameter, and into the micro-component groups of Definition 24.13.

Suppose now that we are in the setting of Theorem 10.10, and that (π, δ) is an irreducible canonical projective representation of type z of a strong real form δ of type J. Theorem 10.10 associates to (π, δ) a complete Langlands parameter $(\phi(\pi, \delta), \tau(\pi, \delta))$, with $\tau(\pi, \delta)$ an irreducible representation of $A_\phi^{loc, Q}$. Define

$$\tau_{univ}(\pi, \delta) = \tau(\pi, \delta) \circ i_{univ}^{loc} : A_{univ}^Q({}^\vee G^\Gamma) \to \mathbb{C}^\times, \qquad (26.10)(d)$$

a one-dimensional character of A_{univ}^Q. In terms of geometric parameters, $\tau_{univ}(\pi, \delta)$ is the character by which $Z({}^\vee G)^{\theta_z, Q}$ acts on the stalks of the equivariant local system (or the equivariant perverse sheaf) corresponding to (π, δ).

We have already considered several cases of this definition. The element $z(\rho)$ belongs to $Z({}^\vee G)^{\theta_z, Q}$, and Definition 15.8 says that

$$\tau_{univ}(\pi, \delta)(\overline{z(\rho)}) = e(\delta), \qquad (26.11)(a)$$

the Kottwitz invariant of the real form. As another example, let $z_0 \in Z(G)^{\sigma_z, fin}$ be the second invariant of the extended group $(G^\Gamma, \mathcal{W})$ (cf. (3.5)). Then Lemma 10.9(b) says (in the notation defined there)

$$\tau_{univ}(\pi, \delta)(\overline{z^Q}) = \chi_{z_0 \delta^2}(z^Q) \qquad (z^Q \in Q). \qquad (26.11)(b)$$

These examples are special cases of the following general fact.

Lemma 26.12. *In the setting of Definition 26.10, the character $\tau_{univ}(\pi, \delta)$ of $Z(^\vee G)^{\theta_z, Q}$ depends only on the G-conjugacy class of δ, and not on the representation π. We may therefore write it as $\tau_{univ}(\delta)$.*

We have not found a particularly compelling proof of this result, but one can proceed along the following lines. The group $Z(^\vee G)^{\theta_z, Q}$ must act by scalars on an irreducible $^\vee G^Q$-equivariant perverse sheaf on the geometric parameter space. By Theorem 15.12, it follows that $\tau_{univ}(\pi, \delta)$ is unchanged if π is replaced by any composition factor of the standard limit representation having π as a quotient. In this way (using the subquotient theorem) we can reduce to the case of ordinary principal series representations of $G(\mathbb{R}, \delta)^{can}$. Since all of these may be regarded as associated to the same Cartan subgroup and set of positive imaginary roots, it is fairly easy to apply the definition of $\tau(\pi, \delta)$ directly (see Chapter 13) and verify the lemma. We leave the details to the reader.

Definition 26.13. Suppose we are in the setting of Definition 26.10, and $\sigma \in A_{univ}^Q(^\vee G^\Gamma)$. Using the notation of (18.3), we now define a function ζ_σ on $G(\mathbb{R}, *)_{SR}^{can}$ by

$$\zeta_\sigma(g, \delta) = \tau_{univ}(\delta)(\sigma) \qquad (26.13)(a)$$

The locally finite formal complex virtual representation $\eta \in \overline{K}_{\mathbb{C}, f}\Pi^z(G/\mathbb{R})_J$ is said to be *strongly σ-stable* if the function

$$\Theta_{SR}(\eta, *)(\zeta_\sigma)^{-1} \qquad (26.13)(b)$$

is constant on the fibers of the first projection $p_1 : G(\mathbb{R}, *)_{SR}^Q \to G^Q$ (Definition 18.6). Define

$$\overline{K}_{\mathbb{C}, f}\Pi^z(G/\mathbb{R})_J^{\sigma-st} \qquad (26.13)(c)$$

to be the vector space of these virtual representations.

Theorem 26.14. *Suppose we are in the setting of Theorem 10.11, and $\sigma \in A_{univ}^Q(^\vee G^\Gamma)$ (Definition 26.10).*

a) The restriction of a strongly σ-stable formal complex virtual representation to single strong real form is stable (Definition 18.2).

b) Suppose δ_0 is a strong real form of G of type J, and η_0 is a stable virtual representation of $G(\mathbb{R}, \delta_0)^{can}$ (Definition 18.2). Then there is a strongly σ-stable locally finite formal complex virtual representation η of G with the property that

$$\Theta_{SR}(\eta, *)(g, \delta_0) = \Theta_{SR}(\eta_0)(g)$$

for every strongly regular element $g \in G(\mathbb{R}, \delta_0)^{can}$. *If δ_0 is quasisplit, then η is unique.*

c) *The vector space* $\overline{K}_{\mathbb{C},f} \Pi^z(G/\mathbb{R})_J^{\sigma-st}$ *has as a basis the set* $\{\eta_\phi^{loc,Q}(i_{univ}^{loc}(\sigma))\}$ *(Definitions 26.5 and 26.10), as ϕ varies over* $\Phi(^\vee G^\Gamma)$.

d) *The vector space* $\overline{K}_{\mathbb{C},f} \Pi^z(G/\mathbb{R})_J^{\sigma-st}$ *has as a basis the set* $\{\eta_\phi^{mic,Q}(i_{univ}^{mic}(\sigma))\}$ *(Definitions 26.8 and 26.10), as ϕ varies over* $\Phi(^\vee G^\Gamma)$.

Proof. Part(a) follows from Lemma 18.5. The function ζ_σ is constant on strong real forms and nowhere zero, so (b) is a formal consequence of Theorem 18.7. Similarly (c) follows from Theorem 18.14. For (d), Theorem 1.31 shows that this set is related to the basis of (c) by an upper triangular matrix with ± 1 in each diagonal entry. Q.E.D.

In the setting of Example 19.17, it is not difficult to see that the function ζ_σ is nothing but the character of the complex formal virtual representation $\eta_{\phi_0}^{mic,Q}(\sigma)$.

Definition 26.15. Suppose $^\vee G^\Gamma$ is a weak E-group (Definition 4.3), and Q is any quotient of $\pi_1(^\vee G)^{alg}$ (Definition 10.10). A *weak endoscopic datum* for $^\vee G^\Gamma$ is a pair $(s^Q, {}^\vee H^\Gamma)$, subject to the following conditions.

i) $s^Q \in {}^\vee G^Q$ is a semisimple element, and $^\vee H^\Gamma \subset {}^\vee G^\Gamma$ is a weak E-group. Write s for the image of s^Q in $^\vee G$.

ii) $^\vee H^\Gamma$ is open in the centralizer of s in $^\vee G^\Gamma$.

In (i), it is understood that the weak E-group structure $^\vee H^\Gamma \to \Gamma$ is the restriction of that for G; that is, that the inclusion

$$\epsilon : {}^\vee H^\Gamma \to {}^\vee G^\Gamma$$

is an L-homomorphism (Definition 5.1). An *endoscopic datum* is a weak endoscopic datum endowed with an E-group structure on $^\vee H^\Gamma$ (Definitions 4.6, 4.14). That is, it is a triple $(s^Q, {}^\vee H^\Gamma, \mathcal{D}_H)$, subject to the conditions above and

iii) $\mathcal{D}_H$ is a $^\vee H$-conjugacy class of elements of finite order in $^\vee H^\Gamma - {}^\vee H$, each acting by conjugation on $^\vee H$ as a distinguished involution.

Endoscopic data $(s^Q, {}^\vee H^\Gamma, \mathcal{D}_H)$ and $((s^Q)', ({}^\vee H^\Gamma)', (\mathcal{D}_H)')$ are *equivalent* if there is an element $g^Q \in {}^\vee G^Q$ with the property that

$$g(^\vee H^\Gamma)' g^{-1} = {}^\vee H^\Gamma, \; g(\mathcal{D}_H)' g^{-1} = \mathcal{D}_H, \; g^Q (s^Q)' (g^Q)^{-1} \in s^Q (Z(^\vee H)^{\theta_z, Q})_0.$$

Here g is the image of g^Q in $^\vee G$. The last condition says that s^Q and $g^Q(s^Q)'(g^Q)^{-1}$ should have the same image in $A_{univ}^{Q_H}(^\vee H^\Gamma)$.

This definition of endoscopic data is compatible with the one given by Langlands and Shelstad in [36], 1.2 (although of course they do not use a covering group); their element s satisfies slightly weaker conditions, but can obviously be modified within their notion of equivalence to satisfy the conditions here. What is more serious is that the notion of equivalence in [36] is substantially less stringent: they replace our coset $s^Q Z(^\vee H^Q)_0$ by $sZ(^\vee H^\Gamma)_0 Z(^\vee G)$. This is part of the reason that the endoscopic lifting of [48] is defined only up to a sign. In any case, we find a natural surjective correspondence from equivalence classes of endoscopic data in our sense to those of [36].

Notice first of all that every element $s^Q \in Z(^\vee G)^{\theta_z, Q}$ (notation as in Proposition 4.4) defines an endoscopic datum $(s^Q, {}^\vee G^\Gamma, \mathcal{D})$. Two of these are equivalent if and only if s^Q and $(s^Q)'$ have the same image in A^Q_{univ} (Definition 26.10). In general, suppose s^Q is any semisimple element of $^\vee G^Q$; write s for its image in $^\vee G$. Define

$$S = \text{centralizer of } s \text{ in } {}^\vee G^\Gamma, \qquad S_1 = \text{centralizer of } s \text{ in } {}^\vee G,$$

and write S_0 for the identity component of S. Then the number of weak endoscopic data corresponding to s^Q is the number of elements of order 2 in the finite group S/S_0 not belonging to S_1/S_0.

Here is an important construction for weak endoscopic data. Suppose $\phi \in P(^\vee G^\Gamma)$ is a Langlands parameter, and $s^Q \in {}^\vee G^Q_\phi$ is a semisimple element (Definition 5.11). (One should think of s^Q as representing an element $\sigma \in A^{loc,Q}_\phi$.) Let $^\vee H$ be the identity component of the centralizer of s in $^\vee G$. Then

$$^\vee H^\Gamma = (^\vee H)\phi(W_\mathbb{R}) \qquad (26.16)$$

is a weak E-group, and the pair $(s^Q, {}^\vee H^\Gamma)$ is a set of weak endoscopic data.

We can now describe the setting for endoscopic lifting. We begin in the setting of Theorem 10.11, with an extended group $(G^\Gamma, \mathcal{W})$ a corresponding E-group $(^\vee G^\Gamma, \mathcal{D})$ with second invariant z, and a quotient Q of $\pi_1(^\vee G)^{alg}$. Fix a set of endoscopic data

$$(s^Q, {}^\vee H^\Gamma, \mathcal{D}_H) \qquad (26.17)(a)$$

for $^\vee G^\Gamma$ as in Definition 26.15, say with second invariant z_H. (We may also call this endoscopic data for G.) An *extended endoscopic group* attached to this set of data is an extended group

$$(H^\Gamma, \mathcal{W}_H) \qquad (26.17)(b)$$

26. Endoscopic lifting

with E-group $^\vee H^\Gamma$ (Definition 4.3) and second invariant 1 (Proposition 3.6). Recall from Definition 1.12 that the extended group structure $\mathcal{W}_\mathcal{H}$ includes an equivalence class of quasisplit strong real forms of H. Fix one of these δ_0. In the terminology of [36], $H(\mathbb{R}, \delta_0)$ is an endoscopic group attached to the endoscopic data. The L-homomorphism

$$\epsilon : {}^\vee H^\Gamma \hookrightarrow {}^\vee G^\Gamma \qquad (26.17)(c)$$

now places us in the setting (26.1), so we can make use of all the constructions introduced there. (It is easy to check that in the present situation the group Q_H of (26.1)(c) is actually equal to Q; but this is something of an accident that disappears in the case of twisted endoscopy, so we will make no use of it.) In particular, we have a restriction homomorphism

$$\epsilon^* : KX({}^\vee G^\Gamma)^Q \to KX({}^\vee H^\Gamma)^{Q_H} \qquad (26.17)(d)$$

and Langlands functoriality

$$\epsilon_* : \overline{K}_\mathbb{C}\Pi^{z_H}(H/\mathbb{R})^{Q_H} \to \overline{K}_\mathbb{C}\Pi^z(G/\mathbb{R})^Q. \qquad (26.17)(e)$$

With these definitions, it is easy to check that the element s^Q from the endoscopic datum belongs to $Z({}^\vee H)^{\theta_z, Q_H}$, and therefore defines a class

$$\sigma_H \in A_{univ}^{Q_H}({}^\vee H^\Gamma) \qquad (26.17)(f)$$

(Definition 26.10).

Definition 26.18. Suppose we are in the setting (26.17). *Endoscopic lifting* is the restriction of Langlands functoriality to σ_H-stable virtual representations. That is,

$$\text{Lift} : \overline{K}_{\mathbb{C},f}\Pi^{z_H}(H/\mathbb{R})^{\sigma_H-st}_{J_H} \to \overline{K}_{\mathbb{C},f}\Pi^z(G/\mathbb{R})_J.$$

Fix a quasisplit strong real form δ_0 of H from the distinguished class given by $\mathcal{W}_\mathcal{H}$, and write

$$\overline{K}_{\mathbb{C},f}\Pi^{z_H}(H(\mathbb{R}, \delta_0))^{st}$$

for the Grothendieck group of stable complex virtual representations (projective of type z_H) of the group $H(\mathbb{R}, \delta_0)$. By Theorem 26.14 (which relies on Shelstad's Theorem 18.7), restriction to $H(\mathbb{R}, \delta_0)$ defines an isomorphism

$$\overline{K}_{\mathbb{C},f}\Pi^{z_H}(H/\mathbb{R})^{\sigma_H-st}_{J_H} \simeq K_\mathbb{C}\Pi^{z_H}(H(\mathbb{R}, \delta_0))^{st}.$$

Using this identification, we may regard lifting as defined on stable virtual (canonical projective) representations of the endoscopic group $H(\mathbb{R}, \delta_0)$:

$$\text{Lift}_0 : K_\mathbb{C}\Pi^{z_H}(H(\mathbb{R}, \delta_0))^{st} \to \overline{K}_{\mathbb{C},f}\Pi^z(G/\mathbb{R})_J.$$

If we fix a strong real form δ of G of type J (that is, satisfying $\delta^2 \in Jz_0$, with z_0 the second invariant of the extended group G^Γ), then we may project the image of lift on the complex virtual (projective of type z) representations of $G(\mathbb{R}, \delta)$. This gives a map

$$\text{Lift}_0(\delta) : K_\mathbb{C}\Pi^{z_H}(H(\mathbb{R}, \delta_0))^{st} \to K_\mathbb{C}\Pi^z(G(\mathbb{R}, \delta)).$$

It is essentially this restricted map that Langlands and Shelstad use.

Lemma 26.19. *Endoscopic lifting depends only on the equivalence class of endoscopic data. More precisely, suppose $(s^Q, {}^\vee H^\Gamma, \mathcal{D}_H)$ and $((s^Q)', ({}^\vee H^\Gamma)', (\mathcal{D}_H)')$ are endoscopic data for G (Definition 26.15), say with second invariants z_H and z'_H. Fix associated extended endoscopic groups $(H^\Gamma, \mathcal{W}_H)$ and $((H^\Gamma)', \mathcal{W}'_H)$ (cf. (26.17)). Fix an equivalence ${}^\vee j$ from $(s^Q, {}^\vee H^\Gamma, \mathcal{D}_H)$ to $((s^Q)', ({}^\vee H^\Gamma)', (\mathcal{D}_H)')$, implemented by conjugation by an element $g^Q \in {}^\vee G^Q$ (Definition 26.15). Finally, choose distinguished (quasisplit) strong real forms δ_0 and δ'_0 for H and H'.*

a) *The isomorphism ${}^\vee j$ carries σ_H to $\sigma_{H'}$ (cf. (26.17)(f)).*

b) *The isomorphism ${}^\vee j$ induces an isomorphism*

$$j : (H^\Gamma, \mathcal{W}_H) \xrightarrow{\sim} ((H^\Gamma)', \mathcal{W}'_H)$$

unique up to an inner automorphism from H.

c) *The isomorphism j allows us to identify equivalence classes of irreducible canonical projective representations of strong real forms of H and H':*

$$j : \Pi^{z_H}(H/\mathbb{R})_{J_H} \xrightarrow{\sim} \Pi^{z_{H'}}(H'/\mathbb{R})_{J_{H'}}.$$

d) *The isomorphism j induces an isomorphism*

$$j_0 : H(\mathbb{R}, \delta_0) \xrightarrow{\sim} H'(\mathbb{R}, \delta'_0)$$

unique up to an inner automorphism from the normalizer of $H'(\mathbb{R}, \delta'_0)$ in H.

e) *The isomorphism j_0 allows us to identify stable virtual representations of $H(\mathbb{R}, \delta_0)$ and $H'(\mathbb{R}, \delta_0')$:*

$$j_0 : K_\mathbb{C}\Pi^{z_H}(H(\mathbb{R},\delta_0))^{st} \xrightarrow{\sim} K_\mathbb{C}\Pi^{z_{H'}}(H'(\mathbb{R},\delta_0'))^{st}.$$

This isomorphism depends only on $^\vee j$.

f) *If η_H is a σ_H-stable formal complex virtual (canonical projective of type z_H) representation of strong real forms of H of type J_H, then*

$$\mathrm{Lift}(\eta_H) = \mathrm{Lift}(j(\eta_H)).$$

g) *If $(\eta_0)_H$ is a complex virtual (canonical projective of type z_H) representation of $H(\mathbb{R}, \delta_0)$, then*

$$\mathrm{Lift}_0((\eta_0)_H) = \mathrm{Lift}_0(j_0((\eta_0)_H)).$$

Proof. Part (a) is immediate from the definition of equivalence and the definition of $A_{univ}^{Q_H}$ (Definition 26.10). Part (b) is Proposition 3.6, and then (c) is immediate. For (d), $j(\delta_0)$ must be a distinguished strong real form coming from $\mathcal{W}_{H'}$. Since these strong real forms constitute a single conjugacy class, conjugation by an element of H' carries $j(\delta_0)$ to δ_0'. The composition of this inner automorphism with j gives j_0. For (e), we need only know that the normalizer of $H'(\mathbb{R}, \delta_0')$ in H' acts trivially on stable virtual representations. This is obvious from the definition of stability (Definition 18.2). Parts (f) and (g) are clear from the definitions. Q.E.D.

The first serious problem we face is showing that Definition 26.18 agrees with the definition of [48], up to a multiplicative constant (which can be chosen more or less arbitrarily in Shelstad's formulation, and which depends on the choice of strong real form δ in ours). To do this, notice that both kinds of lifting can be computed explicitly on the basis elements $\eta_{\phi_H}^{loc, Q_H}(\sigma_H)(\delta_0)$: this is Proposition 26.7 in our case and Theorem 4.1.1 of [48] in hers. In each case the answer is a sum over the final standard limit representations in the L-packet $\Pi^z(G(\mathbb{R}, \delta))_{\phi_G}$, with coefficients given in an explicit combinatorial way in terms of a single Cartan subgroup. Comparing the definitions is a tedious but straightforward exercise (to which Proposition 13.12 is particularly relevant). We omit the details.

Our main theorem on endoscopic lifting concerns the virtual representations of Definition 26.8. In order to apply the results of Chapter 25, we need an element of finite order.

Lemma 26.20. *Suppose G is a (possibly disconnected) complex algebraic group, and $a_0 \in G$ is a semisimple element. Define $H = G_{a_0}$ to be the centralizer of a_0 in G, $Z(H)$ the center of H, and $Z(H)_s$ the group of semisimple elements in $Z(H)$.*

a) The set of elements of finite order is Zariski dense in $Z(H)_s$.

b) The set
$$U = \{\, a \in Z(H)_s \mid \dim G_a = \dim H \,\}$$
is Zariski open in $Z(H)_s$ and contains a_0.

c) There is an element $a' \in a_0(Z(H)_s)_0$ of finite order such that H is open in the centralizer of a' in G.

Proof. The identity component $(Z(H)_s)_0$ is a product of copies of $\mathbb{C}^\times$. The elements of finite order are therefore dense in it. We need only show that they meet every other component. Suppose $b \in Z(H)_s$. Since the group of connected components of $Z(H)$ is finite, $b^n \in (Z(H)_s)_0$ for some positive integer n. Since $\mathbb{C}^\times$ is a divisible group, there is an element $b_0 \in (Z(H)_s)_0$ with $b_0^n = b^n$. Then bb_0^{-1} is an element of order n in the component $b(Z(H)_s)_0$.

For (b), we have $G_a \supset H$ for every $a \in Z(H)_s$. Consequently
$$U = \{\, a \in Z(H)_s \mid \det(\mathrm{Ad}(a)|_{\mathfrak{g}/\mathfrak{h}}) \neq 0 \,\}.$$

This is obviously an open set. Part (c) follows from (a) and (b). Q.E.D.

Suppose we are in the setting (26.17). By Lemma 26.20, we can find an element $s' \in sZ({}^\vee H^\Gamma)_0$ so that s' has finite order, and

$${}^\vee H^\Gamma \text{ is open in the centralizer of } s' \text{ in } {}^\vee G^\Gamma. \qquad (26.21)(a)$$

Now $Z({}^\vee H^\Gamma)_0$ is just $Z({}^\vee H)_0^{\theta_Z}$. We may therefore find a preimage

$$(s^Q)' \in s^Q(Z({}^\vee H)^{\theta_Z, Q_H})_0 \subset {}^\vee G^Q \qquad (26.21)(b)$$

for s'. When the pro-finite group Q is infinite, this preimage may not itself have finite order; but it will have finite order in every algebraic quotient of ${}^\vee H^Q$, and this is all we need to apply the results from Chapter 25. (For example, any particular ${}^\vee G^Q$-equivariant constructible sheaf on $X({}^\vee G^\Gamma)$ will actually be equivariant for an algebraic quotient of ${}^\vee G^Q$.) By Definition 26.15,

$$((s^Q)', {}^\vee H^\Gamma, \mathcal{D}_H) \qquad (26.21)(c)$$

is again an endoscopic datum, equivalent to $(s^Q, {}^\vee H^\Gamma, \mathcal{D}_H)$ by the identity map on ${}^\vee H^\Gamma$. In this setting, we will write σ for the action of s' on $X({}^\vee G^\Gamma)$ or on ${}^\vee G$ (by conjugation); these are compatible automorphisms of finite order. By the choice of s',

$$({}^\vee G^\sigma)_0 \subset {}^\vee H \subset {}^\vee G^\sigma. \qquad (26.21)(d)$$

Lemma 26.22 *In the setting (26.21), $X({}^\vee H^\Gamma)$ is open and closed in $X({}^\vee G^\Gamma)^\sigma$.*

Proof. It is immediate from the definition of the geometric parameter space that

$$X({}^\vee G^\Gamma)^\sigma = \{\, (y, \Lambda) \mid y \in {}^\vee G^\sigma, \operatorname{Ad}(s')\Lambda = \Lambda \,\}.$$

Because a canonical flat Λ is an affine space, and therefore convex, it contains a fixed point of each linear automorphism of finite order preserving it. (Take the center of mass of an orbit.) Hence

$$X({}^\vee G^\Gamma)^\sigma = \{\, (y, \Lambda) \mid y \in {}^\vee G^\sigma, \Lambda \cap \mathfrak{h} \neq 0 \,\}.$$

On the other hand, the image of the closed immersion $X(\epsilon)$ (Corollary 6.21) is

$$\{\, (y, \Lambda) \mid y \in {}^\vee H, \Lambda \cap \mathfrak{h} \neq 0 \,\}.$$

Since ${}^\vee H$ is open and closed in ${}^\vee G^\sigma$, the lemma follows. Q.E.D.

Proposition 26.23. *In the setting (26.17), suppose $x \in X({}^\vee H^\Gamma)$; write $\epsilon(x)$ for the corresponding point of $X({}^\vee G^\Gamma)$.*

a) *The element $s \in {}^\vee G$ from the endoscopic datum defines an automorphism σ of the conormal space $T^*_{{}^\vee G, \epsilon(x)}(X({}^\vee G^\Gamma))$ to the ${}^\vee G$-orbit of $\epsilon(x)$.*

b) *There is a natural isomorphism*

$$\epsilon : T^*_{{}^\vee H, x}(X({}^\vee H^\Gamma)) \xrightarrow{\sim} \left(T^*_{{}^\vee G, \epsilon(x)}(X({}^\vee G^\Gamma))\right)^\sigma.$$

c) *In the isomorphism of (b),*

$$\left(T^*_{{}^\vee G, \epsilon(x)}(X({}^\vee G^\Gamma))_{reg}\right)^\sigma \subset \epsilon\left(T^*_{{}^\vee H, x}(X({}^\vee H^\Gamma))_{reg}\right).$$

d) Suppose $(x,\nu) \in T^*_{{}^\vee H, x}(X({}^\vee H^\Gamma))$, so that $(\epsilon(x), \epsilon(\nu)) \in T^*_{{}^\vee G, \epsilon(x)}(X({}^\vee G^\Gamma))$. With notation as in Definition 24.7, assume that

$$(x, \nu) \in U_{{}^\vee H \cdot x}, \qquad (\epsilon(x), \epsilon(\nu)) \in U_{{}^\vee G \cdot \epsilon(x)}.$$

(This is automatic if the orbit ${}^\vee G \cdot (\epsilon(x), \epsilon(\nu))$ is open in the conormal bundle.) Then there is a natural homomorphism of micro-component groups

$$A^{mic}(\epsilon) : A^{mic, Q_H}_{x,\nu} \to A^{mic, Q}_{\epsilon(x), \epsilon(\nu)}.$$

Proof. This result is true, and not difficult to prove, in exactly the form stated. For our applications, however, it is enough to know that it is true after the endoscopic datum is modified in accordance with (26.21). (This will change the automorphism σ, but not its subspace of fixed points.) In that case it follows from Theorem 25.6, which we are allowed to apply by Lemma 26.22. Q.E.D.

Theorem 26.24. *Suppose we are in the setting (26.17). Fix a Langlands parameter $\phi_H \in P({}^\vee H^\Gamma)$, so that $\phi_G = \epsilon \circ \phi_H$ (that is, ϕ_H regarded as a map into the larger group ${}^\vee G^\Gamma$) is a Langlands parameter for G. Write $x_0 \in X({}^\vee H^\Gamma)$ for the point corresponding to ϕ_H, so that $\epsilon(x_0) \in X({}^\vee G^\Gamma)$ corresponds to ϕ_G. Fix a generic conormal covector $\nu_0 \in U_{{}^\vee H \cdot x_0}$ (Definition 24.7), so that the micro-component group for ϕ_H may be computed at (x_0, ν_0):*

$$A^{mic, Q_H}_{\phi_H} = A^{mic, Q_H}_{(x_0, \nu_0)}$$

(Definition 24.13). Using Proposition 26.23, regard $(\epsilon(x_0), \epsilon(\nu_0))$ as a conormal covector to ${}^\vee G \cdot \epsilon(x_0)$. Assume that this covector is also generic, so that

$$A^{mic, Q}_{\phi_G} = A_{(\epsilon(x_0), \epsilon(\nu_0))}^{mic, Q},$$

and Proposition 26.23 provides a homomorphism

$$A^{mic}(\epsilon) : A^{mic, Q_H}_{\phi_H} \to A^{mic, Q}_{\phi_G}.$$

Define $\sigma_H \in A^{Q_H}_{univ}({}^\vee H^\Gamma)$ as in (26.17)(f). We write also

$$\sigma^{mic}_{\phi_H} = i^{mic}_{univ}(\sigma_H) \in A^{mic, Q_H}_{\phi_H}, \qquad \sigma^{mic}_{\phi_G} = A^{mic}(\epsilon)(\sigma^{mic}_{\phi_H}) \in A^{mic, Q}_{\phi_G}.$$

26. Endoscopic lifting

Then endoscopic lifting (Definition 26.18) sends the σ_H-stable formal complex virtual representation attached to ϕ_H and $\sigma_{\phi_H}^{mic}$ to the one attached to ϕ_G and $\sigma_{\phi_G}^{mic}$:

$$\mathrm{Lift}(\eta_{\phi_H}^{mic,Q_H}(\sigma_{\phi_H}^{mic})) = \eta_{\phi_G}^{mic,Q}(\sigma_{\phi_G}^{mic}).$$

In terms of the restricted map of Langlands and Shelstad, we have

$$\mathrm{Lift}_0(\delta)(\eta_{\phi_H}^{mic,Q_H}(\delta_0)) =$$

$$e(\delta) \sum_{\pi \in \Pi^z(G(\mathbb{R},\delta))_{\phi_G}^{mic}} (-1)^{d(\pi)-d(\phi_G)} (\chi_{\phi_G}^{mic,Q}(\pi)(\sigma_{\phi_G}^{mic}))\pi$$

for each strong real form δ of type J.

Proof. By Lemma 26.19, we may replace the endoscopic datum by an equivalent one without changing Lift. By (26.21), we may therefore assume that s has finite order. Now regard $\mathrm{Lift}(\eta_{\phi_H}^{mic,Q_H}(\sigma_{\phi_H}^{mic}))$ as a linear functional on constructible ${}^\vee G^Q$-equivariant sheaves (Theorem 26.2), and evaluate it on such a sheaf C. By Lemma 26.9, the result is $(-1)^{d(\phi_H)}$ times the alternating sum of the traces of $\sigma_{\phi_H}^{mic}$ on the on the stalks at (x_0, ν_0) of $(Q^{mic})^i(\epsilon^*C)$. By Theorem 25.8, this is equal to $(-1)^{d(\phi_G)}$ times the alternating sum of the traces of $\sigma_{\phi_G}^{mic}$ on the stalks at $(\epsilon(x_0), \epsilon(\nu_0))$ of $(Q^{mic})^i(C)$. By Lemma 26.9 again, this alternating sum is $\eta_{\phi_G}^{mic,Q}(\sigma_{\phi_G}^{mic})$ evaluated at C, as we wished to show. Q.E.D.

Here is the solution to Problem D from the introduction.

Theorem 26.25. *Suppose we are in the setting (26.17). Fix an Arthur parameter $\psi_H \in Q({}^\vee H^\Gamma)$ (Definition 22.4), so that $\psi_G = \epsilon \circ \psi_H$ (that is, ψ_H regarded as a map into the larger group ${}^\vee G^\Gamma$) is an Arthur parameter for G. Write*

$$A^{Arthur}(\epsilon) : A_{\psi_H}^{Q_H} \to A_{\psi_G}^Q$$

for the induced map on component groups. Define

$$\sigma_{\psi_H} = i_{univ}^{Arthur}(\sigma_H) \in A_{\psi_H}^{Q_H}, \qquad \sigma_{\psi_G} = A^{Arthur}(\epsilon)(\sigma_{\psi_H}) \in A_{\psi_G}^Q.$$

Then endoscopic lifting satisfies

$$\mathrm{Lift}(\eta_{\psi_H}(\sigma_{\psi_H})) = \eta_{\psi_G}(\sigma_{\psi_G}).$$

In terms of the restricted lifting map of Langlands and Shelstad,

$$\mathrm{Lift}_0(\delta)(\eta_{\psi_H}(\sigma_{\psi_H})(\delta_0)) =$$
$$e(\delta) \sum_{\pi \in \Pi^z(G(\mathbb{R},\delta))_{\psi_G}} (-1)^{d(\pi)-d(\psi_G)} (\chi_{\psi_G}(\pi)(\sigma_{\psi_G}))\pi$$

for each strong real form δ of G of type J.

In light of the discussion at Definition 24.15, this is just a special case of Theorem 26.24.

27. Special unipotent representations

Arthur's parameters are in many respects most interesting when they are as far as possible from being tempered; that is, when the tempered part of the parameter is as trivial as possible. The corresponding representations are the *special unipotent representations*. These were defined already in [7]. (The theorems of that paper are proved only in the complex case, but the basic definition works in general.) In this chapter we will consider the unipotent case in more detail, and prove that our new definition of Arthur's representations agrees with the old one (Corollary 27.13).

The reader may wonder about the qualifying adjective "special" in the terminology. Arthur's representations should include all the unitary representations of real forms of G that can appear in spaces of square-integrable automorphic forms (with respect to congruence subgroups). They do not, however, include all the interesting unitary representations; perhaps the simplest example of one which is omitted is (either irreducible component of) the metaplectic representation of the complex rank two symplectic group. This representation should be an example of a (still undefined) larger class of interesting unitary representations, for certain of which we are optimistically reserving the term "unipotent." A longer discussion (if not a more illuminating one) may be found in [58].

Definition 27.1. Suppose $^\vee G^\Gamma$ is a weak E-group. An Arthur parameter ψ (Definition 22.4) is said to be *unipotent* if its restriction to the identity component $\mathbb{C}^\times$ of $W_\mathbb{R}$ is trivial. Write $Q_{unip}(^\vee G^\Gamma)$ for the set of unipotent Arthur parameters, and $\Psi_{unip}(^\vee G^\Gamma)$ for the corresponding set of equivalence classes. Following Definition 22.4, write

$$P_{unip}(^\vee G^\Gamma) \subset P_{Arthur}(^\vee G^\Gamma) \subset P(^\vee G^\Gamma)$$

for the corresponding Langlands parameters ϕ_ψ, and

$$\Phi_{unip}(^\vee G^\Gamma) \subset \Phi_{Arthur}(^\vee G^\Gamma) \subset \Phi(^\vee G^\Gamma)$$

for their equivalence classes.

Proposition 27.2. *Suppose $^\vee G^\Gamma$ is a weak E-group. The set of unipotent Arthur parameters for $^\vee G^\Gamma$ (Definition 27.1) may be identified* naturally with the set of L-homomorphisms (Definition 5.1) from the L-group of $PGL(2)$ to $^\vee G^\Gamma$. Suppose that ψ corresponds to ϵ. Then the Langlands parameter associated to ψ (Definition 22.4) is the lift by ϵ of the Langlands parameter of the trivial representation of $PGL(2,\mathbb{R})$ (Proposition 5.4).

Proof. The dual group of $PGL(2)$ is $SL(2)$. Since $PGL(2,\mathbb{R})$ is split, the corresponding automorphism a of the based root datum (Corollary 2.16) is trivial. The L-group is therefore the direct product of the dual group and the Galois group Γ (Chapter 4):

$$\text{L-group of } PGL(2) = SL(2) \times \Gamma.$$

The L-homomorphisms in question are therefore certain homomorphisms from $SL(2) \times \Gamma$ to $^\vee G^\Gamma$. On the other hand, the Weil group $W_\mathbb{R}$ modulo its identity component is just Γ, so it is immediate from Definitions 22.4 and 27.1 that unipotent parameters are certain homomorphisms from $\Gamma \times SL(2)$ to $^\vee G^\Gamma$. One checks immediately that the additional conditions imposed on these homomorphisms by Definition 22.4 correspond precisely to those imposed on L-homomorphisms by Definition 5.1.

Now write $\gamma : W_\mathbb{R} \to \Gamma$ for the natural quotient map (Definition 5.2). The Langlands parameter of the trivial representation of $PGL(2,\mathbb{R})$ is represented by the homomorphism

$$\phi : W_\mathbb{R} \to SL(2) \times \Gamma, \qquad \phi(w) = \left(\begin{pmatrix} |w|^{1/2} & 0 \\ 0 & |w|^{-1/2} \end{pmatrix}, \gamma(w) \right).$$

(This is well-known: in addition to the definitions of Chapters 11–14, one must understand exactly which standard representation of $PGL(2,\mathbb{R})$ contains the trivial representation.) Comparing this with the definition of ϕ_ψ in Definition 22.4 gives the last assertion. Q.E.D.

Corollary 27.3. *Suppose* $^\vee G^\Gamma$ *is a weak E-group. Fix an algebraic homomorphism*

$$\psi_1 : SL(2,\mathbb{C}) \to {}^\vee G,$$

and define

$$S = \text{centralizer of } \psi_1(SL(2,\mathbb{C})) \text{ in } {}^\vee G^\Gamma$$

$$S_0 = S \cap {}^\vee G, \qquad S_1 = S - S_0, \qquad S_1^{(2)} = \{\, y_0 \in S_1 \mid y_0^2 = 1 \,\}.$$

27. Special unipotent representations   297

Then the set of unipotent Arthur parameters ψ restricting to ψ_1 on $SL(2,\mathbb{C})$ may be naturally identified with $S_1^{(2)}$. Two such parameters are equivalent if and only if the corresponding elements of $S_1^{(2)}$ are conjugate under S_0. If the Arthur parameter ψ corresponds to $y_0 \in S_1^{(2)}$, then the Arthur component group for ψ is the component group of the centralizer $S_0(y_0)$ of y_0 in S_0:

$$A_\psi = S_0(y_0)/(S_0(y_0))_0, \qquad A_\psi^{alg} = S_0(y_0)^{alg}/(S_0(y_0)^{alg})_0$$

(Definition 22.4).

This is immediate from the definitions.

Corollary 27.4. *Suppose $^\vee G^\Gamma$ is a weak E-group. Then the set $\Psi_{unip}(^\vee G^\Gamma)$ of equivalence classes of unipotent Arthur parameters is finite.*

Proof. According to a classical result of Dynkin, the number of H-conjugacy classes of algebraic homomorphisms of $SL(2,\mathbb{C})$ into a complex reductive algebraic group H is finite. Applying this to $^\vee G$, we find that there are only finitely many possible ψ_1 up to equivalence. Now Corollary 27.3 completes the proof. Q.E.D.

Example 27.5. Suppose $G = GL(n)$, endowed with the inner class of real forms including $GL(n,\mathbb{R})$. Then the L-group is

$$^\vee G^\Gamma = GL(n,\mathbb{C}) \times \Gamma. \qquad (27.5)(a)$$

A homomorphism ψ_1 of $SL(2)$ into $^\vee G$ is therefore an n-dimensional representation of $SL(2)$. Such a representation may be decomposed as a direct sum of irreducible representations of various dimensions $p_1 > p_2 > \cdots > p_r$; say the representation of dimension p_i occurs m_i times, so that $n = \sum m_i p_i$. It follows from Schur's lemma that the centralizer in $GL(n)$ of the image of ψ_1 is isomorphic to the product of the various $GL(m_i)$; so in the notation of Corollary 27.3,

$$\begin{aligned} S_0 &= GL(m_1) \times \cdots \times GL(m_r), \\ S &= S_0 \times \Gamma, \\ S_1^{(2)} &= \text{(elements of order 2)} \times \{\sigma\}; \end{aligned} \qquad (27.5)(b)$$

here σ is the non-trivial element of Γ. The equivalence classes of ψ associated to ψ_1 therefore correspond to conjugacy classes of elements of order 2 in S_0. Now an element of $GL(m)$ has order 2 if and only if it is diagonalizable with all eigenvalues 1 or -1. There are exactly $m+1$ such conjugacy classes (indexed by the multiplicity of the eigenvalue -1); so we find finally $(m_1+1)\cdots(m_r+1)$ different unipotent Arthur parameters associated to ψ_1.

It turns out that the Arthur packet of each of these parameters contains exactly one representation of $GL(n,\mathbb{R})$. We now describe these representations. Let $P=MN$ be the standard parabolic subgroup of $GL(n)$ with Levi factor

$$M = GL(p_1)^{m_1} \times \cdots \times GL(p_r)^{m_r}. \qquad (27.5)(c)$$

For each i, fix an integer k_i between 0 and m_i. Consider the one-dimensional character π_i of $GL(p_i)^{m_i}$ defined by

$$\pi_i(g_1,\ldots,g_{m_i}) = \operatorname{sgn}(\det(g_1 g_2 \ldots g_{k_i})). \qquad (27.5)(d)$$

Set

$$\pi = \operatorname{Ind}_{P(\mathbb{R})}^{GL(n,\mathbb{R})}(\pi_1 \otimes \cdots \otimes \pi_r) \qquad (27.5)(e)$$

Then π is a unipotent representation attached to the Arthur parameter specified by ψ_1 and the various k_i. (This can be verified by computing ϕ_ψ; what is also true, but not so immediately clear, is that there are no other representations of $GL(n,\mathbb{R})$ in the Arthur packet of ψ.)

Example 27.6. Suppose $G = G_1 \times G_1$, endowed with the inner class of real forms including G_1 (regarded as a real group). The L-group of G is

$${}^\vee G^\Gamma = ({}^\vee G_1 \times {}^\vee G_1) \rtimes \{1,\delta\}, \qquad (27.6)(a)$$

with δ acting on the first factor by interchanging the two factors ${}^\vee G_1$. A homomorphism ψ_1 of $SL(2)$ into ${}^\vee G$ is specified by a pair ψ_1^L, ψ_1^R of such homomorphisms into ${}^\vee G_1$; the centralizer S_0 of its image is a product $S_0^L \times S_0^R$. An element $(g^L,g^R)\delta \in {}^\vee G^\Gamma - {}^\vee G$ conjugates (ψ_1^L,ψ_1^R) to $(\operatorname{Ad}(g^L)\psi_1^R, \operatorname{Ad}(g^R)\psi_1^L)$. It follows at once that the set $S_1^{(2)}$ of Corollary 27.3 is empty (and so there are no unipotent Arthur parameters) unless ψ_1^L is conjugate to ψ_1^R. So we assume that they are conjugate; after replacing ψ_1 by a conjugate, we may even assume

$$\psi_1^L = \psi_1^R, \qquad S_0^L = S_0^R. \qquad (27.6)(b)$$

27. Special unipotent representations

An elementary calculation now shows that

$$S_1^{(2)} = \{\,(x, x^{-1})\delta \mid x \in S_0^L\,\}, \qquad (27.6)(c)$$

and that this set is precisely the S_0 conjugacy class of δ. There is therefore exactly one equivalence class of unipotent Arthur parameters associated to ψ_1. The corresponding representations are (as we will see) the ones considered in [7].

Fix now an algebraic homomorphism

$$\psi_1 : SL(2, \mathbb{C}) \to {}^\vee G. \qquad (27.7)(a)$$

Define $\lambda = \lambda_1$ as in (22.8). Write Λ for the canonical flat through λ (Definition 1.7 and (6.4)), and $\mathcal{O}$ for the orbit of λ in ${}^\vee\mathfrak{g}$. Representations in Arthur packets attached to unipotent Arthur parameters extending ψ_1 must have infinitesimal character $\mathcal{O}$, and we wish to apply to them the results of Chapter 21. We therefore use the notation of (21.1) and (21.5). In particular,

$$e(\Lambda) = \psi_1(-I). \qquad (27.7)(b)$$

The automorphism $\operatorname{Ad} e(\lambda)$ acts by $+1$ on the integer eigenspaces of $\operatorname{ad}(\lambda)$, and by -1 on the half-integer eigenspaces. The map ψ_1 is called *even* if $e(\lambda)$ is central; that is, if $\operatorname{ad}(\lambda)$ has no half-integral eigenspaces. Define

$$E = d\psi_1 \begin{pmatrix} 0 & 1 \\ 0 & 0 \end{pmatrix} \in \mathfrak{n}(\Lambda) \subset {}^\vee\mathfrak{g}(\Lambda) \qquad (27.7)(c)$$

as in (22.8). As in Proposition 22.9, we can use an invariant bilinear form to identify $({}^\vee\mathfrak{g}(\Lambda)/\mathfrak{p}(\Lambda))^*$ with $\mathfrak{n}(\Lambda)$, and so regard E as an element of $T^*_{eP(\Lambda)}(\mathcal{P}(\Lambda)^0)$ or of $\mathcal{N}^*_{\mathcal{P}(\Lambda)^0}$ (cf. (21.5)).

Lemma 27.8. *In the setting (27.7), the element E is $\mathcal{P}(\Lambda)^0$-regular (Definition 20.9); that is, it belongs to the Richardson class $\mathcal{Z}_{\mathcal{P}(\Lambda)^0}$ in ${}^\vee\mathfrak{g}(\Lambda)$ associated to Y (Proposition 20.4). The moment mapping μ has degree 1; equivalently, the only conjugate of $\mathfrak{p}(\Lambda)$ to which E belongs is $\mathfrak{p}(\Lambda)$ itself.*

Proof. This is very well-known, but we sketch an argument. We begin with a direct construction of the parabolic subalgebra $\mathfrak{p}(\Lambda)$ of ${}^\vee\mathfrak{g}(\Lambda)$ from E. Now $\mathfrak{p}(\Lambda)$ is the sum of the non-negative eigenspaces of λ on ${}^\vee\mathfrak{g}(\Lambda)$. Suppose (π, W) is any finite-dimensional representation of

$SL(2)$, with $E = d\pi \begin{pmatrix} 0 & 1 \\ 0 & 0 \end{pmatrix}$. Then the sum of the non-negative weight spaces of W is

$$W^E + W^{E^2} \cap E \cdot W + W^{E^3} \cap E^2 \cdot W + \cdots \qquad (27.9)(a)$$

(It suffices to check this in an irreducible representation W, where it is immediate.) Consequently

$$\mathfrak{p}(\Lambda) = {}^\vee\mathfrak{g}(\Lambda)^{\operatorname{Ad} E} + {}^\vee\mathfrak{g}(\Lambda)^{(\operatorname{Ad} E)^2} \cap (\operatorname{Ad} E) {}^\vee\mathfrak{g}(\Lambda) + \cdots \qquad (27.9)(b)$$

The first consequence of this description is that the centralizer of E in ${}^\vee G(\Lambda)$ must normalize $\mathfrak{p}(\Lambda)$, and consequently must belong to $P(\Lambda)$:

$$ {}^\vee G(\Lambda)^E \subset P(\Lambda). \qquad (27.9)(c)$$

In the setting of (27.9)(a), the operator E must carry the sum of the non-negative weight spaces onto the sum of the positive weight spaces. In our case the sum of the positive weight spaces is just the nil radical $\mathfrak{n}(\Lambda)$ of $\mathfrak{p}(\Lambda)$, so

$$\operatorname{ad}(E)\mathfrak{p}(\Lambda) = \mathfrak{n}(\Lambda). \qquad (27.9)(d)$$

This says that the $P(\Lambda)$ orbit of E is open in $\mathfrak{n}(\Lambda)$, which certainly implies that E is in the Richardson class for $P(\Lambda)$. Proposition 20.4 now says that the containment (27.9)(c) is equivalent to the assertion about the degree of μ. Q.E.D.

We can now reorganize the classification of unipotent parameters in Corollary 27.3 to be consistent with Proposition 6.24 and Theorem 21.2. Expressed in terms of the corresponding Langlands parameters $\phi_\psi = \phi(y, \lambda)$ (cf. (5.8) and (22.8)) the difference is this. In Corollary 27.3, we fix ψ_1 (and therefore $\lambda_1 = \lambda$ and y_1) and look for possible elements $y_0 = yy_1^{-1}$ to go with it. In Theorem 21.2, we fix representatives for the possible y, and then look for λ.

Now the equivalence classes of unipotent Arthur parameters ψ of infinitesimal character $\mathcal{O}$ are in one-to-one correspondence with the equivalence classes of associated Langlands parameters ϕ_ψ (Definition 22.4). These in turn correspond to certain ${}^\vee G$ orbits on $X(\mathcal{O}, {}^\vee G^\Gamma)$ (Proposition 1.10) and therefore to certain K_j orbits on $\mathcal{P}(\Lambda)^0$ (Proposition 6.24). It is these latter orbits that we wish to identify.

Theorem 27.10. *Suppose we are in the setting (21.1), and that the orbit $\mathcal{O}$ arises from a homomorphism of $SL(2, \mathbb{C})$ as in (27.7). The following sets are in one-to-one correspondence.*

1) *Equivalence classes of unipotent Arthur parameters supported on* $X_j(\mathcal{O}, {}^\vee G^\Gamma)$.
2) K_j-*conjugacy classes of parabolic subgroups* $P' \in \mathcal{P}(\Lambda)^0$ *with the following two properties:*
   a) $\theta_j P' = P'$; *and*
   b) $\mathfrak{n}' \cap \mathfrak{s}_j \cap \mathcal{Z}_{\mathcal{P}(\Lambda)^0} \neq \emptyset$. *Here* $\mathfrak{n}'$ *is the nil radical of the Lie algebra of* P'.
3) K_j *orbits on* $\mathfrak{s}_j \cap \mathcal{Z}_{\mathcal{P}(\Lambda)^0}$.

The Arthur (respectively canonical) component group for a parameter may be identified with the component group for the stabilizer of a point in the corresponding orbit of K_j *(respectively* K_j^{alg}*) on* $\mathfrak{s}_j \cap \mathcal{Z}_{\mathcal{P}(\Lambda)^0}$.

Notice that the K_j-conjugacy classes in (2) are precisely the "regular orbits" of K_j on $\mathcal{P}(\Lambda)^0$ in the sense of Definition 20.17; the bijection between (2) and (3) is a consequence of Theorem 20.18, although we will prove it again here.

Proof. Suppose S is an orbit on $X_j(\mathcal{O}, {}^\vee G^\Gamma)$ corresponding to a unipotent Arthur parameter. According to Proposition 6.24, we may choose a point (y_j, Λ') in S, with y_j one of the representatives chosen in (21.1)(g) (uniquely determined) and $P' = P(\Lambda') \in \mathcal{P}(\Lambda)^0$ (determined up to conjugation by K_j). We want to show that P' satisfies the conditions in (2). By the assumption on S, we can find a unipotent Arthur parameter ψ' for which the corresponding elements y', λ', and $E' = E_{\psi'}$ (cf. (22.8)(c)) satisfy

$$y' = y_j, \qquad \lambda' \in \Lambda'. \qquad (27.11)(a)$$

Since E' is in the $+1$ eigenspace of $\operatorname{ad} \lambda'$, it follows that $E' \in \mathfrak{n}(\lambda')$. Since $\operatorname{Ad}(y')E' = -E'$, E' must also be in the -1 eigenspace $\mathfrak{s}_j$ of θ_j. By Lemma 27.8, $E' \in \mathcal{Z}_{\mathcal{P}(\Lambda)^0}$. This gives (2)(b). Since y' and λ' commute, θ_j fixes $P(\lambda') = P'$. This is (2)(a).

Next, suppose we are given a parabolic P' as in (2); we associate to it the unique K_j orbit on $\mathfrak{s}_j \cap \mathcal{Z}_{\mathcal{P}(\Lambda)^0}$ that is dense in $\mathfrak{s}_j \cap \mathcal{Z}_{\mathcal{P}(\Lambda)^0}$.

Finally, suppose we are given an orbit of K_j on $\mathfrak{s}_j \cap \mathcal{Z}_{\mathcal{P}(\Lambda)^0}$. Fix a point E' of this orbit, and construct a homomorphism ψ'_1 from $SL(2)$ to ${}^\vee G(\lambda)$ satisfying

$$d\psi'_1 \begin{pmatrix} 0 & 1 \\ 0 & 0 \end{pmatrix} = E', \qquad \psi'_1(\theta_{SL(2)} x) = \theta_j \psi'_1(x). \qquad (27.11)(b)$$

(Here $\theta_{SL(2)}$ is the involutive automorphism conjugation by $\begin{pmatrix} i & 0 \\ 0 & -i \end{pmatrix}$.) That this is possible is proved in [31]. From [31] one can also deduce that

ψ_1' is unique up to conjugation by K_j (more precisely, by the unipotent radical of the stabilizer of E' in K_j). Since E' is conjugate by $^\vee G(\lambda)$ to our original E, it follows that the element λ' attached to ψ_1' must be conjugate to λ. Because $e(\lambda)$ is central in $^\vee G(\lambda)$, it follows that

$$\psi_1'(-I) = e(\lambda). \qquad (27.11)(c)$$

Define

$$y_0 = y_j \psi_1' \begin{pmatrix} -i & 0 \\ 0 & i \end{pmatrix}. \qquad (27.11)(d)$$

Because of (27.11)(b), y_0 commutes with the image of ψ_1'. Because of (27.11)(c), $y_0^2 = 1$. By Corollary 27.3, the pair (y_0, ψ_1') gives rise to a unipotent Arthur parameter; by construction it depends (up to conjugacy by K_j) only on the K_j orbit from which we began.

We leave to the reader the easy verification that performing any three of these correspondences in succession recovers the equivalence class from which one started. Q.E.D.

Theorem 27.12. *Suppose we are in the setting (21.1), and that the orbit $\mathcal{O}$ arises from a homomorphism of $SL(2,\mathbb{C})$ as in (27.7). Fix an equivalence class of unipotent Arthur parameters supported on $X_j(\mathcal{O}, {}^\vee G^\Gamma)$, say corresponding to a $^\vee G$ orbit S on $X_j(\mathcal{O}, {}^\vee G^\Gamma)$. Write S_K for the corresponding orbit of K_j on Y, and $\mathcal{Z}_S$ for the corresponding orbit of K_j on $\mathfrak{s}_j \cap \mathcal{Z}_{P(\Lambda)^\circ}$ (Theorem 27.10). Then the bijections of Theorem 21.2 identify the following sets:*

1) *irreducible $(^\vee\mathfrak{g}(\Lambda), K_j^{alg})$-modules M annihilated by $I_{P(\Lambda)^\circ}$, with the property that $\mathcal{Z}_S \subset \text{Ch}(M)$ (notation (20.5));*
2) *irreducible K_j^{alg}-equivariant $\mathcal{D}_{P(\Lambda)^\circ}$-modules $\mathcal{M}$, with the property that $T_{S_K}^*(\mathcal{P}(\Lambda)^0) \subset \text{Ch}(\mathcal{M})$ (Definition 19.9); and*
3) *irreducible $^\vee G^{alg}$-equivariant $\mathcal{D}$-modules $\mathcal{N}$ on $X_j(\mathcal{O}, {}^\vee G^\Gamma)$, with the property that $T_S^*(X_j) \subset \text{Ch}(\mathcal{N})$.*

Suppose in addition we are in the setting of Theorem 10.4. Then each of these sets is in bijection with

4) *$\Pi^z(G/\mathbb{R})_\psi$; that is, equivalence classes of irreducible canonical projective representations of type z of strong real forms of G, belonging to the Arthur packet for ψ.*

In cases (1)–(3), the bijections identify the corresponding multiplicities in characteristic cycles. Their common value (a positive integer) is equal to the weight $\chi_\psi(\pi)$ of Theorem 22.7 for the representation in (4).

27. Special unipotent representations

Proof. The equivalence between (1) and (2) is given by Theorem 20.18; between (2) and (3) by Proposition 21.4; and between (3) and (4) by Definition 22.6 (see Definitions 19.13 and 19.15). Q.E.D.

One can extend Theorem 27.12 from multiplicities to the component group representations discussed in Definition 24.15. For the relation between (3) and (4) this is a matter of definition again, and for (2) and (3) it is straightforward. For the relation between (1) and (2) there is a problem, however. In the setting of (1) we have referred to [59] for a definition of the component group representation. That definition is most easily related to the microdifferential operator definition in the setting of (2), which we elected in Chapter 24 not to discuss. For this reason we will not formulate (or use) such an extension.

The bijection between (1) and (4) in Theorem 27.12 can be formulated using just the results of [56]. In this form it was proposed several years ago by Barbasch and Vogan (in various lectures) as the definition of the Arthur packet attached to a unipotent parameter. They also observed the following corollary, which appeared in [7], Definition 1.17 as the definition of special unipotent representations in the complex case.

Corollary 27.13. *Suppose that the infinitesimal character $\mathcal{O}$ (cf. Lemma 15.4) arises from a homomorphism of $SL(2,\mathbb{C})$ into the dual group as in (27.7). Write $J_{max}(\mathcal{O})$ for the maximal primitive ideal in $U(\mathfrak{g})$ of infinitesimal character $\mathcal{O}$. Suppose π is an irreducible canonical projective representation of type z of a strong real form of G, having infinitesimal character $\mathcal{O}$. Then π is a special unipotent representation — that is, π belongs to an Arthur packet $\Pi^z(G/\mathbb{R})_\psi$ for some unipotent parameter ψ — if and only if π is annihilated by $J_{max}(\mathcal{O})$.*

Proof. This follows from Theorem 27.12, Theorem 21.6, and Theorem 21.8. Q.E.D.

To say that a representation has a large annihilator amounts to saying that the representation is small (see [52]). The condition in Corollary 27.13 is therefore equivalent to requiring the Gelfand-Kirillov dimension of π to be minimal, or to requiring the associated variety of the Harish-Chandra module of π to be contained in the closure of a certain complex nilpotent orbit in $\mathfrak{g}^*$ (namely the associated variety of $J_{max}(\mathcal{O})$). The techniques of [6] make it possible to compute exactly how many such representations exist; that is, to find the order of the union of the various unipotent Arthur packets for a single real form and infinitesimal character. Different Arthur packets may overlap, however, and we do not know in general how to calculate the cardinality of an individual one.

Example 27.14. Suppose $G = Sp(4)$ (by which we mean a group of rank 2), endowed with the inner class of real forms including $Sp(4,\mathbb{R})$. The L-group is

$$^\vee G^\Gamma = SO(5,\mathbb{C}) \times \Gamma. \qquad (27.14)(a)$$

Henceforth we will drop the $\mathbb{C}$; since we consider only complex orthogonal groups, no confusion should arise. The group $SL(2)$ is a double cover of $SO(3)$; composing the covering map with the inclusion of $SO(3)$ in $SO(5)$ gives a homomorphism

$$\psi_1 : SL(2) \to SO(5) \simeq {}^\vee G. \qquad (27.14)(b)$$

The centralizer in $SO(5)$ of the image of ψ_1 is evidently

$$S_0 = \left\{ \begin{pmatrix} \epsilon I_3 & 0 \\ 0 & A \end{pmatrix} \mid A \in O(2), \det A = \epsilon \right\} \simeq O(2). \qquad (27.14)(c)$$

Here I_3 is the 3 by 3 identity matrix, and $\epsilon = \pm 1$. Just as in Example 27.5, we find that the equivalence classes of unipotent Arthur parameters ψ associated to ψ_1 correspond to conjugacy classes of elements of order 2 in S_0. There are exactly three such classes, represented by the elements

$$s_{++} = \begin{pmatrix} 1 & 0 \\ 0 & 1 \end{pmatrix}, \quad s_{+-} = \begin{pmatrix} 1 & 0 \\ 0 & -1 \end{pmatrix}, \quad s_{--} = \begin{pmatrix} -1 & 0 \\ 0 & -1 \end{pmatrix}$$
$$(27.14)(d)$$

in $O(2) \simeq S_0$. Write ψ_{++}, etc., for the corresponding Arthur parameters. The corresponding centralizers are just the centralizers in $O(2)$ of these elements, namely

$$S_{\psi_{++}} = O(2), \quad S_{\psi_{+-}} = O(1) \times O(1), \quad S_{\psi_{--}} = O(2). \quad (27.14)(e)$$

The component groups are therefore

$$A_{\psi_{++}} = \mathbb{Z}/2\mathbb{Z}, \quad A_{\psi_{+-}} = \mathbb{Z}/2\mathbb{Z} \times \mathbb{Z}/2\mathbb{Z}, \quad A_{\psi_{--}} = \mathbb{Z}/2\mathbb{Z}.$$
$$(27.14)(f)$$

We also want to calculate the various A_ψ^{alg}. The algebraic universal cover $^\vee G^{alg}$ is the double cover $Spin(5)$, which is also isomorphic to $Sp(4)$. Perhaps the easiest way to calculate S_0^{alg} is to identify the lift $\psi_1^{alg} : SL(2) \to Sp(4)$. This turns out to be the diagonal map of $Sp(2)$ (which is isomorphic to $SL(2)$) into $Sp(2) \times Sp(2)$, followed by the obvious embedding of $Sp(2) \times Sp(2)$ into $Sp(4)$. From this one can deduce that $S_0^{alg} \simeq O(2)$; the covering map

$$O(2) \simeq S_0^{alg} \to S_0 \simeq O(2) \qquad (27.15)(a)$$

is the double cover of $SO(2)$ on the identity component. It follows that

$$S_{\psi_{++}}^{alg} = O(2), \quad S_{\psi_{+-}}^{alg} = \text{quaternion group of order 8}, \quad S_{\psi_{--}}^{alg} = O(2) \qquad (27.15)(b)$$

$$A_{\psi_{++}}^{alg} = A_{\psi_{++}}, \quad A_{\psi_{+-}}^{alg} = \text{quaternion group of order 8}, \quad A_{\psi_{--}}^{alg} = A_{\psi_{--}}. \qquad (27.15)(c)$$

We do not know any entirely elementary way to calculate the Arthur packets, but (using Theorem 27.12) this is not too difficult once one knows something about Harish-Chandra modules for $SO(5)$. We will simply record some of the results. It turns out that there are exactly four equivalence classes of strong real forms of G, having representatives that we denote δ_s, $\delta_{(1,1)}$, $\delta_{(2,0)}$, and $\delta_{(0,2)}$. The last two of these represent the compact real form, which can have no representations of the (singular) infinitesimal character associated to ψ_1. The other two represent the inequivalent real forms $Sp(4,\mathbb{R})$ and $Sp(1,1)$. It turns out that $\delta_{(1,1)}^2 = -I$, so representations of $Sp(1,1)$ correspond to $^\vee G^{alg}$-equivariant perverse sheaves on which $\pi_1(^\vee G)^{alg}$ acts non-trivially (Lemma 10.9(b)). It follows from (27.15)(c) that the Arthur packets for ψ_{++} and ψ_{--} can contain no representations of $Sp(1,1)$. Furthermore, the representation $\tau_{\psi_{+-}}(\pi)$ of $A_{\psi_{+-}}^{alg}$ attached to any irreducible representation $\pi \in \Pi(G/\mathbb{R})_{\psi_{+-}}$ (Definition 24.15) must satisfy

$$\tau_{\psi_{+-}}(\pi)(-I) = \begin{cases} 1, & \text{if } \pi \text{ is a representation of } Sp(4,\mathbb{R}); \\ -1, & \text{if } \pi \text{ is a representation of } Sp(1,1). \end{cases} \qquad (27.16)$$

We mentioned after Corollary 27.13 that one can count unipotent representations using the ideas of [6]. In this way (or any of several others) one finds that $Sp(4,\mathbb{R})$ has exactly eight unipotent representations attached to ψ_1. They are the three irreducible constituents of $\text{Ind}_{GL(2,\mathbb{R})}^{Sp(4,\mathbb{R})}(|\det|)$; the three irreducible constituents of $\text{Ind}_{GL(2,\mathbb{R})}^{Sp(4,\mathbb{R})}(\det)$; and the two irreducible constituents of $\text{Ind}_{GL(1,\mathbb{R}) \times Sp(2,\mathbb{R})}^{Sp(4,\mathbb{R})}(\text{sgn}(\det) \otimes 1)$. These representations may be distinguished by their lowest K-types; the maximal compact subgroup is $U(2)$, so its representations are parametrized by decreasing pairs of integers. The lowest K-types of the three sets of representations described above are $(0,0)$, $(2,2)$, and

$(-2,-2)$; $(1,1)$, $(-1,-1)$, and $(1,-1)$; and $(1,0)$ and $(0,-1)$. We denote the representation of lowest K-type (m,n) by $\pi_s(m,n)$.

An even simpler calculation shows that $Sp(1,1)$ has a unique irreducible unipotent representation attached to ψ_1, namely the unique spherical representation of the correct infinitesimal character. (It lies one-third of the way along the complementary series.) We denote this representation $\pi_{(1,1)}(0)$.

Here are the Arthur packets.

$$\Pi(G/\mathbb{R})_{\psi_{++}} = \{\pi_s(0,0), \pi_s(1,-1)\} \qquad (27.17)(a)$$

$$\Pi(G/\mathbb{R})_{\psi_{--}} = \{\pi_s(1,0), \pi_s(0,-1)\} \qquad (27.17)(b)$$

$$\Pi(G/\mathbb{R})_{\psi_{+-}} = \{\pi_s(1,1), \pi_s(-1,-1), \pi_s(2,2), \pi_s(-2,-2), \pi_{(1,1)}(0)\}$$
$$(27.17)(c)$$

In each case the map $\pi \mapsto \tau_\psi(\pi)$ is a bijection from the packet onto the set of equivalence classes of irreducible representations of A_ψ^{alg}; we have listed the representations so that the one with $\tau_\psi(\pi)$ trivial is first. In particular, $\tau_{\psi_{+-}}(\pi_{(1,1)}(0))$ is the irreducible two-dimensional representation of the quaternion group (cf. (27.16)).

We conclude with a careful study of the case of one-dimensional representations. One of the points of this is to understand the failure of injectivity of the map $\pi \mapsto \tau_\psi(\pi)$. (Failure of surjectivity appears already in the complex case; it corresponds to the fact that we had to use Lusztig's quotient of the component group in [7], rather than just a component group.) All of the examples that we know of this failure seem to arise finally from this very special case. (Part (e) of the following theorem is taken from [34], Lemma 2.11.)

Theorem 27.18. *Suppose $(G^\Gamma, \mathcal{W})$ is an extended group, and $(^\vee G^\Gamma, \mathcal{D})$ is an L-group for G^Γ. Fix a principal three-dimensional subgroup*

$$\psi_1 : SL(2) \to {^\vee G}.$$

a) *The centralizer S_0 of ψ_1 in $^\vee G$ is $Z(^\vee G)$.*
b) *The set of equivalence classes of unipotent Arthur parameters attached to ψ_1 may be identified with*

$$\{z \in Z(^\vee G) \mid z\theta_Z(z) = 1\}/\{w\theta_Z^{-1}(w) \mid w \in Z(^\vee G)\} = H^1(\Gamma, Z(^\vee G)).$$

27. Special unipotent representations

c) If ψ is any unipotent Arthur parameter attached to ψ_1, then

$$A_\psi^{alg} = Z({}^\vee G)^{\theta_Z, alg}/(Z({}^\vee G)^{\theta_Z, alg})_0.$$

This is precisely the universal component group $A_{univ}^{alg}({}^\vee G^\Gamma)$ of Definition 26.10.

d) The unipotent representations (of some real form $G(\mathbb{R}, \delta)$) attached to ψ_1 are precisely the representations trivial on the identity component $G(\mathbb{R}, \delta)_0$.

e) Suppose δ is any strong real form of G^Γ. Then there is a natural surjection

$$H^1(\Gamma, Z({}^\vee G)) \twoheadrightarrow \mathrm{Hom}(G(\mathbb{R}, \delta)/G(\mathbb{R}, \delta)_0, \mathbb{C}^\times).$$

If $G(\mathbb{R}, \delta)$ is quasisplit, this is an isomorphism.

Suppose ψ is a unipotent Arthur parameter attached to ψ_1, and δ is a strong real form of G^Γ. Write $\pi(\psi, \delta)$ for the character of $G(\mathbb{R}, \delta)/G(\mathbb{R}, \delta)_0$ attached to ψ by composing the bijection of (b) with the surjection of (e).

f) Suppose π is an irreducible unipotent representation (attached to ψ_1) of a strong real form $G(\mathbb{R}, \delta)$. Then the integer $\chi_\psi(\pi)$ of Theorem 22.7 is given by

$$\chi_\psi(\pi) = \begin{cases} 1 & \text{if } \pi \simeq \pi(\psi, \delta) \\ 0 & \text{otherwise.} \end{cases}$$

The character $\tau_\psi(\pi)$ of $A_\psi^{alg} \simeq A_{univ}^{alg}({}^\vee G^\Gamma)$ (see (c) above) is

$$\tau_\psi(\pi, \delta) = \chi_\psi(\pi)\tau_{univ}(\delta).$$

If we fix the unipotent parameter ψ attached to ψ_1, we therefore find exactly one unipotent representation for each strong real form. The correspondence $\pi \mapsto \tau_\psi(\pi)$ therefore fails to be injective as soon as the number of strong real forms exceeds the cardinality of $A_\psi^{alg} = Z({}^\vee G)^{\theta_Z, alg}/(Z({}^\vee G)^{\theta_Z, alg})_0$. The simplest example of this is perhaps for $PGL(n)$ endowed with the class of inner forms including the projective unitary groups $PU(p, q)$. There are $[(n+2)/2]$ real forms. The dual group is $SL(n)$, with θ_Z acting on the center by inversion. Consequently A_ψ^{alg} consists of the elements of order 2 in the cyclic group of order n; injectivity fails for $n \geq 3$. For simple groups not of type A the situation

308   The Langlands Classification and Irreducible Characters

is even clearer: typically the number of real forms is on the order of the rank, but A_ψ^{alg} has order at most four.

Proof of Theorem 27.18. Define E, F, and λ for ψ_1 as in (22.8). Because ψ_1 is principal, λ is a regular element with integral eigenvalues. It follows that $P(\lambda)$ is a Borel subgroup of $^\vee G$; write it as $^d B$, and write $^d T$ for the unique maximal torus containing λ in its Lie algebra. Then the centralizer of λ in $^\vee G$ is $^d T$, so $S_0 \subset {}^d T$. The element E is the sum of certain root vectors $\{X_\alpha\}$ for the simple roots, so the centralizer of E in $^d T$ is the intersection of the kernels of the simple roots, which is $Z(^\vee G)$. This proves (a). For (b), write S for the centralizer of ψ_1 in $^\vee G^\Gamma$. By Definition 4.6, we can find an element $^\vee \delta \in \mathcal{D}$ so that

$$^\vee \delta \in \mathrm{Aut}(^\vee G, {}^d B, {}^d T, \{X_\alpha\}) \qquad (27.19)(a)$$

(see Proposition 2.11). In particular, $^\vee \delta$ centralizes E. Now λ belongs to $[^\vee\mathfrak{g}, {}^\vee\mathfrak{g}]$, and $\alpha(\lambda) = 1$ for each simple root α. These properties characterize λ, and they are preserved by $^\vee \delta$. Consequently $^\vee \delta$ also centralizes λ. Finally, it follows from the representation theory of $SL(2)$ that F is characterized by the property $[E, F] = 2\lambda$, so $^\vee \delta$ also centralizes F. Therefore

$$^\vee \delta \in S, \qquad S = Z(^\vee G) \cup Z(^\vee G)^\vee \delta \qquad (27.19)(b)$$

Since $^\vee G^\Gamma$ is an L-group, $^\vee \delta^2 = 1$. Part (b) therefore follows from Corollary 27.3 and (27.19)(b). Part (c) follows from (b) and Corollary 27.3.

For (d), we have already observed that λ belongs to the derived algebra and takes the value 1 on each simple root. The corresponding infinitesimal character is therefore ρ, half the sum of a set of positive roots for G. This is the infinitesimal character of the trivial representation, so the corresponding maximal ideal in $U(\mathfrak{g})$ is the augmentation ideal, the annihilator of the trivial representation. Now (d) follows from Corollary 27.13.

For (e), we begin with the isomorphism

$$Z(^\vee G) \simeq \mathrm{Hom}(\pi_1(G), \mathbb{C}^\times) \qquad (27.20)(a)$$

of Lemma 10.2(a). Along the lines of Lemma 10.2(d), we deduce that

$$H^1(\Gamma, Z(^\vee G)) \simeq \mathrm{Hom}(H^1(\Gamma, \pi_1(G)), \mathbb{C}^\times) \qquad (27.20)(b)$$

To prove (e), it is therefore enough to exhibit an injection

$$G(\mathbb{R}, \delta)/G(\mathbb{R}, \delta)_0 \hookrightarrow H^1(\Gamma, \pi_1(G)). \qquad (27.20)(c)$$

27. Special unipotent representations

For this, we begin with the short exact sequence of groups with Γ actions

$$1 \to \pi_1(G) \to G^{sc} \to G \to 1. \tag{27.20}(d)$$

(Here we make complex conjugation act by δ on G, and pull this automorphism back to G^{sc}.) The corresponding exact sequence in cohomology includes

$$H^0(\Gamma, G^{sc}) \to H^0(\Gamma, G) \to H^1(\Gamma, \pi_1(G)). \tag{27.20}(e)$$

Now H^0 is just the Γ-invariant subgroup; that is (for G and G^{sc}) the group of real points. Hence

$$G^{sc}(\mathbb{R}, \delta) \to G(\mathbb{R}, \delta) \to H^1(\Gamma, \pi_1(G)). \tag{27.20}(f)$$

Since G^{sc} is simply connected, its group of real points is connected. The image of the first map is therefore $G(\mathbb{R}, \delta)_0$. The second map provides (by the exactness of (27.20)(f)) the inclusion in (27.20)(c).

To see that the map of (e) is surjective when $G(\mathbb{R}, \delta)$ is quasisplit, it suffices to find $|H^1(\Gamma, Z({}^\vee G))|$ linearly independent functions on $G(\mathbb{R}, \delta)/G(\mathbb{R}, \delta)_0$. The functions we use are the stable characters $\Theta(\eta_\psi(\delta))$ (notation as in Theorem 22.7 and (18.1)), with ψ a unipotent Arthur parameter attached to ψ_1. The strongly stable characters η_ψ are part of the basis $\{\eta_\phi^{mic}\}$ of strongly stable virtual characters (Corollary 19.16), and are therefore linearly independent. By Shelstad's theorem 18.7, the stable characters $\Theta(\eta_\psi(\delta))$ are also linearly independent. By (d), these characters may be regarded as functions on $G(\mathbb{R}, \delta)/G(\mathbb{R}, \delta)_0$. By (b), there are $|H^1(\Gamma, Z({}^\vee G))|$ such functions, as we wished to show.

For (f), suppose first that $G(\mathbb{R}, \delta)$ is quasisplit. The L-packets attached by the Langlands classification to the parameters ϕ_ψ (with ψ unipotent attached to ψ_1) are disjoint, and each is a non-empty set of representations of the component group of $G(\mathbb{R}, \delta)$. It follows from (e) that the L-packet Π_{ϕ_ψ} contains exactly one representation, and that these exhaust the representations of the component group. We leave to the reader the verification from the definitions that this single representation is actually $\pi(\psi, \delta)$. Because the integer $d(\pi)$ (the dimension of the corresponding orbit on the geometric parameter space) is the same for all finite-dimensional irreducible representations of a fixed group, it follows from Corollary 19.16 that the only representation of $G(\mathbb{R}, \delta)$ in the Arthur packet Π_ψ is $\pi(\psi, \delta)$. Now the two assertions of (f) are immediate.

Finally we consider the general case of (f). Because of the naturality of the surjection in (e), one can check easily that the virtual representation

$$\eta'_\psi = \sum_\delta \pi(\psi, \delta)$$

is strongly stable. By the preceding paragraph, it agrees with the Arthur character η_ψ on each quasisplit real form. By Theorem 18.7, $\eta'_\psi = \eta_\psi$. This is the first claim of (f). The second follows immediately. Q.E.D.

We leave to the reader the formulation of an analogous result in the case of a general E-group. What happens there is that some real forms (possibly all) will have no projective representations of type z trivial on the identity component. The set of equivalence classes of unipotent Arthur parameters attached to ψ_1 may be empty; if not, then it is a principal homogeneous space for $H^1(\Gamma, Z(^\vee G))$.

Bibliography

[1] J. Adams and D. Vogan, "L-groups, projective representations, and the Langlands classification," Amer. J. Math. **114**(1992), 45–138.

[2] J. Arthur, "On some problems suggested by the trace formula," in *Proceedings of the Special Year in Harmonic Analysis University of Maryland*, R. Herb, R. Lipsman, and J. Rosenberg, eds. Lecture Notes in Mathematics **1024**, Springer-Verlag, Berlin-Heidelberg-New York, 1983.

[3] J. Arthur, "Unipotent automorphic representations: conjectures," 13–71 in *Orbites Unipotentes et Représentations II. Groupes p-adiques et Réels*, Astérisque **171–172** (1989).

[4] M. Artin, "Néron models," 213–230 in Arithmetic Geometry, G. Cornell and J. Silverman, eds. Springer-Verlag, New York-Berlin-Heidelberg, 1986.

[5] D. Barbasch, "The unitary dual for complex classical Lie groups," Invent. Math. **96** (1989), 103–176

[6] D. Barbasch and D. Vogan, "Weyl group representations and nilpotent orbits," 21-33 in *Representation Theory of Reductive Groups*, P. Trombi, editor. Birkhäuser, Boston-Basel-Stuttgart, 1983.

[7] D. Barbasch and D. Vogan, "Unipotent representations of complex semisimple Lie groups," Ann. of Math. **121** (1985), 41–110.

[8] A. Beilinson and J. Bernstein, "Localisation de $\mathfrak{g}$-modules," C. R. Acad. Sci. Paris **292** (1981), 15–18.

[9] A. A. Beilinson, J. Bernstein, and P. Deligne, "Faisceaux pervers," 5–171 in *Analyse et topologie sur les espaces singuliers*, volume 1, Astérisque **100** (1982).

[10] A. Borel, "Automorphic L-functions," in *Automorphic Forms Representations and L-functions*, Proceedings of Symposia in Pure Mathematics **33**, part 2, 27–61. American Mathematical Society, Providence, Rhode Island, 1979.

[11] A. Borel et al., *Intersection Cohomology*, Birkhäuser, Boston-Basel-Stuttgart, 1984.

[12] A. Borel et al., *Algebraic D-Modules*, Perspectives in Mathematics **2**. Academic Press, Boston, 1987.

[13] W. Borho and J.-L. Brylinski, "Differential operators on homogeneous spaces I," Invent. Math. **69** (1982), 437–476.

[14] W. Borho and J.-L. Brylinski, "Differential operators on homogeneous spaces III," Invent. Math. **80** (1985), 1–68.

[15] M. Demazure, A. Grothendieck, et al., *Schémas en Groupes I (SGA 3)*, Lecture Notes in Mathematics **151**. Springer-Verlag, Berlin-Heidelberg-New York, 1970.

[16] V. Ginsburg, "Characteristic varieties and vanishing cycles," Invent. Math. **84** (1986), 327–402.

[17] M. Goresky and R. MacPherson, *Stratified Morse Theory*. Springer-Verlag, Berlin-Heidelberg-New York (1983).

[18] M. Goresky and R. MacPherson, "Local contribution to the Lefschetz fixed point formula," preprint.

[19] Harish-Chandra, "The Plancherel formula for complex semisimple Lie groups," Trans. Amer. Math. Soc. **76** (1954), 485–528.

[20] R. Hartshorne, *Algebraic Geometry*. Springer-Verlag, New York, Heidelberg, Berlin, 1977.

[21] M. Hashizume, "Whittaker models for real reductive groups," Japan. J. Math. **5** (1979), 349–401.

[22] J. E. Humphreys, *Introduction to Lie Algebras and Representation Theory*. Springer-Verlag, Berlin-Heidelberg-New York, 1972.

[23] M. Kashiwara, "Index theorem for a maximally overdetermined system of linear differential equations," Proc. Japan Acad. **49** (1973), 803–804.

[24] M. Kashiwara, *Systems of Microdifferential Equations*, Progress in Mathematics **34**. Birkhäuser, Boston-Basel-Stuttgart, 1983.

[25] M. Kashiwara and T. Kawai, "On holonomic systems of microdifferential equations. III –Systems with regular singularities–," Publ. RIMS, Kyoto Univ. **17** (1981), 813–979.

[26] M. Kashiwara and P. Schapira, *Microlocal Study of Sheaves*, Astérisque **128** (1985).

[27] D. Kazhdan and G. Lusztig, "Representations of Coxeter groups and Hecke algebras," Invent. Math. **53** (1979), 165–184.

[28] D. Kazhdan and G. Lusztig, "Schubert varieties and Poincaré duality," in *Geometry of the Laplace operator*, Proceedings of Symposia in Pure Mathematics **33**, part 2, 27–61. American Mathematical Society, Providence, Rhode Island, 1979.

[29] A. Knapp and G. Zuckerman, "Classification of irreducible tempered representations of semisimple Lie groups," Ann. of Math. **116** (1982), 389–501. (See also Ann. of Math. **119** (1984), 639.)

[30] B. Kostant, "On Whittaker vectors and representation theory," Invent. Math. **48** (1979), 101–184.

[31] B. Kostant and S. Rallis, "Orbits and representations associated with symmetric spaces," Amer. J. Math. **93** (1971), 753–809.

[32] R. Kottwitz, "Sign changes in harmonic analysis on reductive groups," Trans. Amer. Math. Soc. **278** (1983), 289–297.

[33] R. Kottwitz, "Stable trace formula: cuspidal tempered terms," Duke Math. J. **51** (1984), 611–650.

[34] R. P. Langlands, "On the classification of representations of real algebraic groups," in *Representation Theory and Harmonic Analysis on Semisimple Lie Groups*, Mathematical Surveys and Monographs **31**, 101–170. American Mathematical Society, Providence, Rhode Island, 1989.

[35] R. P. Langlands, "Stable conjugacy: definitions and lemmas," Canad. J. Math. **31** (1979), 700–725.

[36] R. P. Langlands and D. Shelstad, "On the definition of transfer factors," Math. Ann. **278** (1987), 219–271.

[37] G. Lusztig, *Characters of Reductive Groups over a Finite Field*. Annals of Mathematics Studies **107**. Princeton University Press, Princeton, New Jersey, 1984.

[38] G. Lusztig, "Intersection cohomology complexes on a reductive group," Invent. Math. **75** (1984), 205–272.

[39] G. Lusztig, "Character sheaves I," Adv. in Math. **56** (1985), 193–237.

[40] G. Lusztig and D. Vogan, "Singularities of closures of K-orbits on flag manifolds," Invent. Math. **71** (1983), 365–379.

[41] R. MacPherson, "Chern classes for singular algebraic varieties," Ann. of Math. **100** (1974), 423–432.

[42] T. Matsuki, "The orbits of affine symmetric spaces under the action of minimal parabolic subgroups," J. Math. Soc. Japan **31** (1979), 331–357.

[43] R. Richardson, "Conjugacy classes in parabolic subgroups of semisimple algebraic groups," Bull. London Math. Soc. **6** (1974), 21–24.

[44] M. Sato, M. Kashiwara, and T. Kawai, "Hyperfunctions and pseudodifferential equations," 265–529 in Lecture Notes in Mathematics **287**. Springer-Verlag, Berlin-Heidelberg-New York, 1973.

[45] W. Schmid, "Two character identities for semisimple Lie groups," 196–225 in *Non-commutative Harmonic Analysis and Lie groups*, J. Carmona and M. Vergne, eds., Lecture Notes in Mathematics **587**. Springer-Verlag, Berlin-Heidelberg-New York, 1977.

[46] I. Shafarevich, *Basic Algebraic Geometry*, translated by K. Hirsch. Springer-Verlag, Berlin, Heidelberg, New York, 1977.

[47] D. Shelstad, "Characters and inner forms of a quasi-split group over $\mathbb{R}$," Compositio Math. **39** (1979), 11–45.

[48] D. Shelstad, "L-indistinguishability for real groups," Math. Ann. **259** (1982), 385–430.

[49] B. Speh and D. Vogan, "Reducibility of generalized principal series representatons," Acta Math. **145** (1980), 227–299.

[50] T. A. Springer, "Reductive groups," in *Automorphic Forms Representations and L-functions*, Proceedings of Symposia in Pure Mathematics **33**, part 1, 3–28. American Mathematical Society, Providence, Rhode Island, 1979.

[51] J. Tate, "Number theoretic background," in *Automorphic Forms Representations and L-functions*, Proceedings of Symposia in Pure Mathematics **33**, part 2, 3–26. American Mathematical Society, Providence, Rhode Island, 1979.

[52] D. Vogan, "Gelfand-Kirillov dimension for Harish-Chandra modules," Invent. Math. **48** (1978), 75–98.

[53] D. Vogan, "Irreducible characters of semisimple Lie groups I," Duke Math. J. **46** (1979), 61–108.

[54] D. Vogan, *Representations of Real Reductive Lie Groups*. Birkhauser, Boston-Basel-Stuttgart, 1981.

[55] D. Vogan, "Irreducible characters of semisimple Lie groups III. Proof of the Kazhdan-Lusztig conjectures in the integral case," Invent. Math. **71** (1983), 381–417.

[56] D. Vogan, "Irreducible characters of semisimple Lie groups IV. Character-multiplicity duality," Duke Math. J. **49** (1982), 943–1073.

[57] D. Vogan, "Understanding the unitary dual," in *Proceedings of the Special Year in Harmonic Analysis, University of Maryland*, R. Herb, R. Lipsman and J. Rosenberg, eds. Lecture Notes in Mathematics **1024**. Springer-Verlag, Berlin-Heidelberg-New York, 1983.

[58] D. Vogan, *Unitary Representations of Reductive Lie Groups*. Annals of Mathematics Studies, Princeton University Press, Princeton, New Jersey, 1987.

[59] D. Vogan, "Associated varieties and unipotent representations," in *Harmonic Analysis on Reductive Groups*, W. Barker and P. Sally, eds. Birkhäuser, Boston-Basel-Berlin, 1991.

[60] T. Vust, "Opération de groupes réductifs dans un type de cônes presques homogènes," Bull. Soc. Math. France **102** (1974), 317–333.

[61] D. Vogan, "Dixmier algebras, sheets, and representation theory," in *Operator Algebras, Unitary Representations, Enveloping Algebras and Invariant Theory*, A. Connes, M. Duflo, A. Joseph, and R. Rentschler, eds. Birkhäuser, Boston-Basel-Berlin, 1990.

[62] N. Wallach, *Real Reductive Groups I*. Academic Press, San Diego, 1988.

Index

Arthur packet 2, 20, 242
Arthur parameter 2, 240–241
    associated Langlands parameter 20, 240–241
    unipotent 295, 300–301
Arthur's conjectures 1–3, 242, 263, 293–294
associated cycle 215
    of a Harish-Chandra module 226, 228, 231
associated variety 215
    of a Harish-Chandra module 2226
based root datum 32–33
Beilinson-Bernstein localization 99–100, 192, 235
Bruhat order 90–91, 169
canonical covering 113–114, 117–118
canonical flat 5, 65–66
canonical pairing 13, 172–174, 187–188
Cartan subgroup 120
    of an extended group 139
    of an extended group, based 1150–151
    of an E-group 140–141
    of an E-group, based 151–152
Cayley transform 199–202
character (of a representation) 205–206, 207–208
character identity, Hecht-Schmid 127, 130
character matrix 168
character matrix, geometric 90, 92
characteristic cycle 17, 216, 217, 222–225, 231
characteristic variety 216, 226
complete geometric parameter 8, 86–87
component group 82–84
    for an Arthur parameter 241–242, 244, 300–301
    generic 254, 257
    Langlands 60, 86, 108
    type Q 62, 87
    universal 60–61, 86
conormal bundle 212

constructible sheaves 87
$\mathcal{D}$-modules 88
    regular holonomic 88
derived categories 91
distinguished automorphism 34
dual group 3, 47
    Harish-Chandra modules for 235
E-group 47–50
elliptic elements 239
endoscopic datum 24–25, 285
endoscopic lifting 287–289, 292–294
    of Langlands and Shelstad 288
equivariant fundamental group 9, 82, 84
equivariant micro-fundamental group 19, 258
Euler characteristic, local 14, 16, 17, 90, 249–250
Euler characteristic, microlocal 17, 217
extended group 6, 36, 44–46
fiber product 71–72, 92–93, 223–225
filtration, good 213–215, 229, 236
fixed point theorem 272
fixed point variety 268, 270, 272
fundamental group 84
Gelfand-Kirillov dimension 229–230
geometric parameters 5, 68, 72–73, 80
    complete 8, 86–87
    complete, of type $\mathcal{Q}$ 86–87
    and Langlands parameters 73, 86–87
geometrically stable 16
Goresky-MacPherson fixed point theorem 27, 272
Hecke algebra 180–181
infinitesimal character 169
inner form 30
integral (co)roots 180
Kazhdan-Lusztig algorithm 104, 185, 186
Kazhdan-Lusztig conjectures 10, 173, 185, 186–187
Kazhdan-Lusztig polynomial 104, 185, 286

Index

Kottwitz invariant 13, 171–172, 203–204
L-group 3, 47–50
L-homomorphism 22, 55, 57, 77–78
L-packet 1
Langlands classification 1, 9, 115, 119, 125, 131, 155
    for tori 106, 112, 116
Langlands functoriality . . . 21–23, 57, 77–78, 96–97, 276–277, 280
Langlands parameters 3, 56
    and geometric parameters 73
    complete 61
    complete of type Q 62
Langlands quotient 9, 122
lattice of characters 33, 105–106
lattice of one-parameter subgroups 33, 105
Lefschetz fixed point theorem 272
Lefschetz number, local 249–250
Lefschetz number, microlocal 259, 262–264, 267
length function (on limit characters) 184
lifting 287–289, 292–294
    of Langlands and Shelstad 288
limit character 121–122
    final 130–131
micro-component group 262
micro-packet (of geometric parameters) 218–221
mixed Grothendieck group 181–182
moment map 225
multiplicity matrix 10, 12, 14, 168
    geometric 90, 92
one-parameter subgroups 33, 105
pairing between Cartan subgroups 148–149, 152–153
parity condition 129–130
perverse sheaves 87, 89
Pontriagin duality 107
pro-algebraic group 83, 108–109
projective representation 115, 117–118

real form 28–31
    strong 7, 36, 118
representation (of strong real form) 36–37
Richardson orbit 226, 299
Riemann-Hilbert correspondence 88–89
root subgroup 128–129
smooth morphism 94, 222–223
special unipotent representations 295, 303
stable character 206
    strongly 15, 16, 208–210, 219–221
    strongly σ- 284
standard representation 9, 122, 131
    strongly stable 210–211
strong real forms 7, 36, 118
    and the dual group 284, 306–307
    for tori 110–112
strongly stable 15, 16, 208–210
τ-invariant 128, 132
tempered parameter 1, 6, 239–240
translation datum 101–103, 176
translation functor 102–103, 175
translation principle 102–103, 132–135, 177, 178
Verdier duality 183–185, 196–197, 202
virtual character 9–10
wall crossing 194–196
Weil group 3, 55–56
Whittaker model 41–43, 161, 163

Progress in Mathematics

Edited by:

J. Oesterlé
Département de Mathématiques
Université de Paris VI
4, Place Jussieu
75230 Paris Cedex 05, France

A. Weinstein
Department of Mathematics
University of California
Berkeley, CA 94720
U.S.A.

Progress in Mathematics is a series of books intended for professional mathematicians and scientists, encompassing all areas of pure mathematics. This distinguished series, which began in 1979, includes authored monographs and edited collections of papers on important research developments as well as expositions of particular subject areas.

We encourage preparation of manuscripts in some form of TeX for delivery in camera-ready copy which leads to rapid publication, or in electronic form for interfacing with laser printers or typesetters.

Proposals should be sent directly to the editors or to: Birkhäuser Boston, 675 Massachusetts Avenue, Cambridge, MA 02139, U. S. A.

A complete list of titles in this series is available from the publisher.

53 LAURENT. Théorie de la Deuxième Microlocalisation dans le Domaine Complexe.
54 VERDIER/LE POTIER. Module des Fibres Stables sur les Courbes Algébriques: Notes de l'Ecole Normale Supérieure, Printemps, 1983
55 EICHLER/ZAGIER. The Theory of Jacobi Forms
56 SHIFFMAN /SOMMESE. Vanishing Theorems on Complex Manifolds
57 RIESEL. Prime Numbers and Computer Methods for Factorization
58 HELFFER/NOURRIGAT. Hypoellipticité Maximale pour des Opérateurs Polynomes de Champs de Vecteurs
59 GOLDSTEIN. Séminaire de Théorie des Nombres, Paris 1983–84
60 PROCESI. Geometry Today: Giornate Di Geometria, Roma. 1984
61 BALLMANN/GROMOV/SCHROEDER. Manifolds of Nonpositive Curvature
62 GUILLOU/MARIN. A la Recherche de la Topologie Perdue
63 GOLDSTEIN. Séminaire de Théorie des Nombres, Paris 1984–85
64 MYUNG. Malcev-Admissible Algebras
65 GRUBB. Functional Calculus of Pseudo-Differential Boundary Problems
66 CASSOU-NOGUES/TAYLOR. Elliptic Functions and Rings and Integers
67 HOWE. Discrete Groups in Geometry and Analysis: Papers in Honor of G.D. Mostow on His Sixtieth Birthday
68 ROBERT. Autour de L'Approximation Semi-Classique
69 FARAUT/HARZALLAH. Deux Cours d'Analyse Harmonique
70 ADOLPHSON/CONREY/GHOSH/YAGER. Analytic Number Theory and Diophantine Problems: Proceedings of a Conference at Oklahoma State University

71 GOLDSTEIN. Séminaire de Théorie des Nombres, Paris 1985–86
72 VAISMAN. Symplectic Geometry and Secondary Characteristic Classes
73 MOLINO. Riemannian Foliations
74 HENKIN/LEITERER. Andreotti-Grauert Theory by Integral Formulas
75 GOLDSTEIN. Séminaire de Théorie des Nombres, Paris 1986–87
76 COSSEC/DOLGACHEV. Enriques Surfaces I
77 REYSSAT. Quelques Aspects des Surfaces de Riemann
78 BORHO/BRYLINSKI/MACPHERSON. Nilpotent Orbits, Primitive Ideals, and Characteristic Classes
79 MCKENZIE/VALERIOTE. The Structure of Decidable Locally Finite Varieties
80 KRAFT/PETRIE/SCHWARZ. Topological Methods in Algebraic Transformation Groups
81 GOLDSTEIN. Séminaire de Théorie des Nombres, Paris 1987–88
82 DUFLO/PEDERSEN/VERGNE. The Orbit Method in Representation Theory: Proceedings of a Conference held in Copenhagen, August to September 1988
83 GHYS/DE LA HARPE. Sur les Groupes Hyperboliques d'après Mikhael Gromov
84 ARAKI/KADISON. Mappings of Operator Algebras: Proceedings of the Japan-U.S. Joint Seminar, University of Pennsylvania, Philadelphia, Pennsylvania, 1988
85 BERNDT/DIAMOND/HALBERSTAM/HILDEBRAND. Analytic Number Theory: Proceedings of a Conference in Honor of Paul T. Bateman
86 CARTIER/ILLUSIE/KATZ/LAUMON/MANIN/RIBET. The Grothendieck Festschrift: A Collection of Articles Written in Honor of the 60th Birthday of Alexander Grothendieck. Vol. I
87 CARTIER/ILLUSIE/KATZ/LAUMON/MANIN/RIBET. The Grothendieck Festschrift: A Collection of Articles Written in Honor of the 60th Birthday of Alexander Grothendieck. Volume II
88 CARTIER/ILLUSIE/KATZ/LAUMON/MANIN/RIBET. The Grothendieck Festschrift: A Collection of Articles Written in Honor of the 60th Birthday of Alexander Grothendieck. Volume III
89 VAN DER GEER/OORT/STEENBRINK. Arithmetic Algebraic Geometry
90 SRINIVAS. Algebraic K-Theory
91 GOLDSTEIN. Séminaire de Théorie des Nombres, Paris 1988–89
92 CONNES/DUFLO/JOSEPH/RENTSCHLER. Operator Algebras, Unitary Representations, Envelopping Algebras, and Invariant Theory. A Collection of Articles in Honor of the 65th Birthday of Jacques Dixmier
93 AUDIN. The Topology of Torus Actions on Symplectic Manifolds
94 MORA/TRAVERSO (eds.) Effective Methods in Algebraic Geometry
95 MICHLER/RINGEL (eds.) Representation Theory of Finite Groups and Finite Dimensional Algebras
96 MALGRANGE. Equations Différentielles à Coefficients Polynomiaux
97 MUMFORD/NORI/NORMAN. Tata Lectures on Theta III
98 GODBILLON. Feuilletages, Etudes géométriques
99 DONATO/DUVAL/ELHADAD/TUYNMAN. Symplectic Geometry and Mathematical Physics. A Collection of Articles in Honor of J.-M. Souriau
100 TAYLOR. Pseudodifferential Operators and Nonlinear PDE
101 BARKER/SALLY. Harmonic Analysis on Reductive Groups
102 DAVID. Séminaire de Théorie des Nombres, Paris 1989-90
103 ANGER/PORTENIER. Radon Integrals
104 ADAMS/BARBASCH/VOGAN. The Langlands Classification and Irreducible Characters for Real Reductive Groups